SpringerWienNewYork

Cornelius Lütz
Editor

# Plants in Alpine Regions

## Cell Physiology of Adaption and Survival Strategies

SpringerWienNewYork

*Editor*
Prof. Dr. Cornelius Lütz
University of Innsbruck
Faculty of Life Sciences
Institute of Botany
Sternwartestr. 15
6020 Innsbruck
Austria
cornelius.luetz@uibk.ac.at

This work is subject to copyright.
All rights are reserved, whether the whole or part of the material is concerned, specifically those of translation, reprinting, re-use of illustrations, broadcasting, reproduction by photocopying machines or similar means, and storage in data banks.

The use of registered names, trademarks, etc. in this publication does not imply, even in the absence of a specific statement, that such names are exempt from the relevant protective laws and regulations and therefore free for general use.

© 2012 Springer-Verlag/Wien

SpringerWienNewYork is a part of Springer Science+Business Media
springer.at/

*Cover illustrations*: upper left: *alpine Oxyria digyna*; lower left: *antarctic Deschampsia antarctica* prepared for measurements; right: *Central Alps, Tyrol.*

Typesetting: SPi Publisher Services, Pondicherry, India

Printed on acid-free and chlorine-free bleached paper
SPIN: 12767885

With 112 Figures

Library of Congress Control Number: 2011938122

ISBN 978-3-7091-0135-3 e-ISBN 978-3-7091-0136-0
DOI 10.1007/978-3-7091-0136-0
SpringerWienNewYork

# Preface

Plants inhabiting high alpine and nival zones are considered as living in an extreme environment.

Extreme environments have been attractive for explorers for centuries, and nowadays they also attract tourists. Fortunately biological science is becoming increasingly aware that these remote habitats provide challenging questions that will help to understand the limits of life functions. Biota of cold, extreme environments have been brought closer to the scientific community and to the public by international activities such as the International Year of the Mountains (2002) or the International Polar Year (2007–2009).

While it is indispensable to use model plants such as *Chlorella, Physicomitrella, Hordeum* or *Arabidopsis* to follow single metabolic processes and pathways or fluxes, species from remote locations are difficult to use as model organisms: often they do not grow in culture or they change their metabolism completely under artificial growth.

Plants at the margins of life developed a broad range of adaptation and survival strategies during evolution. These are best studied with species growing in extreme environments – extreme firstly for the researcher, who tries to measure life functions in the field and to harvest samples for later studies in the home laboratory. This has been experienced by an increasing number of scientists working in the fields of geobotany, plant ecophysiology and ecology. Their work has prepared the conditions that allow different aspects in cell physiology of plants from cold environments (the same holds for high temperature biota e.g. of deserts, volcanoes) to be studied by state-of-the-art methods and the results to be interpreted in order to understand the entire organism, and not merely an isolated function.

This book is devoted to the presentation of a collection of articles on adaptation and survival strategies at the level of cell physiology. The plants have partially been investigated in the field and were partially taken directly from alpine or polar habitats for experiments under lab conditions.

The book contains 14 chapters, written by experts from different research areas. Most of the contributions are from scientists from the University of Innsbruck. This is not surprising since the "Alpenuniversität" looks back on more than 150 years of research in alpine regions in many fields, including biology, medicine, weather and climate, geology, geography and others. Surrounded by high mountains, the university still is a center of alpine research today. Several topics have been taken up by colleagues from other universities to integrate their challenging work for a better description of alpine plant cell physiology.

If physiology and cell structural research are not to lose the connection to the organism, at least partial knowledge of the environmental conditions as determinants of most life functions should be considered. Therefore, the first three chapters deal with aspects of the physical environment of alpine plants.

In Chap. 1, *M. Kuhn* explains the conditions of water input in the form of snow or rain in the High Alps. Seasonal variations, regional differences and altitudinal effects or wind exposure strongly determine water and snow situations for the plant communities. The physical characterization of snow cover will help to understand plant survival in winter. Equally important for plant life and development is solar radiation, characterized by *M. Blumthaler* in Chap. 2 for the European Alps. Radiation physics is not an easy issue for plant scientists, but it is well explained in this context. The variation in solar radiation input in the Alps comes from atmospheric factors such as aerosols, dust, clouds, ozone, and further depends on altitude, solar angle and exposure angle of a plant surface to the sun. The biologically effective UV radiation is at the center of this contribution. Chapter 3 by *W. Larcher* deals with the bioclimatic temperatures of mountain plants and connects the first two chapters with the microclimate which is closer to the plants than general weather descriptions allow. Macro- and microclimate temperatures show large differences. Less often taken into account, but of enormous influence are soil temperatures in mountain regions. Soils buffer the large diurnal temperature changes in high altitudes, thus influencing root growth. Recording actual temperatures at the plant body or in the canopy provides important data for understanding plant growth forms and the physiology of temperature adaptation.

The following Chap. 4 by *C. Lütz* and *H.K. Seidlitz* describes effects of anthropogenic increases in UV radiation and tropospheric ozone. Sophisticated climate simulations demonstrate that alpine vegetation as well as one of the two Antarctic higher plants will probably not suffer as a result of the expected increases in UV. In contrast, ozone, which accumulates at higher levels in European mountains than in urban environments, may threaten alpine vegetation by inducing earlier senescence. In a combination of physiological and ultrastructural studies, *Lütz* et al. (Chap. 5) describe cellular adaptations in alpine and polar plants. Chloroplasts show structural adaptations, only rarely found in plants from temperate regions that allow them to use the short vegetation period in a better way. Possible control by the cytoskeleton is discussed. These observations reflect high photosynthetic activities; and development of membranes under snow in some species is documented. By contrast, the dynamic of high temperature resistance in alpine plant species is presented by *G. Neuner* and *O. Buchner* (Chap. 6). Tissue heat tolerance of a large number of alpine species is reported. Heat hardening and developmental aspects are compared. As high temperatures in the Alps normally occur under high irradiation, the authors look more closely at the thermotolerance of photosystem II. Acclimation of photosynthesis and related physiological processes in a broader view are discussed by *P. Streb* and *G. Cornic* in Chap. 7. Aspects of acclimation in alpine plant photosynthesis include C4 and CAM mechanisms and the PTOX electron shuttle. The protection of photosynthesis by energy dissipation and antioxidants is also considered.

*R. Bligny* and *S. Aubert* (Chap. 8) investigate metabolites and describe high amounts of ascorbic acid in some Primulaceae. By using sophisticated NMR methods they also identify methylglucopyranoside in *Geum montanum* leaves, which may play a part in methanol detoxification, and finally study metabolites in *Xanthoria* lichens during desiccation and hydration. In Chap. 9 *F. Baptist* and *I. Aranjuelo* describe metabolisms of N and C in alpine plants – often overlooked by physiologists. Plant development depends greatly on carbon fixation and a balanced N uptake by the roots. Snow cover and time of snow melt determine N uptake for the metabolism. Storage of C and N in alpine plants under the expected climate changes is described and discussed.

The high mountain flora shows that flowering and seed formation function despite the often harsh environmental conditions. In Chap. 10 *J. Wagner* et al. explain how flower formation and anthesis are regulated by species-specific timing based on the plant organ temperatures. Snow melt – again – and day length control reproductive development and seed maturity.

The next two chapters report on recent findings describing adaptation to subzero temperatures. As *S. Mayr* et al. show (Chap. 11), alpine conifers are endangered in winter by limited access to soil water, ice blockages of stored water in several organs, and frost drought in the needles. Embolism and refilling of xylem vessels are studied by biophysical methods and microscopy, resulting in a better understanding of the complex hydraulics in wooden alpine plants. Thematically related, *G. Neuner and J. Hacker* discuss freezing stress and mechanisms of ice propagation in plant tissues (Chap. 12), using alpine dwarf shrubs and herbs. Resistance to freezing stress depends greatly on plant life forms and developmental stages. The capacity of supercooling is studied in some species. By means of digital imaging they describe ice propagation in leaves and discuss the structural and thermal barriers in tissues that are developed to avoid ice propagation.

*D. Remias* continues with snow and ice (Chap. 13), now as a habitat, and reports on recent findings in the cell physiology of snow and ice algae from the Alps and polar regions. The extreme growth conditions require special metabolic and cell structural adaptations, such as accumulation of secondary carotenoids ("red snow") in the cytoplasm, or vacuolar polyphenols as a protection against high PAR and UV radiation (glacier ice algae). Photosynthesis is not inhibited by zero temperatures and not photoinhibited under high irradiation – comparable to many high alpine species. Even smaller in size, but best acclimated to cold temperatures are microorganisms in alpine soils, presented by *R. Margesin* in Chap. 14. These organisms serve as ideal study objects to characterize cold active enzymes, cold shock proteins and cryoprotectans. Microbial activity in alpine soils at low temperatures has an important influence on litter decomposition and nutrition availability, which connects to higher plant root activities.

After many years of studying alpine and polar plants under different aspects, it was a pleasure for me to edit this collection of research contributions; I thank all colleagues for their participation and effort in presenting their data.

I hope that this book expands the information on cell physiology of alpine/polar plants including the connection to the physical environment they are exposed to. The different contributions should encourage more scientists to incorporate plants from extreme environments in their studies in order to understand the limits of cellular adaptation and survival strategies.

Finally I would like to thank the Springer team for their support and valuable suggestions on editing this book, especially Dr. A.D. Strehl and Mag. E.M. Oberhauser.

Innsbruck, spring 2011 Cornelius Lütz

# Contents

# Contributors

**Iker Aranjuelo** Departamento de Biologia Vegetall, Universitat de Barcelona, Avericla Diagonal, Barcelona, Spain

**Serge Aubert** Station Alpine Joseph Fourier, Unité Mixte de Service, Université Joseph Fourier, Grenoble cedex 9, France

**Florence Baptist** Departamento de Biologia Vegetal, Universitat de Barcelona, Avenida Diagonal, Barcelona, Spain

**Barbara Beikircher** Institute of Botany, University of Innsbruck, Innsbruck, Austria

**Paul Bergweiler** German Aerospace Center Project Management Agency, Bonn, Germany

**Richard Bligny** Laboratoire de Physiologie Cellulaire & Végétale, Unité Mixte de Recherche, Institut de Recherche en Technologies et Sciences pour le Vivant, Grenoble cedex 9, France

**Mario Blumthaler** Division for Biomedical Physics, Medical University of Innsbruck, Innsbruck, Austria

**Othmar Buchner** Institute of Botany, University of Innsbruck, Innsbruck, Austria

**Gabriel Cornic** Laboratoire Ecologie Systématique et Evolution, University of Paris-Sud, Orsay, France; CNRS, Orsay, France; AgroParisTech, Paris, France

**Jürgen Hacker** Institute of Botany, University of Innsbruck, Innsbruck, Austria

**Andreas Holzinger** Institute of Botany, University of Innsbruck, Innsbruck, Austria

**Michael Kuhn** Institute of Meteorology and Geophysics, University of Innsbruck, Innsbruck, Austria

**Ursula Ladinig** Institute of Botany, University of Innsbruck, Innsbruck, Austria

**Walter Larcher** Institute of Botany, LTUI, Innsbruck, Austria

**Ilse Larl** Institute of Botany, University of Innsbruck, Innsbruck, Austria

**Cornelius Lütz** University of Innsbruck, Institute of Botany, Innsbruck, Austria

**Rosa Margesin** Institute of Microbiology, University of Innsbruck, Innsbruck, Austria

**Stefan Mayr** Institute of Botany, University of Innsbruck, Innsbruck, Austria

**Gilbert Neuner** Institute of Botany, University of Innsbruck, Innsbruck, Austria

**Lavinia Di Piazza** Institute of Botany, University of Innsbruck, Innsbruck, Austria

**Daniel Remias** Institute of Pharmacy/Pharmacognosy, University of Innsbruck, Innsbruck, Austria

**Peter Schmid** Institute of Botany, University of Innsbruck, Innsbruck, Austria

**Harald K. Seidlitz** Department Environmental Engineering, Institute of Biochemical Plant Pathology, Helmholtz Zentrum München, Neuherberg, Germany

**Gerlinde Steinacher** Institute of Botany, University of Innsbruck, Innsbruck, Austria

**Peter Streb** Laboratoire Ecologie Systématique et Evolution, University of Paris-Sud, Orsay, France; CNRS, Orsay, France; AgroParisTech, Paris, France

**Johanna Wagner** Institute of Botany, University of Innsbruck, Innsbruck, Austria

# Rain and Snow at High Elevation

## The Interaction of Water, Energy and Trace Substances

# 1

Michael Kuhn

## 1.1 Introduction

Plants are major players in the alpine biogeochemical cycles, using water, energy and nutrients from both the atmosphere and the ground for their primary production. They are exposed to rain and snowfall, may be covered by snow for considerable periods, absorb solar radiation and transpire water vapour back to the atmosphere. While the supply of energy, water and nutrients from the atmosphere is the boundary condition for the plants' existence, they significantly determine the return of all three quantities back to the air.

Bioclimatic temperatures in the high Alps are treated in the chapter by Larcher, the supply of solar radiation by Blumthaler. This chapter deals with the significance of rain and snow for high alpine plants. It describes the regional and local distribution of precipitation, its change with elevation and its seasonal course. It emphasizes the importance of snow as a place of water storage, thermal insulation and concentrated release of ions.

## 1.2 Regional, Vertical and Seasonal Distribution of Precipitation

### 1.2.1 Annual Precipitation

The regional distribution of annual precipitation in the Alps has been analysed repeatedly, I recommend reading Fliri (1975), Baumgartner and Reichel (1983), Frei and Schär (1998) and Efthymiadis et al. (2006) for that purpose. All of these authors agree that the distribution of alpine precipitation is dominated by two independent variables: altitude and windward situation (or distance from the northern and southern margins toward the interior Alps).

In the Eastern Alps, the majority of annual precipitation arrives either from the SW or NW, with the passage of a trough on its eastward way from the Atlantic. This explains the frequent succession of south-westerly flow with precipitation at the southern alpine chains followed by north-westerly currents wetting the northern part of the Eastern Alps, a pattern that was described by Nickus et al. (1998): "A trough moving in the Westerlies and moving near the Alps will cause a south-westerly to southerly flow over the Eastern Alps. Precipitation south of the central ridge and Föhn winds in the north are the most frequent weather situation at this stage. With increasing cyclonality air flow will become more westerly, often bringing moist air from the Atlantic. Precipitation will then shift to the central and northern parts of the Alps, starting in the west and continuing to the east. As the flow turns to more northerly directions, a passing cold front may bring precipitation mainly in the northern parts of the Alps and at least an interruption in precipitation in the south."

A consequent rule of thumb is that stations at the northern and southern margins experience three times as much annual precipitation as those in the dry interior valleys (Fliri 1975), in rough figures 1,500–500 mm, with maxima at some stations exceeding 3,000 mm and minima of less than 500 mm per year. In either case, the highest of the central chains experience a secondary maximum as the Glockner Group in Fig. 1.1 (at 47°N and 12.5–13.5°E).

M. Kuhn (✉)
Institute of Meteorology and Geophysics, University of Innsbruck, Innsbruck, Austria
e-mail: michael.kuhn@uibk.ac.at

C. Lütz (ed.), *Plants in Alpine Regions*,
DOI 10.1007/978-3-7091-0136-0_1, © Springer-Verlag/Wien 2012

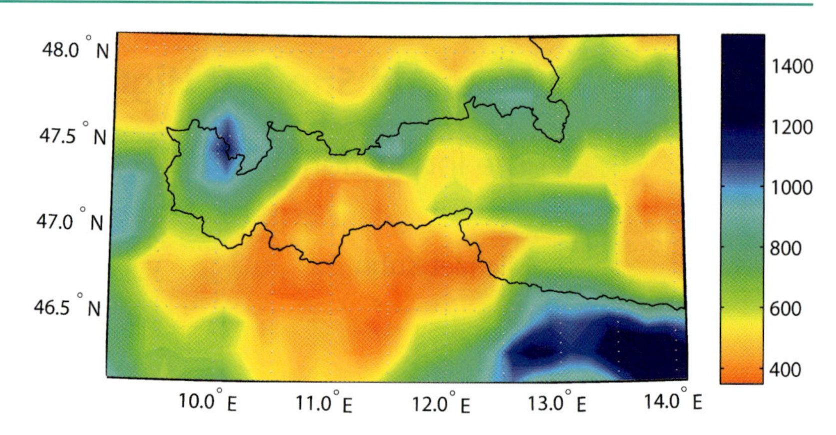


**Fig. 1.1** Winter precipitation (December, January, February) in the Eastern Alps according to the HISTALP analysis, in mm. The HISTALP precipitation data set is described by Eftymiadis et al. (2006)

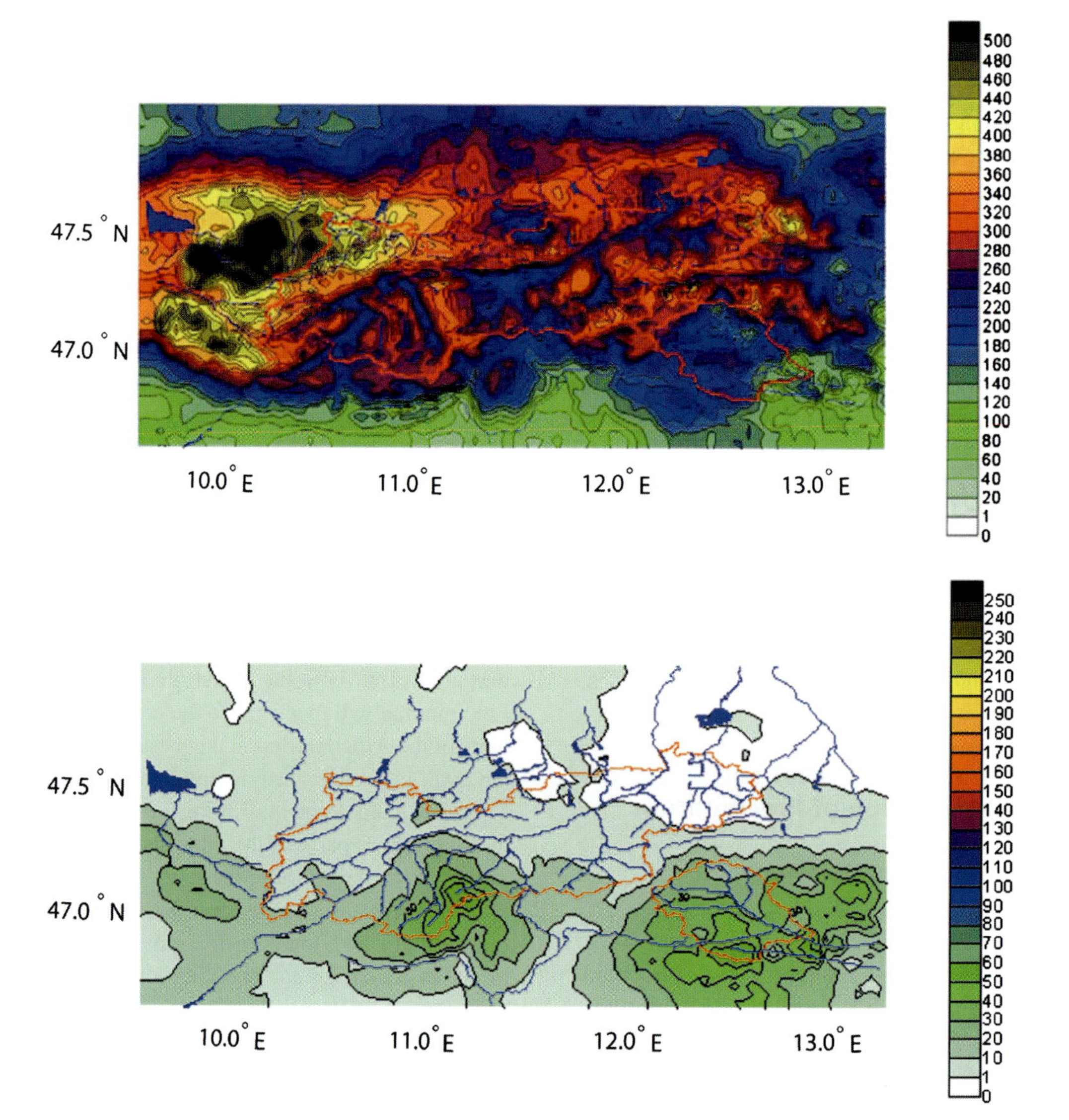


**Fig. 1.2** Total precipitation of the month of August 2010 (*above*) is dominated by north-westerly flow with values exceeding 400 mm in the west and less than 200 in the south-east. The daily sums from 14-08-2010, 07:00 to 15-08-2010, 07:00 given in the *lower* panel describe a situation of southerly flow. From Gattermayr (2010)

The annual values given above are the sums of many individual events. Two examples of these are given in Fig. 1.2. Be aware, however, that many of the details on these precipitation maps are interpolated with algorithms that use elevation and distance from actual meteorological observations as independent variables.

### 1.2.2 Effects of Elevation

Several effects contribute to the higher incidence of precipitation at higher elevations: temperature is lower at higher elevation, the water vapour is thus closer to saturation, and condensation is more likely at higher elevation; when moist air is advected towards a mountain chain, it is forced to ascend and thereby cools; wind speed, and thus horizontal advection of moisture, increases with elevation, the decisive quantity being the product of horizontal wind speed and water vapour density.

Altogether an increase of about 5% per 100 m elevation is observed in various valleys of the Eastern Alps as shown in Fig. 1.3. The annual course of the increase of precipitation with elevation shown in Fig. 1.3 reflects the varying frequencies of advective and convective precipitation. In winter and spring the advective type, associated with the passage of fronts, dominates and leads to high values of the vertical gradient of precipitation. In summer convective precipitation prevails; it depends on local heat sources which are by and large independent of elevation. Convective precipitation most likely occurs over sunlit slopes with vegetation. As both energy supply and vegetation decrease with elevation, convective precipitation may even have an upper limit and thus a lesser increase with elevation; it is certainly more influenced by exposition than by elevation and decreases the mean monthly values of vertical gradients of precipitation given in Fig. 1.3.

### 1.2.3 Seasonal Variation of Precipitation

The seasonal course of precipitation in the Eastern Alps (e.g. 10°E) has a marked change with latitude. In the north, there is a clear dominance of summer rainfalls, monthly sums may be three times as high as those of fall and winter. A summary presentation by Fliri (1975, his Figs. 69–75) shows this summer maximum to extend southward into the dry, central region. Farther south the Mediterranean dry summers split the precipitation curve into two maxima in spring and fall,

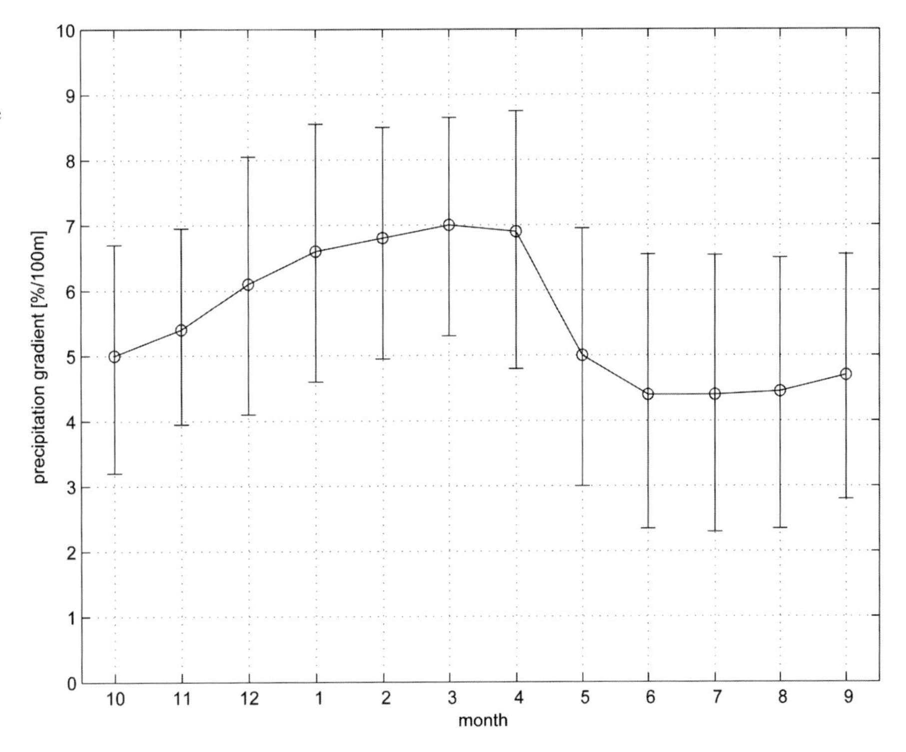


Fig. 1.3 The increase of annual precipitation with elevation, expressed as % per 100 m elevation. Bars indicate upper and lower limits of 10 hydrological basins (Kuhn 2010)

the latter often dominating. This is locally diverse, but generally evident in Fig. 1.4.

An analysis of recent years has shown that the climatological means in Fig. 1.4 have a high interannual variance and that deviations from the mean have a tendency to come in groups of several years. This is true in particular for the occurrence of October and November maxima.

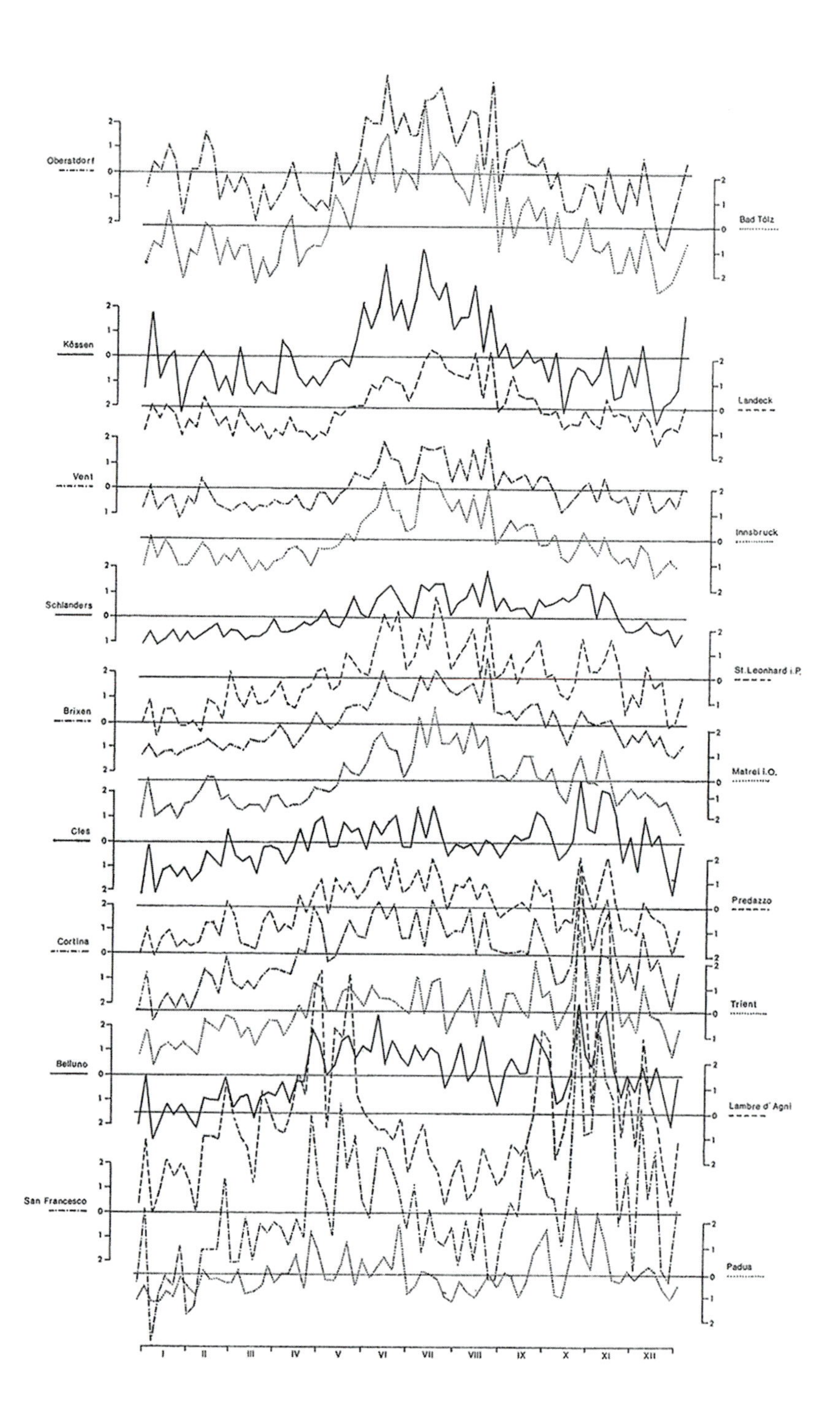


**Fig. 1.4** Annual course of monthly precipitation in a cross section from N to S, approximately 10–12°E, in mm per day. Means of 1931–1960, from Fliri (1975)

## 1.3 Snow Cover

### 1.3.1 The Transition from Rain to Snow

The transition from snow to rain is expected to depend on the 0°C limit. Even at the level of formation of snow flakes it is not exactly the air temperature that determines the freezing of precipitation; it is rather the energy balance of the drop or flake, best approximated by the so called wet bulb temperature which is determined by evaporation and sublimation: at a given air temperature, drops would rather turn into snow flakes at low relative humidity. Snow forms at higher and therefore colder levels so that surface temperatures of about 1°C are an alpine-wide useful approximation for snowfall. The probability of snowfall Q may be expressed by $Q = 0.6–0.1\ T$.

The fraction of solid vs. total precipitation depends on elevation as shown in Fig. 1.5. Considering the fairly regular dependence of temperature on elevation in this figure, it is remarkable that the fraction of solid vs. total precipitation has a much higher variance than that of temperature. The absolute values of solid precipitation F and those of total annual precipitation are dominated by their regional distribution more than by altitude.

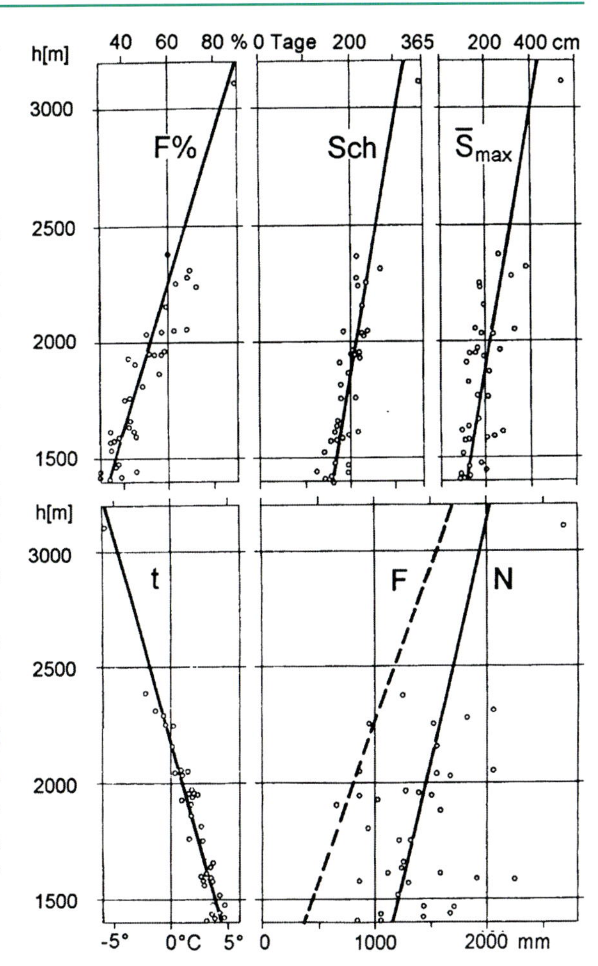


**Fig. 1.5** Dependence on elevation of various quantities describing snow cover: F% fraction of solid vs. total precipitation; Sch duration of the snow cover in days; $\bar{S}_{max}$ annual means of maximum snow height; t temperature in °C; absolute values of solid F and total annual precipitation N in mm (Kuhn 1994 according to data by Lauscher)

### 1.3.2 Accumulation vs. Snowfall

There are several indicators of the amount of snow on the ground. Snowfall per se is best expressed as water equivalent, i.e. the height of its melt water, in mm (or kg per $m^2$). Fresh snow typically has a density of 100 kg $m^{-3}$ which means that a snow cover of 10 cm has a water equivalent of about 10 mm. With a typical mean density of the winter snow cover of 300–400 kg $m^{-3}$, snow packs of 3 m height may be expected at elevations above the tree line; snow packs of 6 m have occasionally been observed in the Austrian Alps.

Once the snow has fallen it is generally redistributed by wind drift, also by avalanches. The amount finally lying on the ground is called accumulation and is expressed in terms of water equivalent. Wind takes snow away from ridges and crests and deposits it in concave terrain where the total accumulation may be twice as high as the original snow fall.

On a small scale, the redistribution of snow may create long lasting covers that profoundly influence vegetation. There are cornices on crests that survive into summer, and creeks that collect and preserve snow (called Schneetälchen in German literature), and there are, on the other hand, crests that are blown free of snow and may suffer much lower soil temperatures than their snow covered, insulated surroundings.

The duration of snow cover at a given elevation has been averaged for all Austrian stations and compared to the mean winter temperature at these stations. With the proper choice of scales, the two curves match very closely in Fig. 1.6.

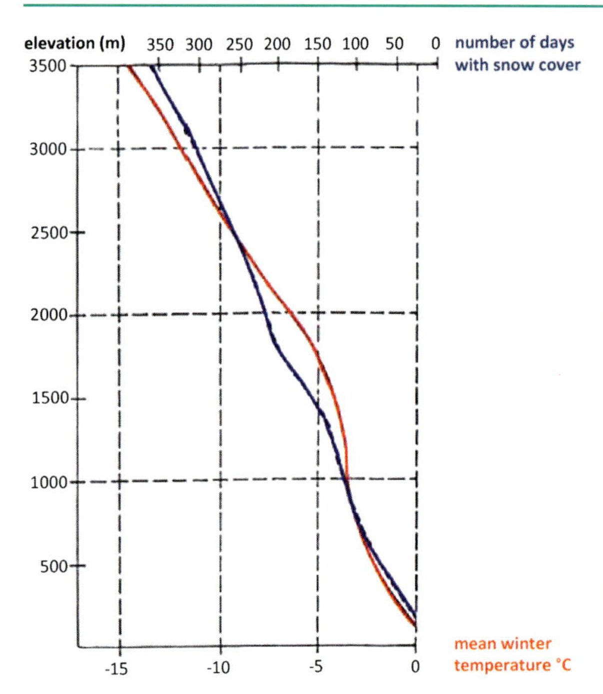


**Fig. 1.6** Duration of snow cover in days and mean winter temperature vs. elevation, from Austrian stations

The seasonal development of the snow cover is determined by both accumulation and ablation. The two graphs in Fig. 1.7 show modelled snow water equivalent vs. elevation in monthly profiles from October to September for the relatively dry basin of the Rofen Valley and for the relatively humid Verwall Valley. In May at 3,050 m elevation, the snow cover in Rofen Valley is about 700 mm w.e., that in Verwall Valley is 1,400 mm w.e. From October through March both valleys have snow covers with low vertical gradients. These are determined mostly by accumulation which in turn increases above the rain/snow limit (compare the fraction of solid precipitation in Fig. 1.5), and in each month these gradients increase with elevation. In April and May, ablation starts at low elevation and diminishes the snow water equivalent, while accumulation keeps adding to the snow cover at high elevation. Thus, strong vertical gradients of snow water equivalent appear in both regions.

These two examples are mean basin values, to which local deviations to either side are caused by exposure and topography. They do, however, clearly show the natural differences in snow line elevation that exist between the dry central regions and the wet margins: Rofen Valley 3,150 m, Verwall Valley 2,550 m. In May the snowline in the early vegetation period is at 2,350 m elevation in Rofen Valley, while it is at only 1,950 m in Verwall Valley during the same period.

### 1.3.3 Energy and Mass Balance of the Snow Pack

The development of the snow pack is influenced by surface temperature, and hence by air temperature in two ways: temperature determines the transition from rain to snowfall; and it determines surface melting and sublimation via the energy balance of the snow (e.g. Kuhn 2008).

In the Eastern Alps, melting is the predominant form of snow ablation, it consumes an amount of 0.33 MJ $kg^{-1}$, subsequent evaporation requires 2.5 MJ $kg^{-1}$. The energy balance of a melting snow cover is

$$S\downarrow + S\uparrow + L\downarrow + L\uparrow + H + C + LS + LM = 0$$

where S is incoming and reflected solar radiation, L long wave (infrared) radiation, H turbulent sensible heat transfer, C is heat conduction in the snow, LS heat required for sublimation and LM for melting. LS includes all water that is first melted and then evaporated while LM represents the melt water that actually runs off. The fluxes S$\downarrow$ and LM are not restricted to the surface but may turn over energy within the snow cover. This is due to solar radiation penetrating into the snow pack and due to melt water percolating and delivering heat to the colder interior.

Snow is a very efficient thermal insulator, which implies that strong vertical temperature gradients may exist in the snow pack. In Fig. 1.8 snow in contact with the soil remains close to 0°C all winter, while a thin top layer may cool down to about –20°C. Associated with these temperature gradients there are gradients of vapour pressure in the snow pack (saturation vapour pressure decreases by a factor of about 2 for each decrease in temperature by 10°C) which in turn lead to a transport of water vapour by diffusion in the pore space, sublimating mass from the lowest snow layers and depositing it above as so called depth hoar. This effectively changes the structure and stability of the lowest snow layers, enabling gas exchange between soil and snow.

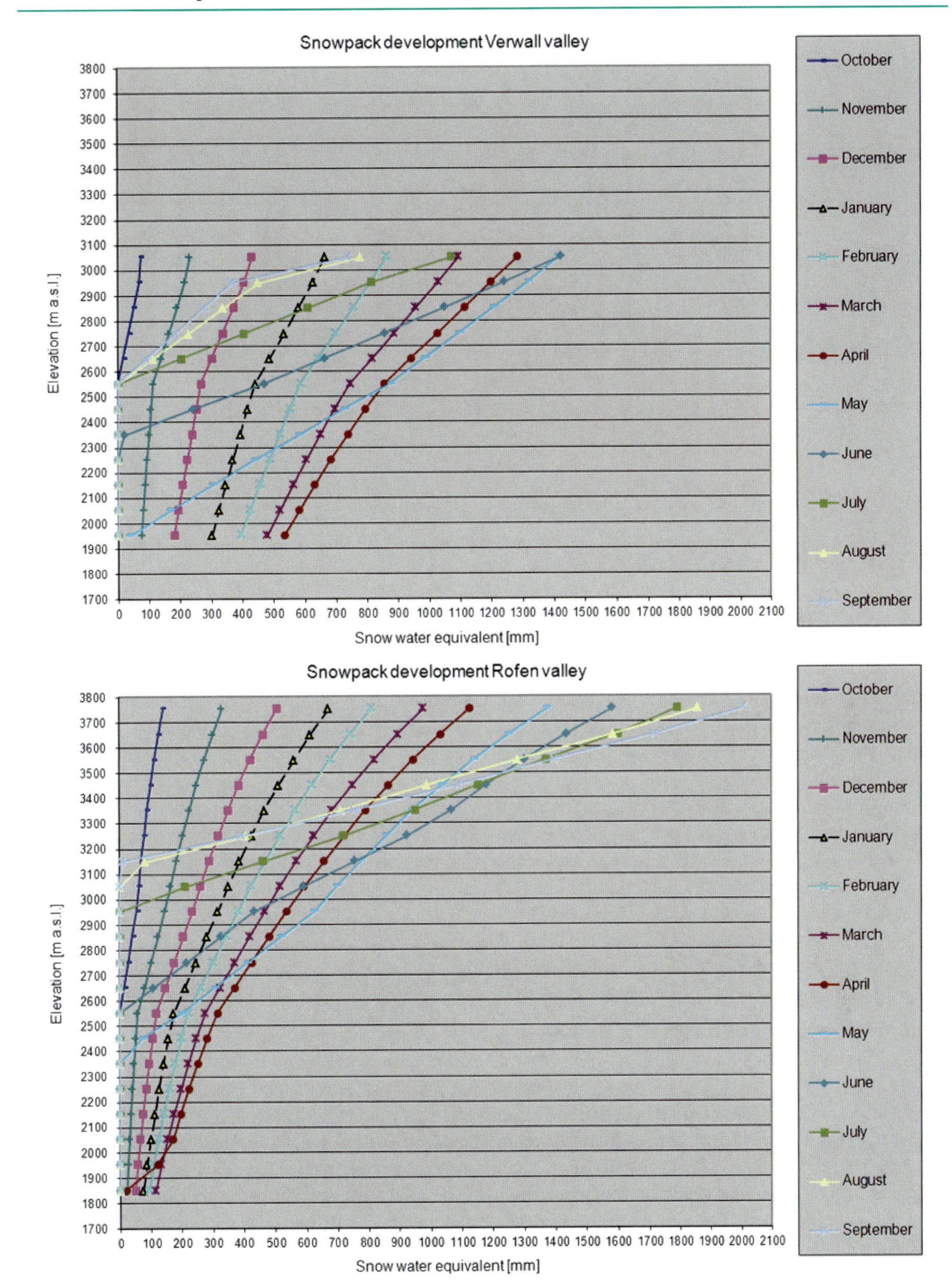


**Fig. 1.7** The seasonal development of the snow cover vs. elevation modelled for two basins. Values are given in mm water equivalent. *Top*: the basin of Verwall (47.1°N, 10.2°E) with abundant precipitation, bottom: the relatively dry basin of Rofen (46.8°N, 10.8°E). Compare the peak values of snow cover at 3,000 m, and the snow line elevation in September

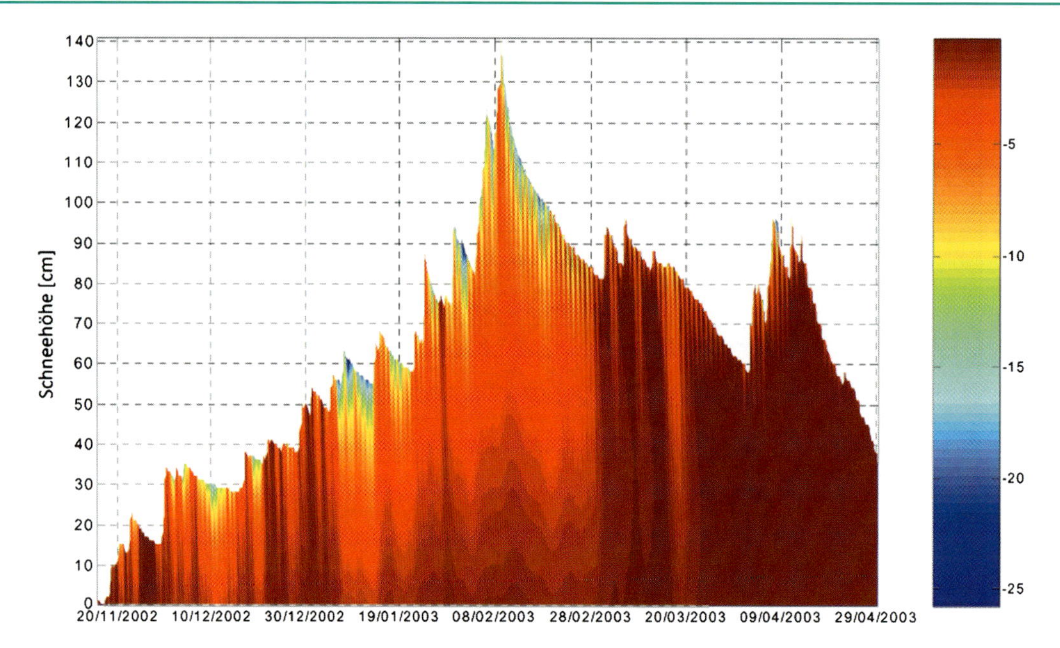


**Fig. 1.8** The distribution of temperature in the snow pack at about 2,000 m elevation in the central Alps. Note the downward penetration of the daily temperature cycle and the associated phase lag. The scale on the right is in °C. From Leichtfried (2005)

### 1.3.4 Percolation of Rain and Melt Water Through the Snow Pack

With occasional rains in winter and with the daily melt cycle in spring, liquid water penetrates into the snow, where it soon refreezes during the accumulation period, forming ice lenses or horizontal layers that may impede vertical gas diffusion. In spring the melt water front will progressively penetrate deeper and will finally reach the soil within a few days. The percolation of melt water was modelled in Fig. 1.9 which is identical to the snow pack in Fig. 1.8.

### 1.3.5 Microscale Contrasts in a Broken Snow Cover

At elevations of 2,000–3,000 m global solar radiation may reach peak values of 500 W $m^{-2}$ in early spring and 1,000 $Wm^{-2}$ in June. This is usually more than sufficient to melt the snow which then has a surface temperature of 0°C. Dark, low albedo objects protruding from the snow, like rocks, trees or patches of bare ground, may then absorb so much solar radiation that in spite of their cold surroundings they may reach exceptionally high temperatures at a small local scale.

An example is given in Fig. 1.10 which displays the record of surface temperature of a rock of 2 m diameter extending half a meter above the snow. With a low albedo of 20% (compared to 70% of the snow), and lacking any energy loss by evaporation, it reached a surface temperature in excess of 37°C in the afternoon. Similar values have been observed on the lower parts of trees. In both examples heat is stored into the night and is transferred to the surrounding snow, soon creating bare patches around the trees or the rocks.

## 1.4 Acid Deposition at High Altitude

Aerosol particles and ions reach the alpine regions by dry deposition, such as dust from local sources or from long distances like the Saharan desert, or by wet deposition in rain and snow. Even if the source strength at far away places remained constant, the deposition in the Alps would always be controlled by both advection and convection, i.e. by synoptic conditions and by local atmospheric stability.

For the Alps, sources of air pollution are in the NW and in the industrial areas of northern Italy. In the typical series of synoptic events that was described in Sect. 1.2.1, it is the Eastern Alps that receive more wet deposition than the Western end of the Alps (Nickus et al. 1998).

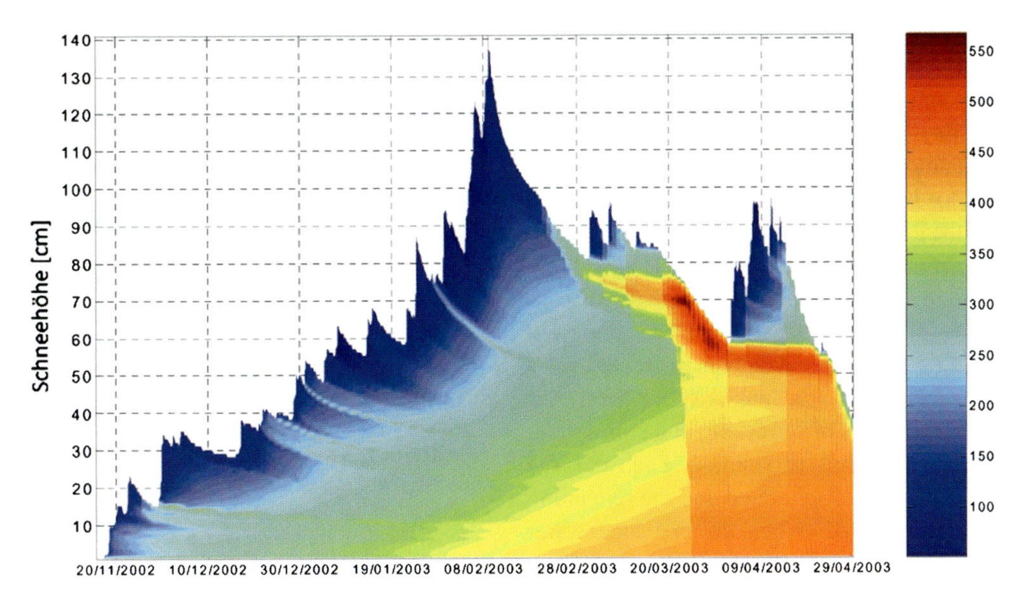


**Fig. 1.9** Liquid water content of the snow pack, given as parts per thousand of the pore space on the right-hand scale. The site is identical to that in Fig. 1.8. From Leichtfried (2005)

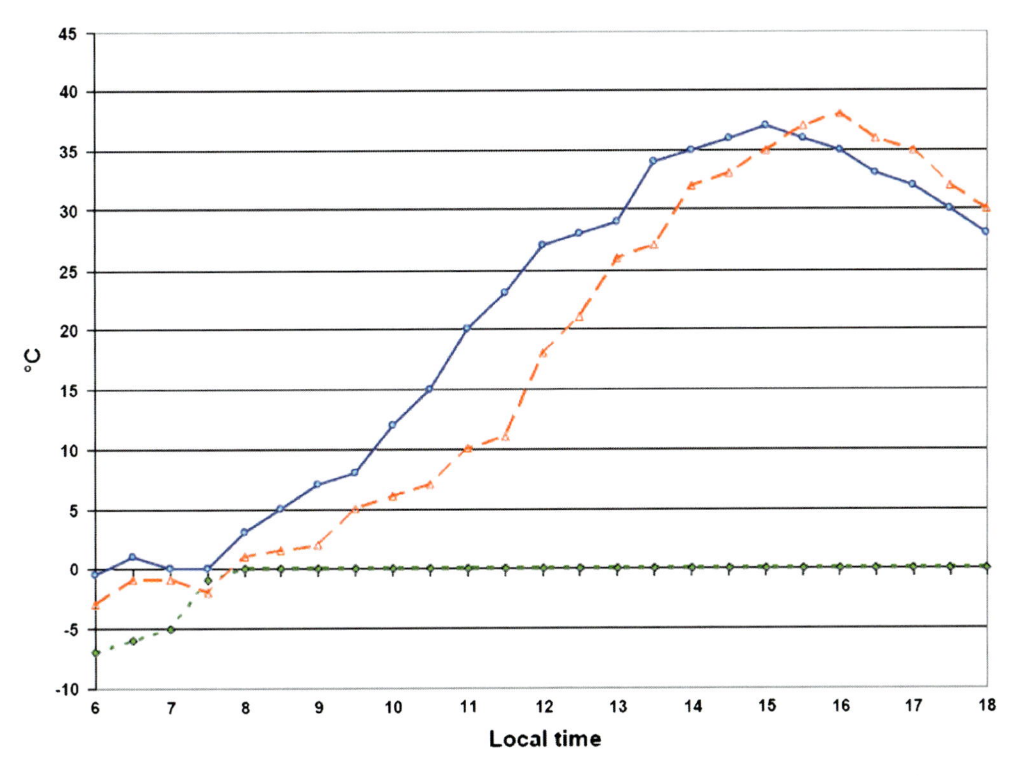


**Fig. 1.10** Records of surface temperature of a rock of 2 m diameter, 0.5 m height, surrounded by snow on a clear summer day at 3,030 m. *Green curve*: snow temperature, blue south face of the rock, red west face

The seasonal development of ion concentration (μequivalents per L) and of total deposition (μequivalents per m$^2$) in the high alpine snow pack is generally characterized by low values in winter due to both low source strength with large areas of Europe being snow-covered, and strong atmospheric stability in the temperature inversions of alpine valleys (Kuhn et al. 1998). With the disappearance of the low land snow cover and the onset of convection in alpine regions in May and June, concentrations of sulfate, nitrate and ammonium at high altitude more than double and total loads increase by more than a factor of five considering the increase in concentration and the simultaneous increase in monthly precipitation (Fig. 1.4).

The physical and chemical processes that redistribute ions in the seasonal snow pack have been reviewed by Kuhn (2001). In the typical daily melt-freeze cycles shown in Figs. 1.8 and 1.9, ions are to a great part excluded in the process of refreezing and concentrate in the intergranular films of solution. Each day the solute becomes more concentrated and the remaining snow grains are purified. When the first melt water penetrates to the ground and runs off, it carries ions at a concentration that may be fivefold the concentration in the unmelted snow pack (Johannessen and Henriksen 1978). In the recent warming, substances of considerable age are being released into the modern alpine ecosystem, a problem that has been reported by Nickus et al. (2010).

## References

Baumgartner A, Reichel E (1983) Der Wasserhaushalt der Alpen. Oldenbourg, München, pp 343

Efthymiadis D, Jones PD, Briffa KR, Auer I, Böhm R, Schöner W, Frei C, Schmidli J (2006) Construction of a 10-min-gridded precipitation data set for the Greater Alpine Region for 1800–2003. J Geophys Res 290(111):1–22. D01105, doi:10.1029/2005JD006120

Fliri F (1975) Das Klima der Alpen im Raum von Tirol. Universitätsverlag Wagner, Innsbruck, pp 179–274

Frei C, Schär C (1998) A precipitation climatology of the Alps from high-resolution rain-gauge observations. Int J Climatol 18:873–900

Gattermayr W (2010) Hydrologische Übersicht August 2010. Hydrographischer Dienst Tirol, S 21

Johannessen M, Henriksen A (1978) Chemistry of snow melt water: changes in concentration during melting. Water Resour Res 14:615–619

Kuhn M (1994) Schnee und Eis im Wasserkreislauf Österreichs. Österreichische Wasser- und Abfallwirtschaft (Wien) 46:76–83

Kuhn M (2001) The nutrient cycle through snow and ice, a review. Aquat Sci 63:150–167

Kuhn M (2008) The climate of snow and ice as boundary condition for microbial life. In: Margesin R (ed) Psychrophiles: from biodiversity to biotechnology. Springer, Berlin, pp 3–15

Kuhn M (2010) The formation and dynamics of glaciers. In: Pellikka P, Rees WG (eds) Remote sensing of glaciers. CRC Press Balkema, Leiden, pp 21–39

Kuhn M, Haslhofer J, Nickus U, Schellander H (1998) Seasonal development of ion concentration in a high alpine snow pack. Atmos Environ 32:4041–4051

Leichtfried A (2005) Schneedeckenmodellierung, Kühtai 2002/2003, Sensitivitätsstudien. Diploma thesis, University of Innsbruck, S 122

Nickus U, Kuhn M, Novo A, Rossi GC (1998) Major element chemistry in alpine snow along a North-South transect in the eastern Alps. Atmos Environ 32:4053–4060

Nickus U, Bishop K, Erlandsson M, Evans CD, Forsius M, Laudon H, Livingstone DM, Monteith D, Thies H (2010) Direct impacts of climate change on freshwater ecosystems. In: Kernan M, Battarbee RW, Moss B (eds) Climate change impacts on freshwater impacts. Wiley-Blackwell, London, pp 38–64

# 2 Solar Radiation of the High Alps

Mario Blumthaler

## 2.1 Introduction

Solar radiation at the Earth's surface varies by orders of magnitude that depend on actual local considerations. Therefore it is crucial to understand the influences of the various factors which determine the actual levels of solar radiation in order to estimate the effects on the whole biosphere. This is especially important for plants in the High Alps, as the levels of solar radiation are highest there due to a combination of several influencing factors.

While the spectrum of solar radiation gives the intensity at each individual wavelength, it is often sufficient to investigate the integral over a certain wavelength range. This is done e.g. for the integral between 280 and 3,000 nm, which is called 'total radiation', and also for the ranges of UVA (315–400 nm), UVB (280–315 nm) and UVC (100–280 nm). For all these integrations the weight of each individual wavelength is the same. This is in contrast to biological reactions, where radiation at different wavelengths usually has a different efficiency in triggering the reaction. Hence, in order to estimate the effects of radiation on the biosphere, one has to know the 'action spectrum' for each individual biological reaction under consideration. The action spectrum is usually a relative quantity in dependence on wavelength and normalised to unity at the wavelength with the maximum of the effect or at a standardised wavelength. The most common biological reaction in the UV wavelength range is the human erythema (McKinlay and Diffey 1987), which is often taken as a general measure for biological effects in the UV. Especially for reactions of UV on plants, a generalized action spectrum was determined (Caldwell et al. 1986), which is not very different from the erythema action spectrum. Therefore many of the findings relating to the variability of erythemally weighted irradiance can be interpreted in the same way for the generalized plant action spectrum. Furthermore, for plants the photosynthetically active radiation (PAR) is of high significance. This is the unweighted integral of radiation in the wavelength range of 400–700 nm. Due to the similarity to the wavelength range of visible radiation (defined by the spectral sensitivity of the human eye), the relations for visible radiation can also be interpreted to be relevant for PAR, and often the total radiation can also be taken as a good approximation for PAR.

The intensity of solar radiation can be measured with different geometries of the detector, corresponding to different applications. The most common type of measurement refers to a horizontal surface, which is called irradiance (intensity per unit area). The intensity of radiation falling on the horizontal surface at an angle to the vertical of the surface (zenith on the sky) is reduced according to the cosine of the zenith angle ('cosine law') due to the change of the projected area. This situation is valid for most types of surfaces, especially for human skin and for plants. For molecules in the atmosphere, which can be dissociated photochemically by radiation, the geometry is different, and in this case the 'actinic flux' (also called 'spherical irradiance') is the relevant quantity.

M. Blumthaler (✉)
Division for Biomedical Physics, Medical University of Innsbruck, Innsbruck, Austria
e-mail: Mario.Blumthaler@i-med.ac.at

C. Lütz (ed.), *Plants in Alpine Regions*,
DOI 10.1007/978-3-7091-0136-0_2, © Springer-Verlag/Wien 2012

Radiation from the whole sphere (4 pi) is received by the object or by the detector without any cosine-weighting, and radiation from all directions from the upper and lower hemisphere has the same weight.

## 2.2 The Sun

The solar spectrum outside of the Earth's atmosphere is close to the spectrum of a black body with a temperature of about 5,800 K. The spectral distribution follows the continuous spectral distribution of a thermal source with maximum irradiance at about 450 nm. However, the spectrum is significantly modified by absorption lines due to atomic absorption in the outer layers of the sun, the so-called Fraunhofer lines. These lines have a very high spectral structure far below 0.1 nm, so that the structure of spectral measurements depends very much on the spectral bandwidth of the instrument.

The total energy as the integral over all wavelengths from the ultraviolet through the visible to the infrared, which is received at the top of the Earth's atmosphere on a surface perpendicular to the radiation, corresponds to 1,367 $Wm^{-2}$. This value holds for the average distance between sun and Earth, it varies by $\pm 3.2\%$ during the year due to the elliptic path of the Earth around the sun, with the maximum value at beginning of January (minimal sun-earth distance) and the minimum value at beginning of July. In the UV wavelength range (100–400 nm) about 7% of the total energy is emitted, in the visible range (400–750 nm) about 46%, and in the infrared range (>750 nm) about 47%. However, the extra-terrestrial spectrum is significantly modified by the processes in the atmosphere, as discussed in detail in the next section, leading especially to a smaller contribution of the UV range.

The intensity of the extraterrestrial solar radiation has some slight temporal variability on different time scales: 27-day solar rotation, 11-year cycle of sunspot activity and occasional solar flares. Mostly the UVC range is affected by up to a few percent (Lean 1987), whereas at longer wavelengths this variability can be neglected. Only an indirect effect may affect UVB levels at the Earth's surface due to effects on the strength of photochemical ozone production in the higher altitudes of the atmosphere.

## 2.3 The Atmosphere

The Earth's atmosphere mainly consists of nitrogen (78%) and oxygen (21%). Many more gases are present but only in very small amounts; however, they can still have a significant influence on the radiative properties (i.e. ozone, see discussion later). The molecules in the atmosphere scatter the light coming from the sun, which means that the direction of the propagation of photons is changed, but the wavelength (and the energy) is – for the purpose discussed here – unchanged. The probability of scattering increases with increasing density of the molecules (i.e. with increasing air pressure). The direction of photons scattered of molecules is mainly forward and backward (with equal amount) and less perpendicular to the propagation. Usually photons are scattered several times in the atmosphere ('multiple scattering'), which leads to a certain amount of 'diffuse' light. Of course, due to backscattering in the atmosphere, photons are also reflected back into space, and thus the total solar energy measured at the Earth's surface is less than the extraterrestrial one.

The probability for scattering of molecules (the so-called 'Rayleigh scattering') strongly increases with decreasing wavelength ($\lambda$) and can be approximated by a function proportional to $\lambda^{-4}$ (Bodhaine et al. 1999). This means that the short blue wavelengths are much more efficiently scattered out of the direct beam of solar radiation than the red ones, and consequently blue dominates in the scattered light ('diffuse' radiation). Therefore the sky looks blue, if the air is clear and clean.

Besides scattering, molecules in the atmosphere can also absorb photons with specific wavelengths. In this case, the energy of the photons is transformed due to chemical reactions and finally transformed into heat, thus reducing the intensity of the radiation. As an example, ozone molecules in the atmosphere can absorb photons very efficiently in the UVC range and to a smaller extent also in the UVB range. This absorption by ozone is responsible for the fact that almost no radiation in the UVC can be observed at the Earth's surface, thus ozone acts as a protection shield against biologically very harmful radiation at these wavelengths.

In the Earth's atmosphere we also find 'aerosols', which are small solid particles of different sizes, shapes and chemical composition, which again can scatter and absorb solar radiation. The size of the aerosols is usually in the range of 0.05–10 μm. Being much larger than molecules, their scattering processes are different ('Mie-scattering'). In this case, the scattering probability depends only slightly on wavelength, as it is proportional to about $\lambda^{-1.3}$ (Angström 1964). Thus the scattered diffuse light due to aerosols is whiter (in contrast to the blue light scattered of molecules), producing a 'milky' sky in the event of large amounts of aerosols. Also, the direction of the scattered photons is different for aerosols compared to molecules; Mie-scattering has a very strong forward peak, so that the sky looks especially 'milky' in the surrounding of the sun. The absorption by aerosols is usually small, only a high concentration of soot (i.e. from forest fires) may lead to significant absorption.

Both together, scattering and absorption, are called 'extinction' and lead to a reduction of the intensity of the direct beam of solar radiation, and due to scattering to an increase in diffuse radiation. The extinction can be described with the 'Beer-Lambert-Law', where for the application relevant here, the scattering by molecules, absorption by ozone and extinction by aerosols are considered:

$$I = I_0 * \exp(-\tau_R(\lambda) * m_R - \tau_O(\lambda) * m_O - \tau_A(\lambda) * m_A)$$

'I' is the intensity at the Earth's surface, '$I_0$' is the intensity of the incident radiation at the top of the atmosphere; 'τ' is the dimensionless 'vertical optical depth', which depends on the amount of molecules or aerosols (counted from the altitude of the observer to the top of the atmosphere) and which strongly depends on wavelength (λ); 'm' is the 'air mass', which characterises the pathlength of the photons through the atmosphere. If the sun is in the zenith, then $m = 1$, otherwise – for the assumption of a plane-parallel atmosphere – $m = 1/\cos(sza)$, where sza is the 'solar zenith angle', which corresponds to (90° – solar elevation). For sza larger than about 85°, a correction of the atmosphere's spherical shape is necessary. The index 'R' marks the quantities for Rayleigh scattering of the molecules, 'O' for ozone absorption and 'A' for aerosol extinction.

As a consequence of the scattering processes in the atmosphere, the radiation field at the Earth's surface can be separated into the direct solar beam and the diffuse sky radiance. The angular distribution of the diffuse radiance is not homogeneous across the sky; it has a maximum around the sun, a minimum around 90° to the sun in the plane through the sun and the zenith, and especially for longer wavelengths it increases significantly towards the horizon (Blumthaler et al. 1996a). Both components together, direct solar beam and diffuse radiation, measured on a horizontal plane, are called 'global irradiance'. The share of diffuse radiation in global radiation is very variable (Blumthaler et al. 1994a). If the sun is completely hidden by clouds, all radiation is diffuse. Under cloudless conditions, the following dependencies can be summarised for the share of diffuse radiation in global radiation: it depends strongly (a) on wavelength with the highest values at the shortest wavelengths in the UVB, (b) on solar elevation with higher values at lower solar elevation (due to the longer pathlength) and (c) on the amount of aerosols with higher values at higher amounts. Furthermore, the share of diffuse radiation in global radiation is smaller at high altitude above sea level, because at higher altitudes the amount of scattering molecules (and aerosols) is lower due to lower air pressure, and thus scattering is less significant. As an example of these relations, measurements taken on a cloudless day at a high Alpine station (Blumthaler and Ambach 1991) indicated that for total irradiance and solar elevation above about 30° only about 10% of global irradiance is diffuse. These values can also be taken for PAR. In contrast, in the UVB-range generally about 50% of global irradiance is diffuse, and this value goes up to 90% and even 100% at very low solar elevations.

The separation of the radiation field into direct and diffuse components must also be considered if quantification of radiation on tilted surfaces is investigated. The contribution of the direct beam of solar radiation on a tilted surface can be calculated if solar elevation and horizontal angle of the position of the sun relative to the orientation of the tilted surface is known. The cosine of the angle between the sun and the vertical of the plane has to be considered in the calculation of the intensity of the direct beam (cosine law). The contribution of the diffuse radiation can be calculated straightforward under the simplified assumption that the diffuse radiance of the sky is homogeneous (Schauberger 1990). Otherwise this requires an extended calculation using a sophisticated radiative transfer model. As a tilted surface receives diffuse radiation from a part of the sky and also from a part of the ground, it is essential for the determination of the diffuse radiation falling on a tilted surface to know how much of the radiation is reflected from the ground (the so-called 'albedo'). This is of particular importance if the ground is covered by snow. The detailed discussion of the albedo follows in the next section.

An example of quantitative results was found in measurements on a vertical surface facing southward at the High Alpine Research Station Jungfraujoch in Switzerland (3,576 m). There the daily total erythemally weighted irradiance varied by a factor of 0.4–1.6 relative to a horizontal surface, depending on the season of the year (Blumthaler et al. 1996b). In winter time, when solar elevation is low, the vertical surface receives much more solar radiation compared to the horizontal one, whereas in summertime with high solar elevation the relation is the opposite. At a measurement campaign in Izana (2,376 m, Tenerife, Spain) the large range of variation of the irradiance ratio for vertical to horizontal surfaces was investigated as a function of wavelength and time of the day (Webb et al. 1999). At 500 nm an up to sixfold increase was observed when the solar elevation was low, whereas the increase was only about 20% at 300 nm under the same conditions. This is a consequence of the different relation between direct and diffuse irradiance, as discussed previously. Furthermore, when comparing the measurements from Izana and Jungfraujoch, the difference in ground albedo with snow at Jungfraujoch and snow-free terrain at Izania is also significant.

## 2.4 Variability of Solar Radiation Under Cloudless Conditions

The most important parameter determining the level of solar radiation under cloudless conditions is solar elevation. Furthermore, altitude above sea level, albedo (reflectivity) of the ground and amount and type of aerosols have a significant influence, and in addition the total amount of atmospheric ozone is of specific importance for the level of UVB radiation. Parameters that have only a minor effect on the level of solar radiation especially at higher altitudes, are the temperature profile in the atmosphere, the vertical distribution of the aerosols and the vertical profile of ozone. In the following sections, the effects of the main parameters are discussed in detail.

### 2.4.1 Effect of Solar Elevation

Solar elevation changes during the day, and the maximum value of solar elevation reached at solar noon depends on the season of the year and on the latitude of the observation site. When latitude, longitude, date and time are given, then solar elevation and azimuthal position of the sun can be calculated exactly. Of course, the higher the solar elevation the higher the level of solar radiation. However, this relation depends strongly on wavelength. UVB radiation is much more strongly absorbed at low solar elevations compared to radiation at longer wavelengths due to the longer pathlength within the ozone layer. Therefore, the diurnal course of UVB radiation is steeper compared to total radiation or PAR (Fig. 2.1).

For the same reason, the ratio between maximum daily values in summer relative to winter is also much higher for UVB radiation. Measurements at the High Alpine Research Station Jungfraujoch in Switzerland (3,576 m) have shown that the ratio of maximum daily totals in summer relative to winter is 18:1 for erythemally weighted UV irradiance and only 5:1 for total radiation and PAR (Blumthaler 1993).

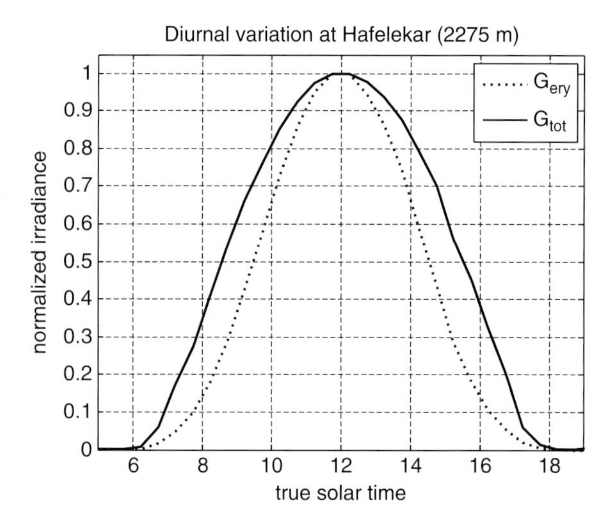


**Fig. 2.1** Diurnal variation of erythemally weighted global irradiance ($G_{ery}$) and total global irradiance ($G_{tot}$) on a cloudless day at Hafelekar near Innsbruck (2,275 m above sea level). Both quantities are normalized to their maximum values at local noon

At high latitudes solar elevation is relatively low even in summertime. Therefore, the intensity of solar radiation is also relatively low, but the daily sum is significantly increased due to the longer length of the day.

### 2.4.2 Effect of Altitude

Generally global solar irradiance increases with increasing altitude above sea level. This increase is mainly due to a pronounced increase of direct irradiance, whereas for altitudes below about 3,000 m the diffuse irradiance is more or less constant (Blumthaler et al. 1997). The attenuation of direct irradiance due to extinction (following the Beer-Lambert-Law) becomes smaller, when the amount of scattering and absorbing molecules and aerosols becomes smaller. Furthermore, as the Beer-Lambert-Law depends strongly on wavelength, the increase of irradiance with altitude also depends strongly on wavelength. This increase with altitude is quantified with the 'altitude effect', which is defined as the increase of global irradiance for an increase in altitude of 1,000 m, relative to the lower site. Measurements of spectral irradiance at stations with different altitudes (Blumthaler et al. 1994b) show the dependence of the altitude effect on wavelength in the UV range (Fig. 2.2).

The strong increase towards the shorter wavelengths is a consequence of the smaller amount of atmospheric ozone at higher altitudes, although only about 9% of the total amount of atmospheric ozone is distributed in the troposphere. Additionally, the scattering on molecules (Rayleigh scattering) also increases strongly with decreasing wavelength. The figure also shows the relatively large range of variability of the altitude effect. This is a consequence of the strongly varying amount of aerosols and tropospheric ozone in the layer between the high and low altitude measurement stations. Under cloudless conditions, measurements in the Alps showed an altitude effect for UVB irradiance of 15–25% and of 10–15% for total irradiance (Blumthaler et al. 1992). In contrast, in the Andean mountains the altitude effect for UVB irradiance was only about 9% (Piazena 1996) or about 7% (Zarrati et al. 2003), because there the amount of aerosols and tropospheric ozone was very small. Furthermore, the altitude effect is additionally enhanced especially at shorter wavelengths if the ground is covered by snow at the mountain station and the ground is free of snow at the station in the valley (Gröbner et al. 2000).

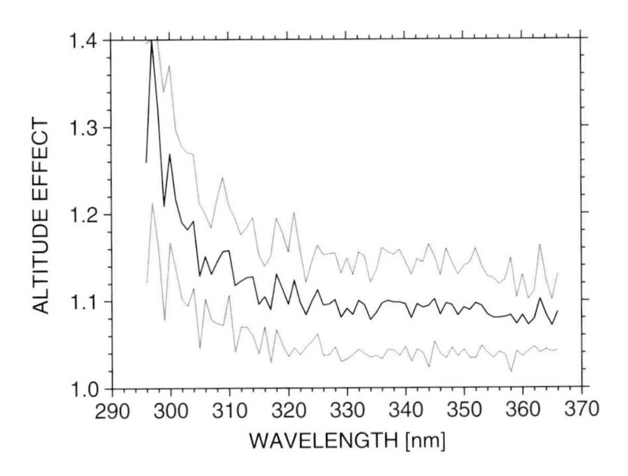


**Fig. 2.2** Increase of spectral irradiance for an increase in altitude of 1,000 m, average (*solid line*) and range of variation (*dashed lines*) (From Blumthaler et al. 1994b)

As an example of the combined effect of solar elevation and altitude as discussed in 1.4.1 and 1.4.2, Fig. 2.3 compares the results of measurements of erythemally weighted UV irradiance ($G_{ery}$) and of total global irradiance ($G_{tot}$) at the High Alpine Research Station Jungfraujoch in Switzerland (3,576 m) and in Innsbruck (577 m).

The envelope of the seasonal course marks the cloud-free days with maximum values, whereas the other days are affected by clouds reducing the daily sum. Comparing the seasonal maximum values at Jungfraujoch and in Innsbruck shows the altitude effect, which is more pronounced in the shorter wavelength range. Comparing the seasonal course of total and erythemally weighted irradiance shows the different effects of solar elevation on the different wavelength ranges.

### 2.4.3 Effect of Albedo

The albedo of a surface is defined as the ratio of reflected irradiance to incoming irradiance. If the albedo is high, then the reflected irradiance is high, and consequently the diffuse irradiance is increased due to multiple reflections between the ground and the atmosphere. As a cloud layer will enhance these multiple reflections, the increase of diffuse irradiance due to albedo is highest under overcast conditions.

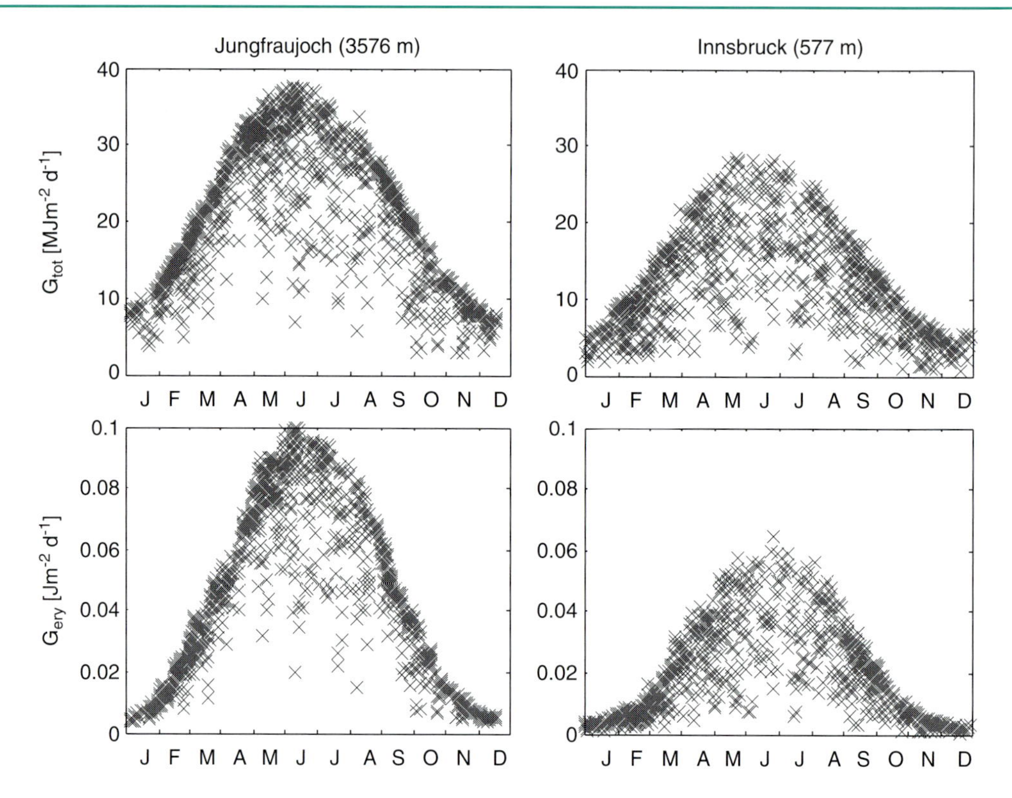


**Fig. 2.3** Seasonal course of daily sums of total ($G_{tot}$) and erythemally weighted global irradiance ($G_{ery}$) at the High Alpine Research Station Jungfraujoch (Switzerland) and in Innsbruck (Austria)

The albedo depends on the type of surface, and also on wavelength for each surface. Spectral measurements of the albedo of various surfaces (Fig. 2.4) show a clear separation of albedo values: only snow-covered surfaces have a high albedo, which can exceed 90% in the case of fresh snow and be somewhat reduced if the snow is getting older and more polluted. For all other types of surfaces the albedo is relatively small and usually increases with increasing wavelength. The smallest albedo values were measured for green grassland, where the albedo was less than 1% in the UV range.

The increase of global irradiance due to a higher albedo can be quantified with an amplification factor, which gives the enhancement for a change in albedo by 10%. This factor generally increases with decreasing wavelengths, even if it is assumed that the albedo itself would be constant for all wavelengths, because the multiple reflections between atmosphere and ground are much more efficient for shorter wavelengths (Raleigh-scattering). However, in the UVB range, where ozone absorption is significant, the amplification decreases with decreasing wavelengths, because the longer pathlength of the photons due to multiple reflections will result in a more pronounced absorption by ozone. Under cloudless conditions the amplification of global irradiance due to a change of albedo by 10% is found to be about 1.03 at 300 nm, 1.035 at 320 nm and then decreasing to about 1.02 at 400 nm and 1.01 at 500 nm. As an example of the maximum effect of albedo, the enhancement of global irradiance for a change from green grassland to fresh snow is estimated to be about 30% at 320 nm and about 8% at 500 nm.

## 2.4.4 Effect of Aerosols

At higher altitude the amount of aerosols is usually relatively low, so that the effect of aerosols on global and diffuse irradiance is also relatively small. Aerosols mainly decrease the direct irradiance and increase the diffuse irradiance due to scattering. If the absorption of aerosols is high (e.g. for aerosols from biomass burning), then diffuse irradiance is less increased. Therefore, in many cases reduction by aerosols is

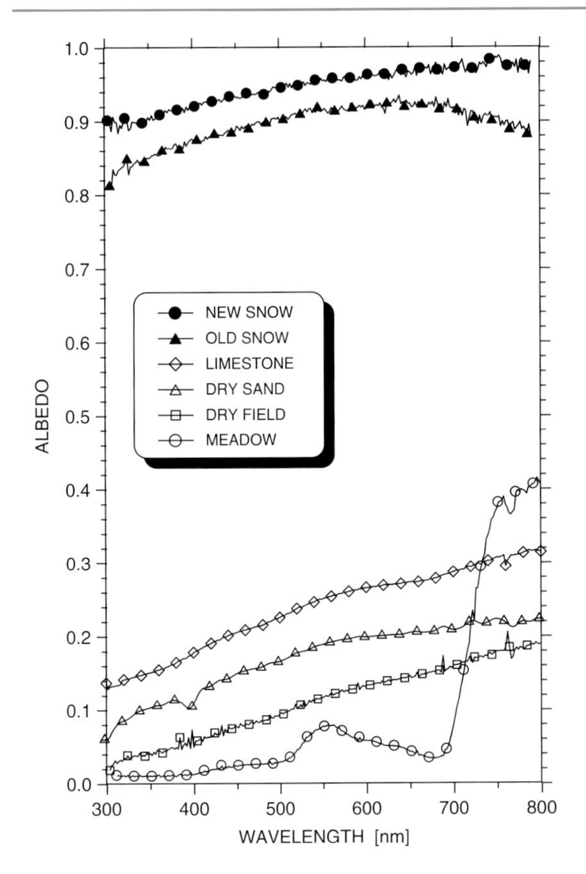


Fig. 2.4 Albedo of various types of surfaces as a function of wavelength (From Blumthaler 2007)

moderate (around 10–20%, significantly less at higher altitudes) for global irradiance, but occasionally it can be 30% or higher (Kylling et al. 1998). A special situation can occur if Saharan dust is transported far north and even up to the Alps. This type of aerosols is usually found at an altitude above the mountains and thus also affects the radiation in the High Alps, predominately scattering and only marginally absorbing, thus significantly increasing diffuse irradiance and only slightly reducing global irradiance.

### 2.4.5 Effect of Ozone

Ozone in the atmosphere mainly affects the UVB wavelength range due to its spectral absorption cross-section. For wavelengths higher than 330 nm its effect can be neglected, for shorter wavelengths it strongly increases with decreasing wavelength. This absorption by ozone in the atmosphere is the reason why almost no radiation with wavelengths below 285 nm can be measured at the earth's surface, although it is emitted by the sun.

At mid and high latitudes the total amount of ozone in the atmosphere varies strongly with time. Maximum values occur in springtime and minimum ones in autumn. Especially in springtime large day-to-day variability can occur with variations even larger than the seasonal ones, which results in significant variations of UVB radiation. To quantify the effect of ozone variations on UVB radiation, the so-called 'radiation amplification factor' is used, which gives the percentage increase of global irradiance for a decrease of ozone by 1%. This form is a simplification, which is in agreement with radiative transfer model calculations for ozone changes up to a few percent. The radiation amplification factor depends strongly on wavelength due to the wavelength dependency of the ozone absorption cross-section. For erythemally weighted irradiance the factor is about 1.1, for the generalized DNA damage it is 1.9, and for the generalized plant action spectrum it is about 1.8 (Madronich et al. 1998).

## 2.5 Variability of Solar Radiation Due to Clouds

In general clouds will of course attenuate solar radiation at the earth's surface, when a significant part of solar radiation is reflected back to space because of the high albedo of the top of clouds. Within a cloud, solar radiation is mainly scattered and usually only marginally absorbed. This scattering process within a cloud is almost independent of wavelength. The bottom of the cloud, as seen from the earth's surface, looks darker when the optical density of the cloud becomes higher, which depends on the density of the water droplets in the cloud. Very high clouds (Cirrus clouds, above about 7 km) usually consist of ice particles and always look relatively bright.

The effect of clouds on solar radiation at the earth's surface varies over a very large range, as does the density of the clouds and their distribution across the sky. In addition, it is of high significance whether clouds cover the sun itself or only the sky beside the sun. Under completely overcast conditions a reduction of global irradiance down to about 20–30% of the clear sky value as a rough average can be observed at low altitude (Josefsson and Landelius 2000), whereas at high altitude this reduction is somewhat less, the level being about 40–50% (Blumthaler et al. 1994a). This is caused by an average smaller thickness of clouds at higher altitudes. Although the scattering process itself within a cloud is almost independent of wavelength, the final attenuation of solar radiation becomes wavelength-dependent (Kylling et al. 1997). Solar UV radiation is attenuated up to 40% less than total radiation (Blumthaler et al. 1994a), which is mainly caused by the higher share of diffuse radiation at shorter wavelengths due to Raleigh-scattering.

However, under certain conditions clouds can also effect radiation enhancement at the earth's surface. If the sun itself is not covered by clouds, but big cumulus-type clouds are near to the sun, then reflections from the sides of the clouds may occur, which will enhance global irradiance at the surface. The degree of enhancement depends strongly on the local conditions, but for short time intervals an enhancement of total global irradiance of more than 20% can be observed (Cede et al. 2002). The enhancement is strongest for total radiation and about half the degree for UV radiation, as in the UV wavelength range the direct component (which is reflected from the sides of the clouds) contributes less to the global irradiance than in the total wavelength range.

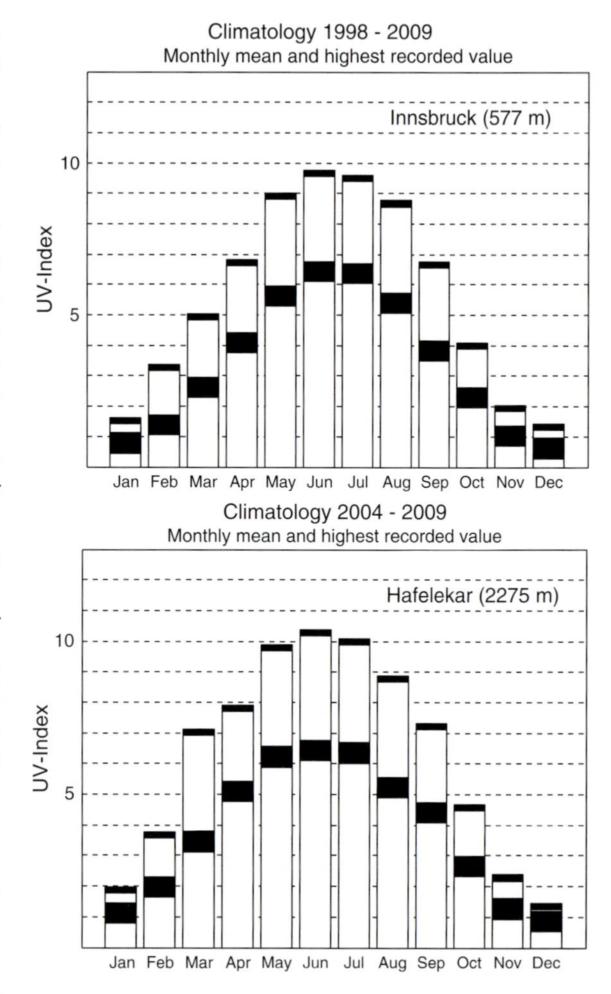


**Fig. 2.5** Seasonal course of the monthly mean values of daily maximum erythemally weighted irradiance (*broad bars*) and the highest recorded value (*thin bar*) in Innsbruck, 577 m, Austria, (*top*) and at Hafelekar near Innsbruck, 2,275 m (*bottom*)

## 2.6 Climatological Aspects

In the previous sections the effects of individual factors on solar radiation were discussed; however, in reality it is always a combination of these factors that determines the actual level of solar radiation. In order to estimate the average radiation levels in the High Alps, longer time series of measurements are necessary. Although several decades of continuous measurements would be desirable to derive a complete climatology of solar radiation at a specific site, it is possible to discuss climatological aspects based on a shorter time series. For solar UV radiation the longest time series of measurements in Alpine regions are not much longer than one decade, but still they provide important characteristics. Figure 2.5 shows two examples taken from the Austrian UV monitoring network, which started to monitor the erythemally weighted UV radiation under different environmental conditions (urban/rural, low/high altitude) at several sites in Austria in 1998 and which includes 12 stations today. The raw data are collected every 10 min, and after conversion to absolute units they are published in near real time on the web site (www.uv-index.at). Following international recommendations, the erythemally weighted solar radiation is presented in units of the so-called 'UV-Index' (Global solar UV Index). This gives the erythemally weighted irradiance, expressed in W $m^{-2}$, multiplied by a scaling factor of 40, thus leading to values up to about 11 under the conditions of the Alpine environment. However, on a worldwide scale, UV index values up to 20 were observed in cities at high altitude in the Andeans.

In Fig. 2.5 the data for daily maximum values (expressed in the units of the UV index) are analysed as monthly averages (broad bars) for all available years of measurements. In addition, the thin bars indicate the maximum value of the UV index in each month, observed in any of the years of measurements. The top graph shows the results for Innsbruck (577 m), while the bottom graph shows the results for the nearby mountain station Hafelekar (2,275 m). As these data are average values, they include all weather conditions from cloudless to heavy rainfall or snowfall. It is quite surprising that the altitude effect as derived for cloudless conditions in 2.4.2 is almost invisible for the average values presented here. Thus on average the higher levels of radiation due to higher altitude can be masked by a higher average frequency of clouds at a mountain station. This might especially be the case for a station like Hafelekar, which is situated on a mountain ridge, where some convective clouds are frequently concentrated. Only the maximum values of the UV index are slightly higher at the mountain station (generally by less than 10%), which is again less than the altitude effect for cloudless conditions. The consequence of these analyses is that the average levels of radiation at higher altitudes depend very much on the local conditions of cloudiness, which can have a more significant influence on the intensity of solar radiation than the higher altitude. Furthermore, as clouds are the dominating parameter for the average values, the climatological results for erythemally weighted UV irradiance presented here can be generalized for PAR too.

Only for the months of March and April one can see significant differences between the measurement data for Innsbruck and Hafelekar as shown in Fig. 2.5. It is obvious that in these months the additional snow cover at the mountain station also leads to a significant increase for the average values presented here.

## References


Angström A (1964) The parameters of atmospheric turbidity. Tellus 16:64–75

Blumthaler M (1993) Solar UV measurements. In: Tevini M (ed) UV-B radiation and ozone depletion – effects on humans, animals, plants, microorganisms, and materials. Lewis Publishers, Boca Raton, pp 71–94

Blumthaler M (2007) Factors, trends and scenarios of UV radiation in arctic-alpine environments. In: Orbak Kallenborn, Tombre Hegseth, Petersen Hoel (eds) Arctic alpine ecosystems and people in a changing environment. Springer, Berlin, pp 181–194

Blumthaler M, Ambach W (1991) Spectral measurements of global and diffuse solar ultraviolet-B radiant exposure and ozone variations. Photochem Photobiol 54:429–432

Blumthaler M, Ambach W, Rehwald W (1992) Solar UV-A and UV-B radiation fluxes at two alpine stations at different altitudes. Theor Appl Climatol 46:39–44

Blumthaler M, Ambach W, Salzgeber M (1994a) Effects of cloudiness on global and diffuse UV irradiance in a high-mountain area. Theor Appl Climatol 50:23–30

Blumthaler M, Webb AR, Seckmeyer G, Bais AF, Huber M, Mayer B (1994b) Simultaneous spectroradiometry: a study of solar UV irradiance at two altitudes. Geophys Res Lett 21 (25):2805–2808

Blumthaler M, Gröbner J, Huber M, Ambach W (1996a) Measuring spectral and spatial variations of UVA and UVB sky radiance. Geophys Res Lett 23(5):547–550

Blumthaler M, Ambach W, Ellinger R (1996b) UV-Bestrahlung von horizontalen und vertikalen Flächen im Hochgebirge. Wetter Leben 48(1–2):25–31

Blumthaler M, Ambach W, Ellinger R (1997) Increase of solar UV radiation with altitude. J Photochem Photobiol B:Biol 39 (2):130–134

Bodhaine BA, Wood NB, Dutton EG, Slusser JR (1999) On Rayleigh optical depth calculations. J Atmos Ocean Tech 16 (11):1854–1861

Caldwell MM, Camp LB, Warner CW, Flint SD (1986) Action spectra and their key role in assessing the biological consequence of solar UV-B radiation change. In: Worrest Caldwell (ed) Stratospheric ozone reduction, solar ultraviolet radiation and plant life. Springer, Berlin, pp 87–111

Cede A, Blumthaler M, Luccini E, Piacentini RD, Nunez L (2002) Effects of clouds on erythemal and total irradiance as derived from data of the Argentine network. Geophys Res Let 29(24):76/1–76/4

Global solar UV Index: a practical guide; WHO, WMO, UNEO, ICNIRP, ISBN 92 4 159007 6.

Gröbner J, Albold A, Blumthaler M, Cabot T, De la Casinieri A, Lenoble J, Martin T, Masserot D, Müller M, Philipona R, Pichler T, Pougatch E, Rengarajan G, Schmucki D, Seckmeyer G, Sergent C, Touré ML, Weihs P (2000) Variability of spectral solar ultraviolet irradiance in an Alpine environment. J Geophys Res 105:26991–27003

Josefsson W, Landelius T (2000) Effect of clouds on UV irradiance: as estimated from cloud amount, cloud type, precipitation, global radiation and sunshine duration. J Geophys Res 105:4927–4935

Kylling A, Albold A, Seckmeyer G (1997) Transmittance of a cloud is wavelength-dependent in the UV-range: physical interpretation. Geophys Res Lett 24(4):397–400

Kylling A, Bais AF, Blumthaler M, Schreder J, Zerefos CS (1998) The effect of aerosols on solar UV irradiances during the PAUR campaign. J Geophys Res 103:26051–26060

Lean J (1987) Solar ultraviolet irradiance variations: a review. J Geophys Res 92:839–868

Madronich S, McKenzie RL, Björn L, Caldwell MM (1998) Changes in biologically-active ultraviolet radiation reaching the Earth's surface. J Photochem Photobiol B: Biol 46(1–3): 5–19

McKinlay AF, Diffey BL (1987) A reference action spectrum for ultraviolet induced erythema in human skin. CIE Journal 6(1):17–22

Piazena H (1996) The effect of altitude upon the solar UV-B and UV-A irradiance in the tropical Chilean Andes. Sol Energy 57(2):133–140

Schauberger G (1990) Model for the global irradiance of the solar biologically-effective ultraviolet-radiation on inclined surfaces. Photochem Photobiol 52(5):1029–1032

Webb AR, Weihs P, Blumthaler M (1999) Spectral UV irradiance on vertical surfaces: a case study. Photochem Photobiol 69(4):464–470

Zarrati F, Forno R, Fuentas J, Andrade M (2003) Erythemally weighted UV variations at two high altitude locations. J Geophys Res 108:ACH5_1–6

# 3 Bioclimatic Temperatures in the High Alps

Walter Larcher

## 3.1 Introduction

Characteristic of a high mountain climate are lower temperatures, frequency and intensity of wind and a more irregular distribution of precipitation. High mountain climate is defined by small-scale, terrain-dependent and short-term changeability.

Macroclimatic data recorded by meteorological stations form the basis for temperature studies. Yet to identify the causal effects of climate on the functions of plants, topoclimatical and microclimatical factors have to be taken into account. The bioclimate is a particular climate: it is more stable, warmer and wetter than the climate of the outer air. These climate factors affect essential processes like metabolism, growth, and reproductive development as well as the environmental limits of survival.

This article concentrates on a few examples of bioclimate temperatures from the life zones in the local area of the Austrian Alps that have been recorded in the context of scientific studies on plants in high mountains: temperature thresholds and heat sums must be reached for vegetative development and reproductive processes (flowering and seed development) of alpine forbs (*Gentianella germanica*: Wagner and Mitterhofer 1998), grasses and graminoids (*Carex*-species: Wagner and Reichegger 1997), subnival cushion plants (*Saxifraga*-species: Ladinig and Wagner 2005, 2009; Larl and Wagner 2006) and nival rosettes (*Ranunculus glacialis:* Wagner et al. 2010). To analyse the eco-physiological influences on the exchange of carbon dioxide in the natural habitat, temperatures have to be recorded on an hourly basis. This permits determining optimal and unfavorable temperature ranges for carbon dioxide exchange. Net photosynthesis and respiration in high mountain plants in the Tyrolean Alps were tested by Cartellieri (1940) for the first time and were later investigated with modern methods in the alpine zones (Grabherr 1977; Wohlfahrt et al. 1999; Wieser and Bahn 2004) reaching as far as the subnival and nival areas (Moser et al. 1977; Körner and Diemer 1987). Absolute and mean temperature extremes are essential to determine heat and frost constraints. Resistance to extreme temperatures was observed in many species of the dwarf shrub heath (e.g. Larcher 1977), in forbs and grasses of the alpine zone (e.g. Larcher and Wagner 1976; Neuner et al. 1999; Taschler and Neuner 2004) and in glacier species (Buchner and Neuner 2003; Larcher 1977).

## 3.2 Climate in Mountain Regions

### 3.2.1 Air Temperatures in Mountain Regions

Air temperatures decrease with increasing elevation. In the Alps, the temperatures of the free atmosphere drop, according to the adiabatic lapse rate, by an annual mean 0.55–0.62°C per 100 m and during the summer ca. 0.60–0.65°C per 100 m from the bottom of the valley to the high mountain regions (Franz 1979).

W. Larcher (✉)
Institute of Botany, LTUI, Innsbruck, Austria
e-mail: walter.larcher@uibk.ac.at

C. Lütz (ed.), *Plants in Alpine Regions*,
DOI 10.1007/978-3-7091-0136-0_3, © Springer-Verlag/Wien 2012

**Table 3.1** Long-term air temperature at 2 m height at the timberline (1,950 m a.s.l.) and the summit (2,247 m a.s.l.) of Mt Patscherkofel, at the glacier foreland of Mittelbergferner (Ötztal Alps 2,850 m a.s.l.) and at the summit of Mt Brunnenkogel (Ötztal Alps 3,440 m a.s.l.) provided by the Central Institute for Meteorology and Geodynamics, Regional Center for the Tyrol and Vorarlberg (zamg.ac.at)

| Altitude (m) | Period | Tm annual (°C) | Tm warmest month (°C) | Tabs max min (°C) | Days frost-free (d) |
|---|---|---|---|---|---|
| 3,440 | 2003–2009 | −5.8 | 1.9 | 16.5–30.8 | 48 |
| 2,850 | 1995–2009 | −2.3 | 5.7 | 17.6–29.0 | 94 |
| 2,247 | 1961–1990 | 0.2 | 8.1 | 23.4–29.5 | 142 |
| 1,950 | 1963–1992 | 2.1 | 9.7 | 26.0–28.0 | 182 |

Tm = annual mean air temperature; Tm mean air temperature mean warmest month; Tabs max and min = temperatures of absolute maximum and minimum; Days frost-free = number of days with minimum > 0°C

In the Central Alps the mean air temperatures from June until the end of August are ca. 10°C at the timberline, ca. 8°C 300 m higher in the alpine zone, ca. 5°C at 2,850 m a.s.l. in the glacier foreland and 1°C at 3,440 m a.s.l. in the nival zone. At the timberline (1,950 m a.s.l.), 182 days of the year were frost-free. At the top of the upper alpine life zone, measurements showed 142 frost-free days in the year. In the glacier area of the Central Alps the length of the frost-free period is considerably shorter, namely 94 and 48 days in the year (Table 3.1). In the alpine zone absolute minimum air temperatures at 2 m of −7°C (June) and −3°C to −4°C (July and August) occur; whereas in midsummer (July, August) frosts of about −7°C in the subnival zone and −8°C to −10°C in glacier regions can be measured.

The absolute lowest air temperatures ever measured in standard weather stations were −36°C to −37°C (Steinhauser 1954; Cappel 1977). These temperatures are of course isolated extremes. Most of the absolute minimum air temperatures of the free atmosphere in winter range from −18°C to −24°C in the alpine zone and reach as low as −30°C in glacier regions. During winter the plants are protected against these low temperatures by a layer of snow. Only several pioneer plants (cushion plants, tussocks, mosses and lichens) growing on snow-free and windy ridges are directly exposed to the low air temperatures.

### 3.2.2 Climatically Potential Growing Season: Snowmelt and Winter Snowfall

The period of the *climatically potential growing season* is defined as the period between snowmelt in spring and daily mean temperatures below freezing or the formation of a continuous snow cover in autumn (Svoboda 1977; Wagner et al. 1995).

In the Central Alps the winter snow cover lasts for an average of 127 days at 1,000 m a.s.l., 167 days at 1,500 m a.s.l. and 214 days at 2,000 m a.s.l. (Lauscher and Lauscher 1980). Over the years, winter snow depth is variable and the timing of melting also changes from year to year: for instance, when observing the very same microsite in a glacier foreland (2,650 m a.s.l.) over a period of 4 years, the earliest snowmelt happened on May 9 in 2003 and the latest on June 5 in 2002 (Larl and Wagner 2006).

The melting of the snow in spring is defined by solar radiation on various expositions and the inclination (Fig. 3.1). Generally, the duration of snow cover decreases from shadowy steep slopes to sunny slopes. The longest duration of snow cover can be measured on SE- and E-facing slopes, the shortest on SW-facing slopes. If the steepness increases, the duration of snow cover decreases. Up to an inclination of 30° 70%, all the winter days showed snow cover. With a slope of more than 60° there is hardly ever a permanent snow cover.

Above the timberline, the duration of snow cover not only depends on the altitude but also on topographic conditions. With incomplete snow cover, characteristic patterns of snow patches and snowmelt areas develop for a given terrain. In these patches, which can be covered by winter snow as late as mid June and August, the growing season is very short for high mountain plants, making it unfavourable for growth and development. Two different annual temperature patterns measured in habitats located at the same altitude show an average growing season (65 days) and a shortened growing season (35 days per year) in a glacier region (Fig. 3.2).

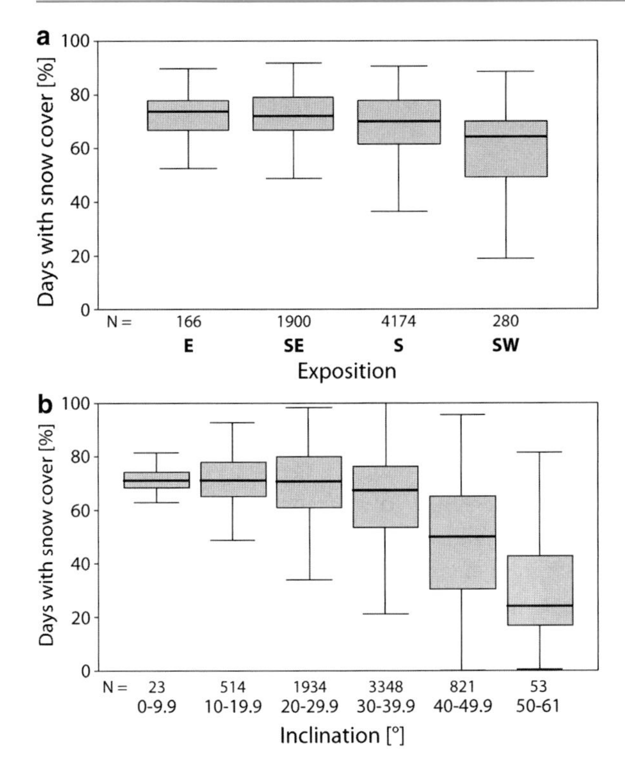


Fig. 3.1 (**a**) Distribution of duration of snow cover in relation to exposition, (**b**) distribution of duration of snow cover in relation to inclination between 1,900 m and 2,100 m a.s.l. (Tasser et al. 2001)

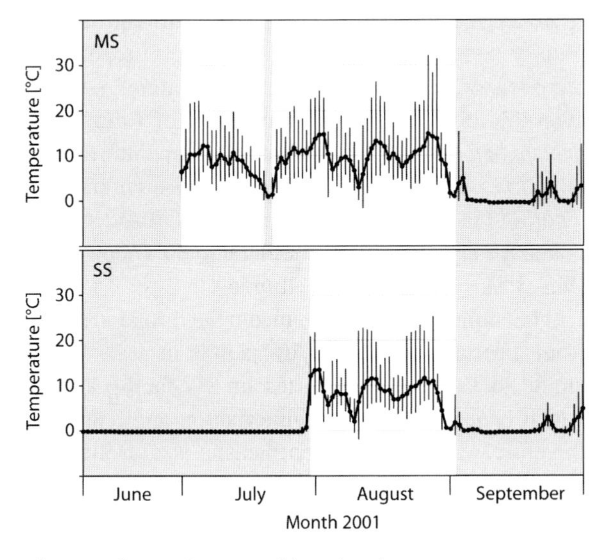


Fig. 3.2 Seasonal course of boundary layer temperatures at a mid-season site (MS) and a short-season (SS) site in the foreland of the Hintertux Glacier (2,650 m a.s.l., Zillertal Alps). *Grey:* snow cover (Ladinig and Wagner 2009)

### 3.2.3 Soil Temperatures in Mountain Regions

Soil temperatures in mountain regions depend on the composition and structure of the soil, plant cover, oxygen content, soil humidity and the presence of snow cover but also on the exposition and inclination of the slope. The soil acts as a thermal buffer by taking up a considerable amount of heat during the day and by releasing it again at night. In contrast to the boundary layer temperatures, which show very high amplitudes, soil temperatures below a depth of 10–15 cm vary only slightly. Furthermore, changes in soil temperatures are delayed when compared to air temperatures.

During the warmest month (July) the root zone temperatures down to 30 cm depth were on average between 14°C and 15°C at 1,500 m a.s.l. and between 11°C and 12°C at 2,000 m a.s.l. At a depth of 50–100 cm the warmest soil temperatures were recorded in August and ranged from 13°C to 14°C (1,500 m a.s.l.) and 10–11°C (2,000 m a.s.l.). The long-term annual means of soil temperatures (Eckel 1960) were 6°C (−10 cm depth) at 1,500 m a.s.l. and 4°C (30–100 cm depth) at 2,000 m a.s.l.

Körner and Paulsen (2004) measured soil temperatures (10 cm below the surface) at the treeline (2,050 ± 20 m a.s.l.) during the growing period. The values ranged from 7.3°C on a N-slope to 6.7°C on an E-slope, and from 6.9°C on a S-slope to 7.4°C on a W-slope. The critical temperature for root growth is ca. 6°C, which equates to the mean soil temperature at climatic treelines (Alvarez-Uria and Körner 2007).

Under a dense *Rhododendron* cover and during the snow-free period, soil temperatures at −10 cm were 8–10°C. On an average of 120–160 days, temperatures above 5°C occur in the root zone of favourable sites; at higher altitudes this time span is reduced to about 80 days (Larcher and Wagner 2004). The soil surface in the alpine grassland is warmer during the growing period than the soil 10 cm below the surface, as there are no trees to shade the ground (Fig. 3.3). In a world-wide study of root zone temperatures (Körner et al. 2003), soil temperatures in alpine grasslands in the Alps were also measured. In July and August the monthly mean soil temperatures (−10 cm depth) of the Curvuletum in the Ötztal Alps (2,520 m a.s.l.) were between 9°C and 11°C with maximum temperatures of up to 15°C; in the East Valais Alps,

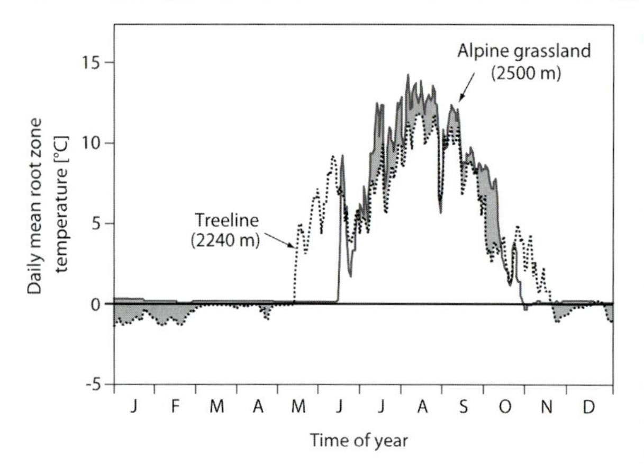


**Fig. 3.3** Comparison of soil temperatures at the treeline and in alpine grasslands of the East Valais Alps, Switzerland. The loggers were positioned under closed vegetation at a depth of −10 cm (Körner et al. 2003)

Switzerland, mean temperatures between 10°C and 13°C with maximum temperatures of 19°C were recorded, and in the Dolomites (2,300 m a.s.l.) mean temperatures ranged from 13°C to 19°C. Across all sites, these temperatures were calculated with data from two seasons and differed by <1 K between the years.

Above ca. 2,500 m a.s.l. and up to the permanent snow-line it is difficult to identify a distinct gradient in altitude. Soil temperatures vary according to the various topographic microsites. In the glacier regions the soil temperatures (−10 cm) pass the sub-zero temperature threshold for only 3 months. During the warmest month (July) average soil temperatures of 2.8°C, mean maximum temperatures of 6.1°C and mean minimum temperatures of 0.2°C were recorded on a ridge site. On isolated sunny days (daily maximum air temperature 7°C) temperatures of about 15°C were measured on a ridge site and about 10°C on an S-facing slope (Moser et al. 1977). In summer at the Jungfraujoch (3,700 m a.s.l., Swiss Alps) the rock temperature drops each night to below 0°C and even down to −5°C (Mathys 1974). During clear days minimum temperatures of 0–5°C and maximum temperatures of 24–28°C occur in crevices (−10 cm); at a depth of −20 cm the temperatures range from 3°C to 8°C (minimum) and 20–22°C (maximum).

## 3.3 Bioclimatic Temperatures in the Phytosphere

Bioclimate is the microclimate from the upper surface of the vegetation down to the deepest roots in the soil (Lowry 1967; Cernusca 1976a). This bioclimate is a characteristic climate: a particular bioclimate, however, is also continuously subjected to influences from all climatic spheres. The influence of the macroclimate and mesoclimate defines the range of radiation and precipitation. Beneath closed canopies, the bioclimate is more stable, warmer and wetter than the climate of the outer air. On sunny days the temperature at the surface of vegetation is higher than that of the air above it.

Bioclimate is determined by the height of the plant cover and growth forms of different plant communities. The small-scale mosaic of vegetation in the high mountain regions shows predictable bioclimatic patterns. Dwarf-shrub heaths and alpine grassland are warmer than the forest and scattered trees at the timberline under sunny conditions (Fig. 3.4).

During the climatically potential growing season the average mean canopy temperatures were about 8°C in *Rhododendron ferrugineum* shrubs at the treeline (1,900–2,000 m a.s.l.). During the main phase of growth (July and August) temperatures of about 10°C were recorded during the warmest month (Larcher and Wagner 2004). In the dwarf shrub community temperatures of 30–32°C were repeatedly recorded on sunny slopes and temperatures reached even 35–38°C. In the prostrate mats of *Loiseleuria procumbens* mean maximum temperatures of 20–30°C were recorded on summer days with strong incoming radiation leaf temperatures of 40–42°C, and boundary layer temperatures of up to 50°C can occur (Fig. 3.5).

The alpine grasslands covers the broad transect from treeline to about 2,200–2,400 m a.s.l. These alpine grasslands occur on S- or SW-facing slopes, which benefit from the steep radiation angle and are therefore warmer and drier than the shady N-facing slopes or hollows. In alpine grasslands (at 2,000 m a.s.l.) during the growing season, canopy temperatures are

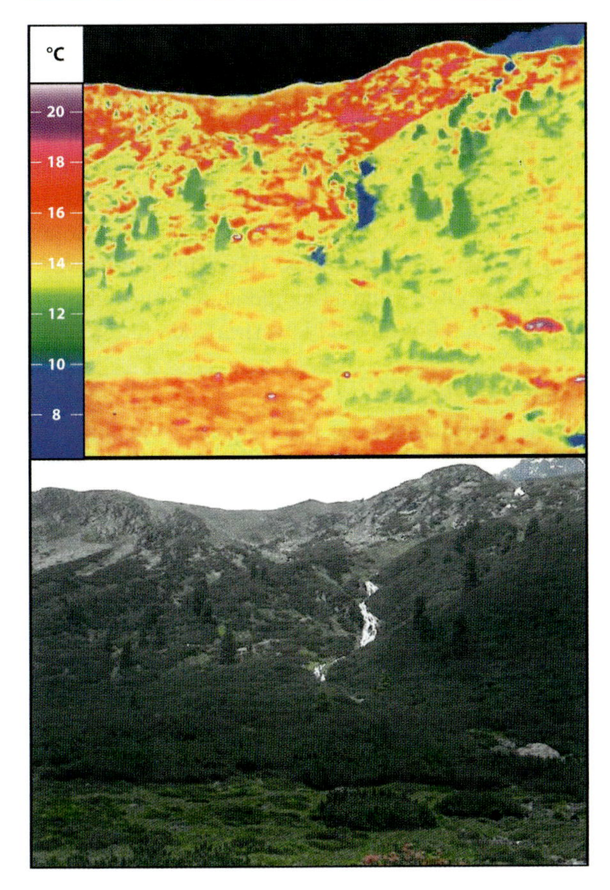


Fig. 3.4 Bioclimatic temperatures across an elevational transect in the Stubai valley (Central Alps) recorded with thermography. The higher and sunnier sites in the alpine plant communities are the warmest regions, whereas the soil surface of the summit (Ruderhofspitze 3,474 m) and the creek are the coldest areas. Picture taken July 3, 2009, 1100 h (Photo: U. Tappeiner)

about 10°C on SW-facing slopes (Tasser et al. 2001). The mean boundary layer temperatures on the NE-facing slope are cooler by about 2 K; the maxima on sunny days, however, are 5–7 K higher on the sunlit site than on the shady northern site (Wagner and Reichegger 1997).

Beyond the upper alpine zone (2,300–2,500 m a.s.l.) begin the prostrate plant life forms like short graminoids, rosettes and cushion plants. In winter prostrate plants growing on microsites sheltered from the wind are mostly covered with snow. These plants experience temperatures between 0°C and −3°C. On sunny slopes snow-free periods start in the first and second week of May.

The duration of the climatically potential growing season is normally 100–120 days. In July and August

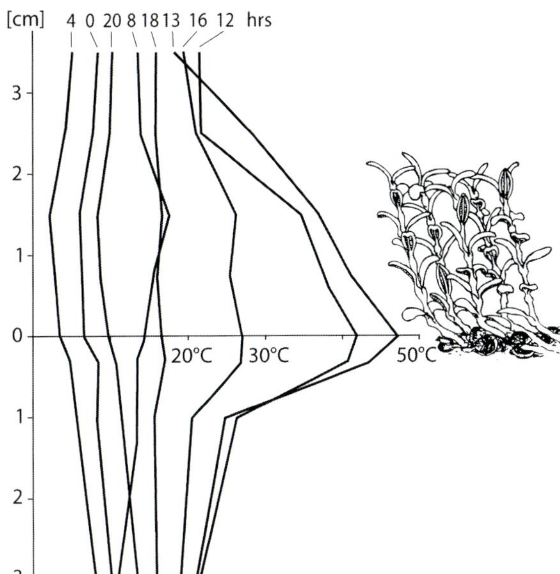


Fig. 3.5 Vertical temperature profile of a *Loiseleuria procumbens* carpet at 2,175 m a.s.l. on a calm and clear mid summer day. At 1300 h the air temperature was 19.4°C (10 cm above the canopy) and 16.6°C (2 m height). The litter was most prone to overheating, reaching 47.5°C (Cernusca 1976b; modified)

the monthly means of boundary layer temperatures were about 9–11°C (Ladinig and Wagner 2005). The daily minimum temperatures during the summer were 3–10°C, daily maximum temperatures on an N-slope about 20–25°C and on a W-slope 25–30°C.

Isolated rosette plants and cushion species were most commonly found in this scant patchy vegetation. Rosette plants are exposed to a broad range of boundary layer temperatures over an entire day. *Primula minima* temperatures provide an example of an extreme temperature range. Due to strong irradiation and shelter from wind, the temperatures may differ by 30 K between sunrise and early afternoon (Fig. 3.6). On clear nights lower temperatures occur in the morning because of emitted thermal re-radiation. Even on sunny winter days, noon temperatures of up to 15–20°C can occur on windy ridges under a thin layer of snow or on snow-free sites. Cushion plants can heat up very much. There is a steep temperature gradient across the cushion. In *Silene acaulis*, the differences between the sunny and shady side reach a maximum at 10 a.m. (9 K) and at 4 p.m. (12 K) and are smaller during the midday hours, when the angle of incidence of solar radiation is higher (Körner and De Moraes 1979; Körner 2003).

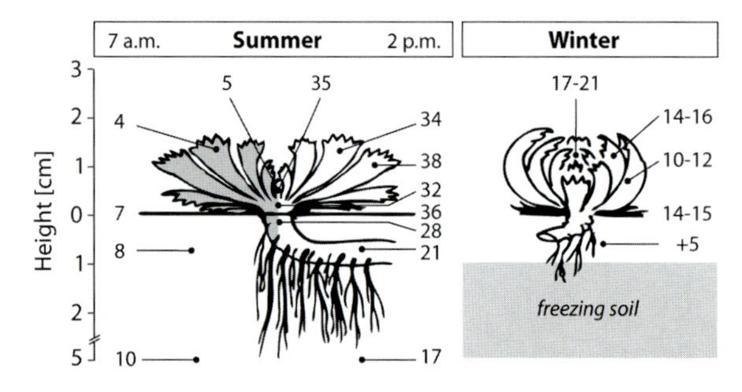


**Fig. 3.6** Vertical temperature profile (°C) of *Primula minima* at 2,200 m a.s.l. *Left*: daily temperature amplitude during the summer. Temperatures were measured with thermocouples 7 h before sunset in the shade (air temperature 8°C) and in the early afternoon at 1400 h (air temperature 16°C, solar radiation 800 W $m^2$). *Right*: snowdrift on a site in winter. Temperature profile at 1300 h (air temperature −2.5°C) of fading and rolled up leaves of the rosette. *Grey area*: ground frost at a depth of 1 cm (Larcher 1980 and unpubl.)

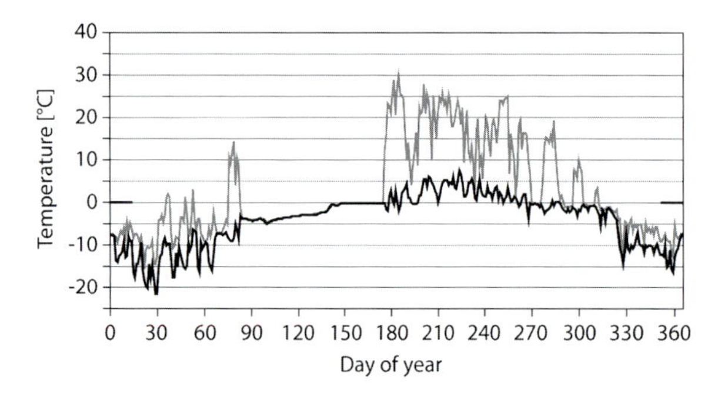


**Fig. 3.7** Annual course of boundary layer temperatures at a subnival site in the glacier foreland of the Schaufelferner (Stubai Alps 2,880 m a.s.l.) in the climatically normal year of 2004. The temperature logger was installed in a cushion of *Saxifraga bryoides*. *Upper line*: daily maximum, *lower line*: daily minimum (Larcher et al. 2010)

Plant temperatures were measured in the glacier foreland at 2,880 m a.s.l. with scattered vegetation (Fig. 3.7). During winter, temperatures under the permanent snow cover fell to −5°C and to −15°C. At microsites of snow-free spots, which are very often covered with cushion plants, temperatures of down to −25°C may occur.

At the glacier foreland, the growing season ranged from about 5 weeks (late-thawing site) to 2 months (early-thawing site). Frosty temperatures were regular in the sparse vegetation and at the soil surface. In July and August temperature minima between −2°C and −3°C were recorded on about 20 days in the subnival ecotone, and temperature minima down to −5°C were recorded on about 50 days on the nival summits. On the other hand, due to high irradiation, cushions and rosettes plants can reach temperatures of about 25°C.

Long-term temperature records (2003–2009) in a cushion of *Saxifraga bryoides* during the main phase of growth and reproductive development showed an average of 9.1 ± 1.4°C for July and August. Mean cushion plant temperatures were 3.2 ± 0.8 K warmer than the free air temperatures. The difference between plant temperatures and air temperatures was highest in the year 2006 (4.8 K) and lowest in 2008 (2.3 K).

## Conclusions

The characteristics of plant growth forms have an important influence on the different temperatures. On sunny days the temperature at the surface of vegetation is higher than that of the air above it. Particularly rosette and cushion plants, which grow close to the ground, heat up much more than erect plants. These microclimatic temperatures decrease the impact of the rough high mountain climate and allow alpine plants to successfully survive and reproduce.

The high mountain regions offer an excellent investigation area as they combine vertical and topographical changes in environmental conditions and thus open up possibilities for comparative studies. In the mountains different ecotypes, which would be hundreds or maybe even thousands of kilometres apart in the valley, can be found right next to each other. This natural experiment offers ideal conditions for plant physiological studies.

**Acknowledgements** Many thanks to the Central Institute for Meteorology and Geodynamics, Regional Center for Tirol and Vorarlberg (Dr. Karl Gabl), and to the Institute of Ecology, University Innsbruck (Prof. Dr. Ulrike Tappeiner), for providing data. Thanks, to "pdl, Dr. Eugen Preuss" Innsbruck, for image processing.

## References

Alvarez-Uria P, Körner C (2007) Low temperature limits of root growth in deciduous and evergreen temperate tree species. Func Ecol 21:211–218

Buchner O, Neuner G (2003) Variability of heat tolerance in alpine plant species measured at different altitudes. Arctic Antarctic Alpine Res 35:411–420

Cappel A (1977) Extremwerte der Lufttemperatur auf der Zugspitze (1900–1976). Jb Sonnblick-Verein 74(75):37–42

Cartellieri E (1940) Über Transpiration und Kohlensäureassimilation an einem hochalpinen Standort. SB Akad Wiss Wien, Math-nat Kl I 149:95–143

Cernusca A (1976a) Bestandes-Struktur, Bioklima und Energiehaushalt von alpinen Zwergstrauchbeständen. Oecolog Plant 11:71–102

Cernusca A (1976b) Energie- und Wasserhaushalt eines alpinen Zwergstrauchbestandes während einer Föhnperiode. Archiv für Meteorologie. Geophys Bioklima Ser B 24:219–241

Eckel O (1960) Bodentemperatur. In: Steinhauser F, Eckel O, Lauscher F (eds) Klimatographie von Österreich, Bd. 3 (2). Denkschrift Österr Akad Wiss. Springer, Wien, pp 207–292

Franz H (1979) Ökologie der Hochgebirge. Ulmer, Stuttgart

Grabherr G (1977) Der $CO_2$-Gaswechsel des immergrünen Zwergstrauches *Loiseleuria procumbens* (L.) Desv. in Abhängigkeit von Strahlung, Temperatur, Wasserstreß und phänologischem Zustand. Photosynthetica 11:302–310

Körner C (2003) Alpine plant life, 2nd edn. Springer, Berlin

Körner C, De Moraes JAPV (1979) Water potential and diffusion resistance in alpine cushion plants on clear summer days. Oecol Plant 14:109–120

Körner C, Diemer M (1987) In situ photosynthetic responses to light, temperature and carbon dioxide in herbaceous plants from low and high altitude. Func Ecol 1:179–194

Körner C, Paulsen J (2004) A world-wide study of high altitude treeline temperatures. J Biogeogr 31:713–732

Körner C, Paulsen J, Pelaez-Riedl S (2003) A bioclimatic characterisation of Europa's alpine areas. In: Nagy L, Grabherr G, Körner C, Thompson DBA (eds) Alpine biodiversity in Europa. Springer, Berlin, pp 13–30

Ladinig U, Wagner J (2005) Sexual reproduction of the high mountain plant *Saxifraga moschata* Wulfen at varying lengths of the growing season. Flora 200:502–515

Ladinig U, Wagner J (2009) Dynamics of flower development and vegetative shoot growth in the high mountain plant *Saxifraga bryoides* L. Flora 204:63–73

Larcher W (1977) Ergebnisse des IBP-Projektes "Zwergstrauchheide Patscherkofel". SB Österr Akad Wiss Math-nat Kl I 186:301–371

Larcher W (1980) Klimastreß im Gebirge. Adaptationstraining und Selektionsfilter für Pflanzen. Rheinisch-Westf Akademie der Wissenschaft N291. Westdeutscher Verlag, Leverkusen, pp 49–88

Larcher W, Wagner J (1976) Temperaturgrenzen der $CO_2$-Aufnahme und Temperaturresistenz der Blätter von Gebirgspflanzen im vegetationsaktiven Zustand. Oecol Plant 11:361–374

Larcher W, Wagner J (2004) Lebensweise der Alpenrosen in ihrer Umwelt: 70 Jahre ökophysiologische Forschung in Innsbruck. Berichte naturwiss-med Verein Innsbruck 91:251–291

Larcher W, Kainmüller C, Wagner J (2010) Survival types of high mountain plants under extreme temperatures. Flora 205:3–18

Larl I, Wagner J (2006) Timing of reproductive and vegetative development in *Saxifraga oppositifolia* in an alpine and a subnival climate. Plant Biol 8:155–166

Lauscher A, Lauscher F (1980) Vom Schneeklima der Ostalpen. Jb Sonnblick Verein 1978–1980:15–23

Lowry WP (1967) Weather and life. An introduction to biometeorology. Academic, New York

Mathys H (1974) Klimatische Aspekte zu der Frostverwitterung in der Hochgebirgsregion. Mitt Naturforschung Ges Bern NF 31:49–62

Moser W, Brzoska W, Zachhuber K, Larcher W (1977) Ergebnisse des IBP-Projekts "Hoher Nebelkogel 3184 m". SB Österr Akad Wiss Math-nat KlI 186:386–419

Neuner G, Braun V, Buchner O, Taschler D (1999) Leaf rosette closure in the alpine rosette plant *Saxifraga paniculata* Mill.: significance for survival of drought and heat under high irradiation. Plant. Cell Environ 22:1539–1548

Steinhauser F (1954) Klimatabelle für den Sonnblick (3106 m) 1901 bis 1950. Jb Sonnblick Verein 49(50):56–60

Svoboda J (1977) Ecology and primary production of raised beach communities, Truelove Lowland. In: Bliss LC (ed) Truelove lowland, Devon Island, Canada: a high arctic ecosystem. University Alberta Press, Edmonton, pp 185–216

Taschler D, Neuner G (2004) Summer frost resistance and freezing patterns measured in situ in leaves of major alpine plant growth forms in relation to their upper distribution boundary. Plant Cell Environ 27:737–746

Tasser E, Tappeiner U, Cernusca A (2001) Südtirol Almen in Wandel. Ökologische Folgen von Landnutzungsänderungen. Europäische Akademie, Bozen

Wagner J, Mitterhofer E (1998) Phenology, seed development, and reproductive success of an alpine population of *Gentianella germanica* in climatically varying years. Botan Acta 111:159–166

Wagner J, Reichegger B (1997) Phenology and seed development of the alpine sedges *Carex curvula* and *Carex firma* in response to contrasting topoclimates. Alpine Arctic Res 29:291–299

Wagner J, Achalkazi M, Mayr St (1995) Anwendung quantitativ embryologischer Methoden in Entwicklungsbiologie und Reproduktionsökologie der Pflanzen. Anzeiger der Österreichischen Akademie der Wissenschaften. Math-nat Kl I 131: 7–18

Wagner J, Steinacher G, Ladinig U (2010) *Ranunculus glacialis* L.: successful reproduction at the altitudinal limits of higher plant life. Protoplasma 243:117–128

Wieser G, Bahn M (2004) Seasonal and spatial variation of woody tissue respiration in a *Pinus cembra* tree at the alpine timberline in the central Austrian Alps. Trees 18:576–580

Wohlfahrt G, Bahn M, Haubner E, Horak I, Michaele W, Rottmar K, Tappeiner U, Cernusca A (1999) Inter-specific variation of the biochemical limitation to photosynthesis and related leaf traits of 30 species from mountain grassland ecosystems under different land use. Plant Cell Environ 22:1281–1296

zamg.ac.at (Regional Centre for Meteorology)

# Physiological and Ultrastructural Changes in Alpine Plants Exposed to High Levels of UV and Ozone

# 4

Cornelius Lütz and Harald K. Seidlitz

## Abbreviations

| | |
|---|---|
| AWI | Alfred Wegener institute for polar research |
| Fv/Fm | Photosynthetic optimum quantum yield |
| MED/h | Minimal erythema dosis per hour |
| PAR | Photosynthetic active radiation |
| TEM | Transmission electron microscopy |

## 4.1 Introduction

Alpine and polar plants resist a variety of unfavourable conditions such as short vegetation periods, rapid changes in temperature and weather conditions, and often high irradiation. These environmental conditions, especially for the European alpine situation, are addressed in the first three chapters of this book. Additional growth site characterizations can be found in Amils et al. (2007), Blumthaler (2007), Crawford (2008), Huiskes et al. (2003), Körner (2003), Körner and Larcher (1988), Larcher and Wagner (2009), Nagy and Grabherr (2009).

C. Lütz (✉)
University of Innsbruck, Institute of Botany, Innsbruck, Austria
e-mail: Cornelius.luetz@uibk.ac.at

H.K. Seidlitz
Department Environmental Engineering, Institute of Biochemical Plant Pathology, Helmholtz Zentrum München, Neuherberg, Germany

Higher plants – the focus in this article – have developed many strategies to cope with these growth conditions. Numerous scientific contributions published over the last decades have shown that most species run individual sets of adaptation processes, indicating that a general rule of how to survive in high mountains or polar regions obviously does not exist. Such adaptation processes are described (as a selection) by Alberdi et al. (2002), Bravo et al. (2001), Crawford (1997), Kappen (1983), Körner (2003), Larcher (2001), Larcher et al. (2010), Lütz (2010), Lütz, Chap. 5 of this volume, Oerbaek et al. (2004), Streb et al. (1997), Wielgolaski and Karlsen (2007), Xiong et al. (1999).

In addition to natural stressors, human activities have developed further stress loads: particularly in the European (Northern) Alps above 1,500 m altitude, tropospheric ozone concentrations reach high values due to ozone transport from traffic dominated and industrialized regions, but low destruction rates in the mountains (Smidt 1993, 1998; Smidt and Englisch 1998; Herman et al. 1998). The area studied for this work is part of the northern limestone Alps (Karwendel, Bavaria), with mean monthly values for tropospheric ozone reaching 40–60 ppb ozone at 1,730 m a.s.l and peak values of up to 140 ppb in summer (continuous measurements of the Tyrolean weather service). These concentrations are considered to be damaging to plants in many studies and to be higher than the plant damage limit expressed as AOT 40 (Langebartels et al. 1997; Herman et al. 1998).

In addition to frequent episodes of high concentrations of tropospheric ozone, human activities – in particular the release of chlorofluorocarbon compounds into the atmosphere – have caused a

C. Lütz (ed.), *Plants in Alpine Regions*,
DOI 10.1007/978-3-7091-0136-0_4, © Springer-Verlag/Wien 2012

seasonally dependent depletion of the stratospheric ozone layer and concomitantly an increased exposure of the biosphere with detrimental short wave UV-irradiation (UV-B radiation 280–315 nm) (Krupa and Kickert 1989; Flint and Caldwell 2003; Kirchhoff and Echer 2001; Xu and Sullivan 2010). International efforts to counteract this trend resulted in the Montreal Protocol which has been successful in limiting the stratospheric ozone depletion. However, it will take several decades until ozone will return to 1980 levels (McKenzie et al. 2011), and episodes with dramatically depleted stratospheric ozone – so-called ozone holes – still occur even in the northern hemisphere (AWI, press releases March 14, 2011: "Arctic on the verge of record ozone loss" and April 5, 2011 Record depletion of the Arctic ozone layer caused increased UV radiation in Scandinavia – Brief episodes of increased UV radiation may also occur over Central Europe; http:/www.awi.de/en/news).

The spring formation of ozone holes over Antarctica has been well documented and surveyed by satellites for years (Marchant 1997; Bargagli 2008). Possible negative impacts on antarctic biota have been widely discussed (e.g. Huiskes et al. 2003, Wiencke et al. 2008); for the northern hemisphere this additional irradiation load seems not to be of biological relevance. Despite an estimated annual increase of about 1% (Blumthaler and Ambach 1990; Frederick 1990) it is a matter of debate whether this small enhancement endangers crop plants (Tevini and Teramura 1989; Rozema et al. 1997) or not (Fiscus and Booker 1995; Johnson and Day 2002).

Unfortunately a precise prediction of future terrestrial UV exposure is not yet possible, as the interaction between ozone depletion and climate change is poorly understood (McKenzie et al. 2011).

The present knowledge of the effects of solar UV-irradiation on terrestrial ecosystems has been summarized by Rozema et al. (2005) and Ballaré et al. (2011). It appears that plant growth has generally been only little affected by ozone decline since 1980. The reason is that plants have developed rather effective defence mechanisms against damage from UV-B radiation, e.g. forming certain UV-protective phenolic compounds (Bornman et al. 1997; Nybakken et al. 2004). However, it should be kept in mind that other stressors, e.g. high levels of PAR and UV-A radiation (Götz et al. 2010) or low temperatures (Oppeneiger 2008; Albert et al. 2009; Lütz 2010) can also upregulate the formation of phenolic metabolites.

For plants growing in extreme habitats, such as high alpine regions, however, the additional stress of UV-irradiation plays an important role although these plants have evolutionarily adapted to extreme irradiation regimes and to the naturally occurring amplitudes of their changes. In the mountains UV-irradiation increases with altitude and it can reach high values particularly on slopes with perpendicular orientation to the sun, even in winter, when back scattering from snow surface increases especially the UV component (Blumthaler et al. (1996, 1997), Blumthaler, Chap. 2, this book).

Taken together, elevated ozone levels and high UV-B irradiation may increase the oxidative stress in plants. Today both "stressors" have been well studied separately especially in lowland and agricultural or forest plants (see below), but not in combination, and not for alpine plant species, which are already exposed to a high external pressure from extreme climate conditions (Lütz 2010).

Therefore we will present the only simulation studies so far performed on alpine plants using elevated ozone and UV irradiation and compare the results with known investigations on the application of theses stressors in single exposures. The results of the simulation studies were published as reports for the Bavarian Ministry for Environment and Health (Lütz, 2001 and 2004).

## 4.2 Current Situation of UV Irradiation and Ozone Pollution in the Tropospere with Respect to Alpine Plant Covers

### 4.2.1 The Research Tool of Climate Simulation

Large exposure chambers and sun simulators specially designed for plant research of the *Helmholtz Research Centre Munich, Institute of Biochemical Plant Pathology, Department of Environmental Simulation* were used for our simulation studies. Technical details of the research chambers are described in Payer et al. (1993), Thiel et al. (1996), Döhring et al. (1996). In addition to natural simulation of a wide spectrum of environmental parameters and air pollutants, such as ozone, acid mist, $CO_2$ etc., they provide a naturally balanced solar-like simulation of daylight throughout the spectral range of terrestrial global radiation from the short wave UV to the infrared. PAR levels in excess of 1,600 μmol $m^{-2}$ $s^{-1}$ and UV-B levels up to 7 MED $h^{-1}$ can be reached even at temperatures down to 5°C.

Experiments were been performed with alpine plants taken from the Bavarian limestone Alps near Mittenwald at 2,250 m a.s.l.. The plant community "Caricetum firmae", typical of this area, was represented in the experiments by *Carex firma* (CF), *Dryas octopetala* (DO) *Ranunculus alpestris* (RA) and sometimes *Salix retusa* (SR). Cushions of the plants with the soil were excised from their natural environment after snow melt and were directly transferred to the plant climate simulation chambers. Mean ozone values in this region are 60 ppb with peaks up to 120 ppb (measurements by the Bavarian weather service), and on sunny days an UV-B irradiation of 4 MED/h is frequent.

The ozone treatment experiments were performed in two consecutive years (vegetation periods), followed by ozone/ozone UV-B treatments in two further years, each lasting at least 10–15 weeks.

*Climatic conditions (for ozone only applications):* 11–16 h light, depending on the simulated season, with max. light (1,400 μmol $m^{-2}$ $s^{-1}$ between 10 a.m. and 14 p.m.) on 4 days a week. Two days had a low irradiation of 375 μmol $m^{-2}$ $s^{-1}$, 1 day with intermediate light values. Temperatures ranged between 16°C and 27°C at day, according to the light regime, and at night between 8°C and 12°C. Relative humidity provided was set according to the light regimes between 50% and 90%.

*Ozone treatments*: 12/10 ppb ozone (day/night) in the controls and 90/80 ppb (day/night) in the treatment chambers, with ozone peaks up to 120 ppb ozone during 4 h a week for 4 days. This setup was chosen because governmental weather services regularly report monthly ozone means of 60 ppb with peaks up to 140 ppb in the Karwendel mountains.

*UV-treatments*: at max. light of 1,400 μmol $m^{-2}$ $s^{-1}$ PAR in the control chamber a UV-B radiation of 4 MED/h was applied, and in the treatment chamber it was increased to 6–7 MED/h at max. 1,600 μmol $m^{-2}$ $s^{-1}$ PAR. Day length for the entire experiment was 5 a.m. to 9 p.m., with max. UV-B increasing to a max. level between 11 a.m. and 3 p.m., then decreasing. Temperatures were set at 15°C during max. light and 5°C in the night.

### 4.2.2 Physiological Investigations

*Photosynthesis and respiration:* Leaf gas exchange was recorded as oxygen turnover in an oxygen electrode (Hansatech, Kings Lynn, UK) according to Lütz (1996). Temperature: 18°C, saturating $CO_2$ was provided and activity measured at PAR of 500 and of 1,500 μmol $m^{-2}$ $s^{-1}$.

PS II activity was measured with PEA from Hansatech as Fv/Fm and calculated according to Strasser et al. (1995).

*Plastid pigments and α-tocopherol*: leaves from the photosynthesis assays were extracted and pigments and tocopherol separated by HPLC according to Wildi and Lütz (1996).

*Antioxidants and flavonoids*: Total glutathione content of leaves was determined with an assay according to Griffiths (1980), and total ascorbic acid by HPLC as described in Wildi and Lütz (1996), except that 0.3% metaphosphoric acid was used for extraction, as for the glutathione assays. – The semi-quantitative measurements of flavonoids based on extraction in methanol followed Schnitzler et al. (1996). Absorption spectra were taken between 250 nm and 450 nm in a photometer (DW 2000 SLM Aminco). For semi-quantification the maximum absorption wavelength in the UV-A region was determined for each species, and data expressed as absorption at this wavelength per fresh weight.

Data analysis: the data shown are means of 3–5 replicates with standard error (SE). For statistical analysis of differences, the *t*-test was used: *: $p < 0.05$; **: $p < 0.01$; ***: $p < 0.001$.

### 4.2.3 Do High Alpine Plants Suffer Under Elevated Ozone or Increased UV irradiation?

The visual appearance of the leaves after an exposure time of up to 15 weeks remained unchanged in case of the UV-treatments or their controls.

However, easily visible yellowing occurred in SR and RA; DA showed a bronzing of the leaf surface,

**Fig. 4.1** Visual pattern of the ozone-treated plants after 2 months of incubation in comparison to the control leaves. *Left group*: control samples, *right group*: after ozone treatment (see text). *From top*: *Dryas octopetala; Salix retusa; Carex firma; Ranunculus alpestris*

**Fig. 4.2** Leaves under field conditions and the sampling area ("Karwendelgrube") near Mittenwald (Bavaria), 2,250 m a.s.l., first week of September

starting at the tips; CF developed necrotic brown leaf tips (the oldest tissues in the leaf) (Fig. 4.1).

Such yellowing occurs also at that time in the field (Fig. 4.2), however, it is not possible to interpret this field yellowing of leaves exclusively as ozone based.

All physiological measurements as well as compound extractions were performed only in the completely green parts of the leaves to get comparable starting situations.

*Photosynthetic activity* (oxygen evolution) as well as *Fv/Fm* as a direct measure of PS II activity were found reduced after 3 months of high ozone treatment, but statistically not significant (data not shown). By contrast, high UV-B led to a significant reduction in Fv/Fm in DO and in CF, and a small reduction in CF (Fig. 4.3). Photosynthesis was clearly reduced, significant ($p < 0.01$) in CF.

*Photosynthetic pigments* (total chlorophylls and total carotenoids) under high ozone were found reduced in RA and DO (Fig. 4.4), in CF there is only a minor reduction in total chlorophylls. Significant changes developed in the pool sizes of the xanthophyll

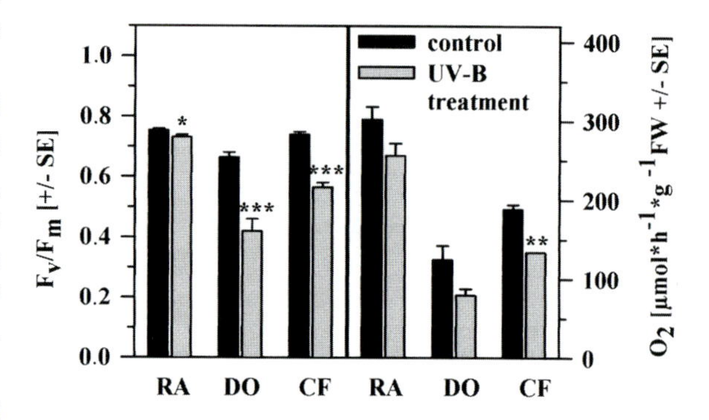


**Fig. 4.3** Effects of enhanced UV-B treatment on photosynthetic oxygen development (*right*) and on PS II activity (*left*) after 12 weeks of: *Ranunculus alpestris* (RA), *Dryas octopetala* (DO), *Carex firma* (CF). FW: fresh weight. Significance levels, see methods

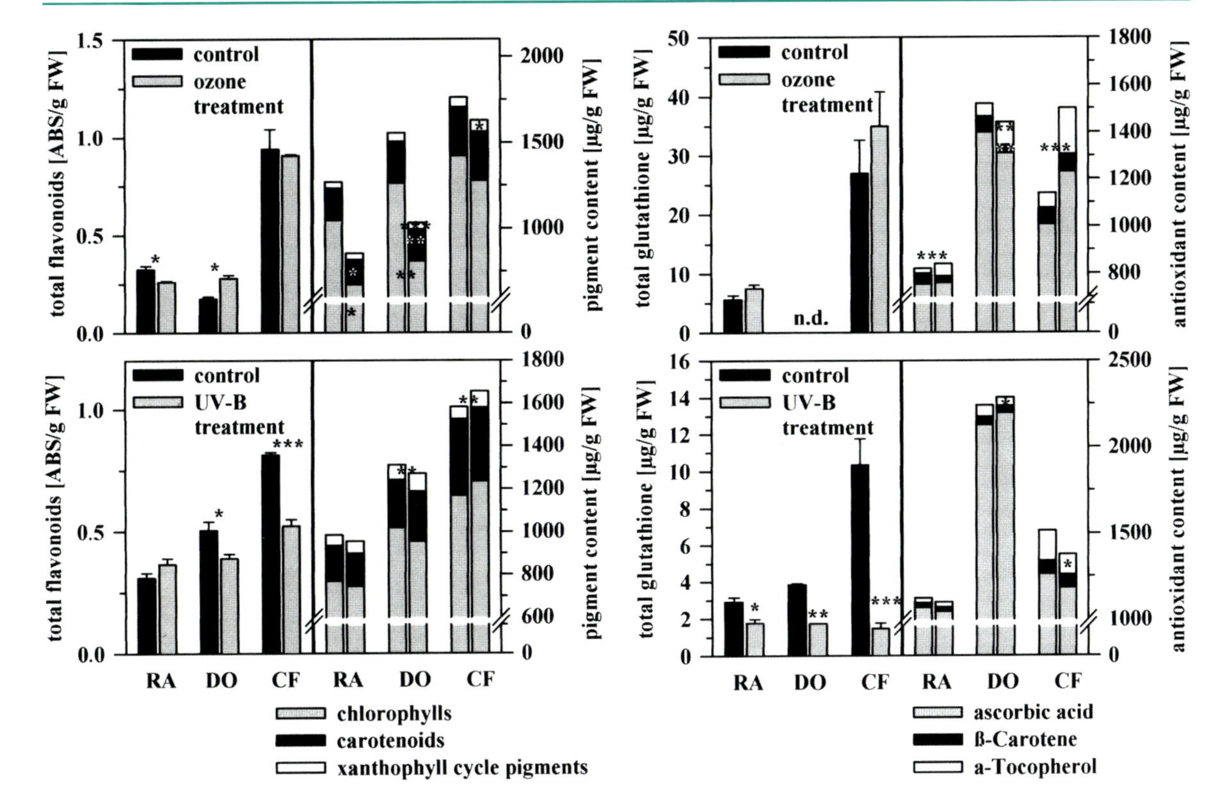


**Fig. 4.4** Upper graph: effects of ozone treatment on total flavonoids (*left*) and on total photosynthetic pigments (*right*) after 15 weeks of simulation. FW: fresh weight; ABS: rel. absorption at the plant-specific peak in the UV-A. Lower graph: effects of enhanced UV-B treatment after 12 weeks of simulation. *Left*: total flavonoids, *right*: total photosynthetic pigments

**Fig. 4.5** Upper graph: effects of ozone treatment on total glutathion (*left*) and on total antioxidants, composed of ascorbic acid, ß-carotene and α-tocopherol, (*right*) after 15 weeks of simulation. FW: fresh weight. Lower graph: effects of enhanced UV-B treatment after 12 weeks of simulation. *Left*: total glutathion, *right*: total antioxidants

cycle pigments, which are important to dissipate excess energy: reduced in DO, but increasing in CF. After UV-B enhancement there are no significant changes in chlorophylls and total carotenoids, but xanthophyll cycle pigments increased significantly in DO and in CF.

*Flavonoids* and additional *antioxidants* should protect against stress-induced (oxygenic) radicals and as epidermal flavonoid screen against most of the UV-irradiation. After ozone treatment each plant behaved differently (Fig. 4.4): there was a slight decrease in RA, an increase in DO and no change in CF. Unexpectedly, enhanced UV-B reduced total flavonoids in DO and in CF, but led to a small, insignificant increase in RA.

*Four antioxidants* were determined to describe plant cell reactions in the simulation scenario in more detail: ß-carotene, α-tocopherol, total ascorbic acid and glutathion (Fig. 4.5). Ozone induced a tocopherol increase in all species, most pronounced in CF, which also showed the highest absolute content of this antioxidant. Apart from this, the only other significant change in other antioxidants due to ozone treatment is the reduction of ß-carotene in DO. In this species total carotenoids also decrease under high ozone.

After UV-B treatment, the most striking change in antioxidant content is a reduction in glutathion content in all three species (Fig. 4.5), most pronounced in CF. Except from a slight decrease in tocopherol contents of DO and CF, other antioxidants seemed to remain unaffected.

In spring and sometimes during summer, alpine plants can be covered by snow for several hours or days. This is a protection against UV-B irradiation, even with a shallow snow cover. Own measurements in 2,500 m altitude in the Central Alps at the end of June in a larger remaining snow bed, which melted away weeks later, resulted in the following data: UV-B irradiation at noon near to the snowfield: 4 MED/h. Measured under a snow cover of 2 cm: 0.4 MED/h only; 4 cm: 0.08 MED/h, 6 cm: 0.02 MED/h; 8 cm: $3*\ 10^{-3}$ MED/h: 10 cm: $1*10^{-3}$ MED/h; 20 cm: $1*10^{-6}$ MED/h. That means that 2 cm of snow will allow for recovery from possible irradiation damage, especially because alpine plants are physiologically adapted to lower temperatures in their metabolism.

The simulation showed that alpine plants, well protected by firmly bound and extractable flavonoids, seem not to be affected by a UV-B load much higher than they would experience in their natural environment. Irrespectively of whether increased UV comes from stratospheric spring ozone holes or from a migration to higher growth sites as an effect of global warming (Nagy and Grabherr 2009), they will not be threatened. The reduction of glutathion contents should be investigated in more detail as it points to an interesting adaptation in the network of reducing compounds, such as ascorbic acid or the tocopherols, driven by the reducing power of photosynthesis. Difficult to understand is the decrease in total flavonoids in DO and in CF. Both plants have not reached their upper limits of alpine distribution; therefore they should manage higher UV irradiation. However, when the set of flavonoids is changed in a way that newly formed soluble or cell wall integrated compounds such as phenylpropane acids etc. developed, we were not able to measure these compounds with the described method. Figure 4.6 shows the *firmly bound flavonoids* and *phenylpropanes* after detection by fluorescence in leaf sections of DO and SR. Especially the outer epidermal cell walls are stained, and this picture did not change with samples from the field or from the simulation studies.

A more detailed view into cellular changes was allowed by electron microscopy studies. While UV-treatments and the comparison with field samples did not show any remarkable response to the different growth conditions, ozone treatment resulted in drastic ultrastructural changes in the chloroplasts (only). In Fig 4.7 the leaf plastids show the expected intact ultrastructure which is equal in field samples and in samples after 2 months under simulated control conditions – a proof of the quality of the simulation, and of the obvious natural adaptation to the ozone concentrations as supplied in the controls. The plastids of RA showed the plastid protrusions, as they have been described by us for most high alpine plants (Lütz and Moser 1977; Lütz and Engel 2007; Lütz 2010, and Chap. 5 of this book).

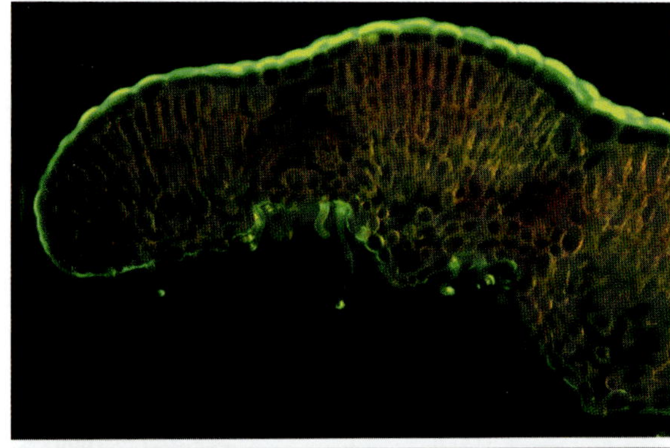

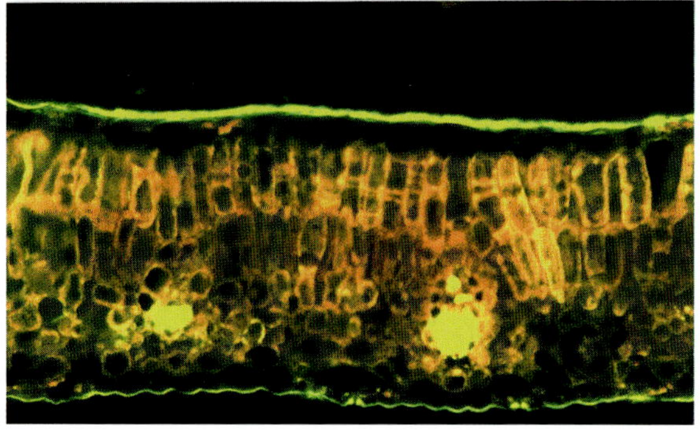

**Fig. 4.6** Localization of firmly bound flavonoids and precursors in the cell walls of *Dryas octopetala* (*top*) and *Salix retusa* (*bottom*). Detection by fluorescence with "Naturstoffreagenz A" in methanol (Hutzler et al. 1998). Especially the outer epidermal walls show the typical greenish-yellow appearance

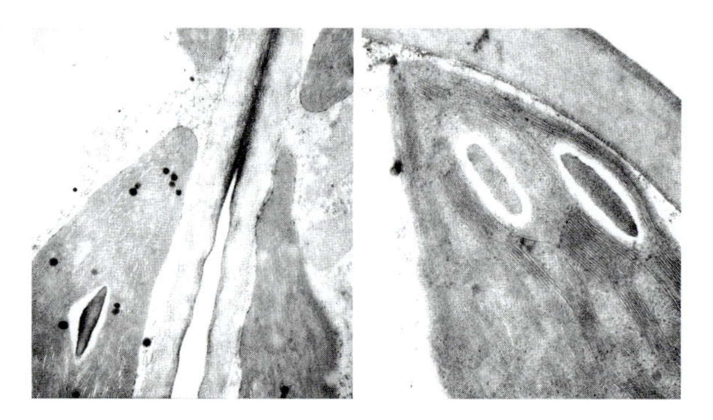

**Fig. 4.7** Examples of plastid ultrastructure in leaves of *Ranunculus alpestris* (*left*, magn. 8,000x), and of *Carex firma* (*right*, magn. 35,000x) after 2 months in climate simulation with low ozone. TEM fixation and embedding procedures followed conventional methods as is described e.g. by Lütz (1996)

High ozone levels, however, changed the thylakoid architecture significantly (Fig. 4.8). Thylakoids

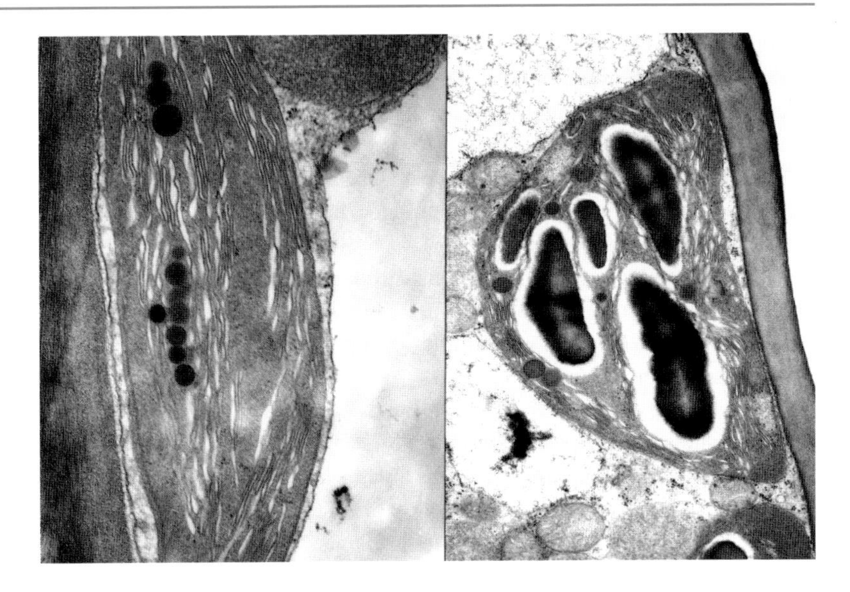

Fig. 4.8 Examples of changes in plastid ultrastructure in leaves of *Ranunculus alpestris* (*left*, magn. 30,000x), and of *Carex firma* (*right*, magn. 20,000 x) after 2 months in the climate simulation with high ozone. TEM fixation and embedding procedures: see Fig. 4.7

developed dilatations, strongest in RA, often observed in CA, but less often seen in DO (not shown). In the dilatations, the membranes undulate; a sign of loosening physiological cooperation between membranes and stroma. The occurrence of plastid protrusions seemed to be strongly reduced. These observations correspond to the reductions found in pigment contents and in photosynthesis. Very similar ozone-induced plastid ultrastructural changes were observed by us in the relatively ozone-resistant spruce trees (Barnes et al. 1995; Schiffgens and Lütz 1992) as well as in other alpine plants taken from the central Alps from up to 3,000 m a.s.l. after similar simulation experiments (Thron 1996).

On a structural cellular level, high ozone, as it can occur under field conditions especially in hot summers in alpine regions, affects thylakoid membranes to a remarkable extent.

High ozone exposures change the vitality of plants considerably. This has been clearly demonstrated in several studies mostly involving trees from subalpine and mountain regions (Blank and Lütz 1990; Herman et al. 1998; Langebartels et al. 1997; Lütz et al. 1998). The typical forest decline symptoms in spruce develop if the trees had Mg deficiency (often measured in mountain spruce needles) *and* additional high ozone in the field. Siefermann-Harms et al. (2004) and Boxler-Baldoma et al. (2006) could clearly show that under these conditions the light harvesting complex is damaged in thylakoids, and that considerable ultrastructural changes in the vascular tissues of the needles occur. If alpine plants above the timberline have imbalanced nutrition and are exposed to higher ozone doses, damage will develop. This effect is stronger if warm weather together with long periods of high insulation prevails. These stress conditions require all defence systems in the cells, which cannot be balanced by photosynthesis in the long run, and finally will result in earlier senescence of the plant. At cooler, humid climate conditions plants better resist ozone (Senser et al. 1990).

These data show that each plant developed individual stress responses and decomposition processes under prolonged oxidative exposures. Ozone remains a problem for herbaceous as well as for forest plants in mountain or alpine regions above 1,500 m a.s.l., UV-B increases only will be tolerated.

### 4.2.4 The Situation for High Alpine Plant Species Under Enhanced UV Plus Enhanced Ozone

Under natural conditions the plants in the Northern Limestone Alps are exposed to UV according to altitude, and the irradiance can increase or decrease as a result of snow cover or cloud formation (see Chap. 1, 2 of this book). The exposure to tropospheric ozone depends strongly on traffic, industry and weather conditions in the alpine environment. In the preceding paragraph we showed that UV-B enhancement did not endanger the physiology or cell structure of the selected alpine plant species. High ozone fumigation, however, could induce several detrimental effects, which are interpreted finally as an earlier onset of senescence.

The combination of both stressors should once more be tested in a simulation setup not far from mountain conditions.

Climate and pollution simulations similar to the chamber studies outlined briefly in Paragraph two were performed for 4 weeks in summer in two consecutive years under the combined stress loads of ozone plus UV-B. The controls were exposed to 3–4 MED/h of UV-B and a background ozone level of 20–30 ppb. Again, plants with soil were taken from the field at 2,250 m a.s.l. and studied as described before.

The simultaneous treatment of alpine plants (RA, DO, CF) with high ozone (90 ppb average) plus high UV-B (6 MED/h) resulted in slightly, but not significantly reduced ozone symptoms for several parameters when compared with the ozone-only treatment in a parallel simulation. In addition, the natural UV resistance of the plant leaves is not diminished by this oxidative load. By contrast, flavonoid pools increased significantly in CF, and slightly in DO and RA. According to HPLC-absorption spectra of separated flavonoids, they represent mostly kaempferol and quercetin derivatives, and in case of CF also luteolin. Their unexpected stability under ozone is supported by studies on lowland or laboratory plants suggesting that UV-B accelerates those defence pathways which were also described for ozone. This includes the huge flavan pool, such as anthocyanins and flavonoids, acting as antioxidants against oxygen radicals in green leaves (Gould et al. 2002; Stratmann 2003; Smirnoff 2005; Fujibe et al. 2004).

Other parameters remained unaffected when comparing the UV/ozone treatment with the control group: this holds for PS II activity (as Fv/Fm), lipid peroxidation levels or content of xanthophyll cycle pigments. Photosynthetic oxygen development seemed to be slightly increased under the double treatment during 4 weeks of observation, and the level of total antioxidants is significantly higher (Fig. 4.9 for RA, similar in DO and CF). In addition, α-tocopherol more than doubles in RA, and by about 30% in DO and CF (data not shown). The assay of total antioxidants includes flavonoids and tocopherols among others.

From these results we can assume that high alpine plants can cope with a simultaneous double load of increased UV and high ozone – as long as high ozone is not present for most of the vegetation period.

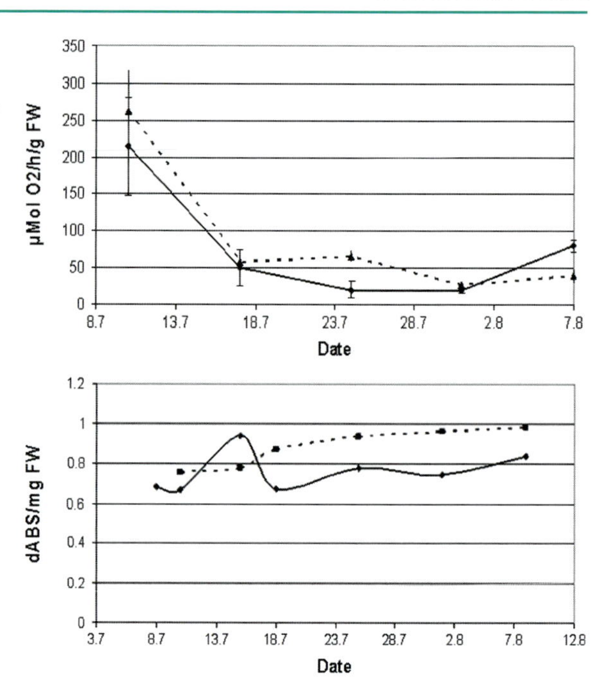


**Fig. 4.9** Changes in photosynthetic oxygen formation (*top*) and in total antioxidants (*bottom*) over 4 weeks of simulation for *Ranunculus alpestris*. Continuous line: control, broken line: UV + $O_3$ treatment

## 4.3 The Situation for Polar Plant Species

Polar regions are not endangered by high ozone emission problems like in highly populated and industrialized regions of the world. Therefore human activities as a possible problem for metabolism of polar plants is currently restricted to enhanced UV irradiation, with a greater short wave input in the southern hemisphere (Kirchhoff and Echer 2001; Huiskes et al. 2003; Bargagli 2008; Xu and Sullivan 2010).

For the northern hemisphere, several experiments, either shielding natural UV with filters, or, less frequently, enhancing UV irradiation e.g. in the field, have been reported (Björn et al. 1999; Caldwell et al. 1982, 2007; Nybakken et al. 2004; Rozema et al. 2005). When experimental setups use a control environment with "no UV", of course the "treatment" with natural UV or UV increments will result in reduced activities of a number of physiological processes (Ruhland and Day 2000; Albert et al. 2008). However, such experimental designs do not respect the natural adaptation to the surrounding light climate, which always contains UV, and therefore cannot be interpreted for negative effects of UV irradiation. Omitting UV in the long run is dangerous for field grown plants, because UV-B and especially UV-A induce pathways such as flavonoid formation, which is of great importance to establish antiradical and antioxidant power as a defence against different pathogens (Smirnoff 2005; Lütz 2010).

A number of studies have shown that arctic plant species are well adapted to higher UV irradiation (Björn et al. 1999; Hessen 2002; Crawford 2008). This can, at least in part, be explained by the occurrence of the same species (DO, *Oxyria digyna*, *Saxifraga oppositifolia*, *Bistorta vivipara*, *Poa alpina* etc.) or closely related species (many Ranunuculales, Saxifragaceae etc.) in the Alps or in the Rocky Mountains, where elevation exposes the plants to much higher UV intensities than under arctic ozone holes. This resistance seems to be genetically fixed and has so far protected arctic plants.

UV as a possible stressor in the only two antarctic phanerogams, *Deschampsia antarctica* (Gramineae) (DA) and *Colobanthus quitensis* (CQ) (Caryophyllaceae), has been studied by several groups, again using UV blocking as the control (Day et al. 1999; Ruhland and Day 2000). Lütz et al. (2008) described the flavonoid composition of both plants, harvested on King George Island, maritime Antarctic, after HPLC separation and identified some compounds which are also found in European alpine plants in considerable amounts.

Apart from flavonoids, polyamines have been described as powerful protecting substances in the plant cell; they are partially located in the thylakoid membranes (Kotzabasis et al. 1999). If polyamines are applied to plants, chilling injury can be avoided to some extent (Kramer and Wang 1989). Their protection has further been studied e.g. in tobacco under high ozone fumigation in the same simulation chambers as described above (Navakoudis et al. 2003). Polyamines play an important role in protecting cucumber plants under exposure to UV-B irradiation (Kramer et al. 1991). Since in a previous study we could confirm that polyamines protect photosynthesis under enhanced UV simulation in different tobacco cultivars (Lütz et al. 2005), we studied the content of polyamines in both antarctic phanerogams with samples taken together with the samples for the flavonoid studies mentioned in Lütz et al. (2008). Polyamine identification was performed in the lab of K. Kotzabasis (Dept Biology, Univ. Heraklion, Greece) as described by Kotzabasis et al. (1993) and Lütz et al. (2005). Figure 4.10 compares the contents of the three polyamins putrescine, spermin and spermidin in both plants. The three data sets per plant represent plant samples from different locations so as to get an impression of the natural variability. DA contains about twice the amount of putrescine compared to CQ, while spermidine and spermine occur in similar amounts in both plants.

The photosynthetic stability of both antarctic plants will additionally be guaranteed by the polyamines; their role in UV- and cold protection has to be evaluated in the future.

In contrast to these setups, we were able to study DA for possible effects of naturally simulated UV irradiation higher than that measured around the antarctic peninsula, with this irradiation as a control. Plants of DA were taken from Robert Island by G. Zuniga (Univ. Conception, Chile), grown further in a normal climate chamber under low PAR without UV for 1 year. In cooperation with G. Zuniga and L.J. Corcuera some plants of DA were sent to our climate simulation group at the Helmholtz Zentrum

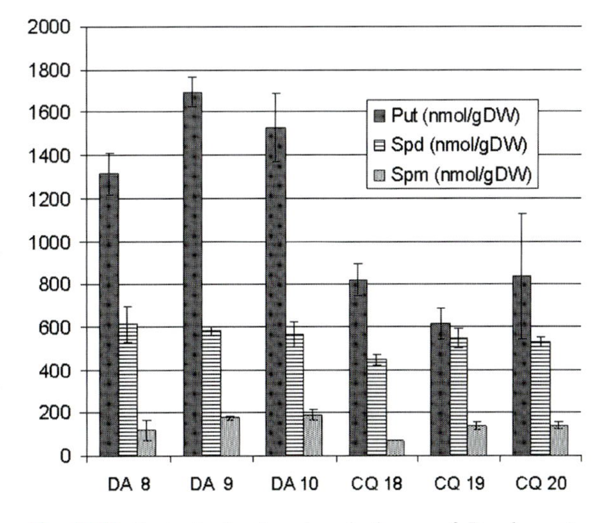


**Fig. 4.10** Content of polyamines in leaves of *Deschampsia antarctica* and *Colobanthus quitensis*. Put: putrescine; Spd: spermidin; Spm: spermin. Courtesy of Prof. K. Kotzabasis, Heraklion, Greece

München, where the plantlets had to be raised to obtain more biomass under similar lab conditions as at the Univ. of Conception. This means that the antarctic hair grass (DA) has not obtained PAR or UV as at the growth sites for more than a year. For the UV-stress experiments, plants were exposed to PAR of 1,000 μmol $m^{-2}$ $s^{-1}$ and UV-B irradiation of 3 MED/h in the aforementioned sun simulators (similar to Palmers station, maritime Antarctic) in the control; the enhanced UV-B treatment had 1,100 μmol $m^{-2}$ $s^{-1}$ PAR and 5 MED/h. Samples were taken during and after the experiment, which lasted for 14 days of simulation.

At the end of the experiment, the control (with "antarctic" UV-B) and the "high UV-B" group were visibly equal. The same holds for the ultrastructural appearance (Fig. 4.11). Cells appear unaffected in all samples, and the plastids contain well developed thylakoids, grana, envelopes. The only difference stems from the shape of the plastids. Earlier results of DA samples prepared by us for TEM directly at the growth site on King George Island (maritime Antarctic) showed plastid protrusions as reported by us for more than 30 polar and alpine plants as an adaptation to the short vegetation period (Lütz and Moser 1977; Lütz 1987; Lütz and Engel 2007; Lütz 2010, and Chap. 5 in this book). These protrusions did not develop in these tested samples of DA (Fig. 4.9) which had been grown for more than a year under artificial – and in this case beneficial – conditions without the pressure to adapt their physiology and ultrastructure to cold temperatures plus occasionally strong light. This observation supports our findings with *Oxyria digyna*, which did not form protrusions after having grown this high alpine plant under mild climate conditions in an institute garden (Lütz and Engel 2007).

Physiological observations of DA plants with tissues that had developed more or less UV–free and under low PAR before the simulation started did not reveal striking differences. Oxygen development of photosynthesis was followed for 13 days (Fig. 4.12). After 2 days down-regulation of activities ends up in stable activities without differences between the treatments. This down-regulation was expected as the plants were transferred from low light and continuous mild temperatures to the simulation conditions. The lower light intensity (PAR) during the assays represents the non-light-saturated condition, and the

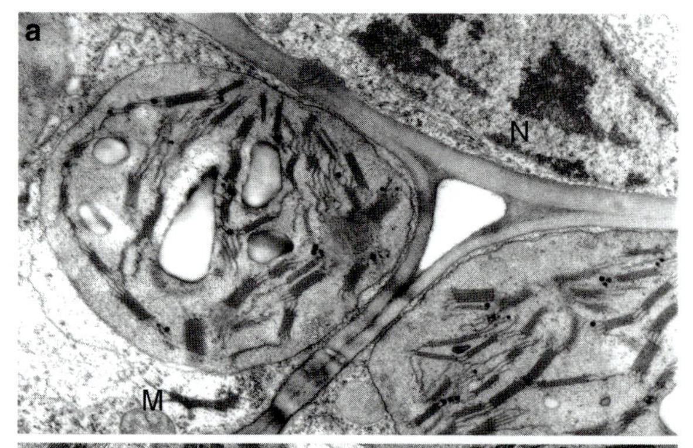


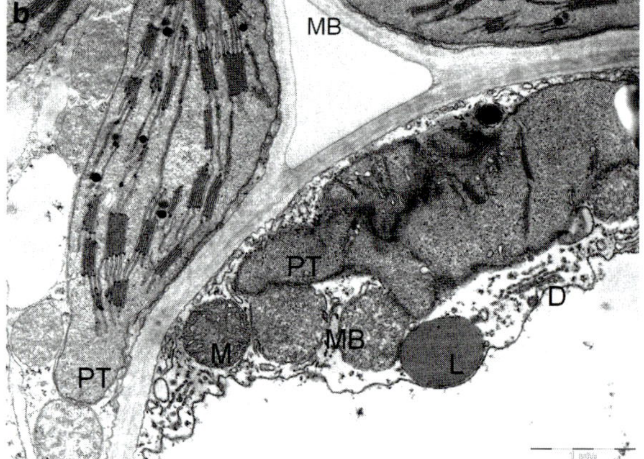


**Fig. 4.11** (**a**) *Deschampsia antarctica* plastids after 2 weeks under high UV-B treatment. These samples were grown before the experiment for more than 1 year under artificial, low PAR-light (no UV) conditions. The control groups did not differ in ultrastructure. (**b**) *Deschampsia antarctica* plastids fixed for TEM at the growth site on King George Island, maritime Antarctic. Only these samples developed protrusions (PT), close to mitochondria (M) and microbodies (MB). L: lipid body, D: part of a dictyosome. N: nucleus

1,500 μmol $m^{-2}$ $s^{-1}$ light intensity represents saturating, but not photoinhibition conditions. At the lower light intensity during measurement a small inhibiting effect of high UV-B became visible, which disappeared with time.

No significant changes were observed in the content or composition of all photosynthetic pigments. Only the thylakoid-located α-tocopherol became significantly reduced over time (Fig. 4.13a). This does not hold for the precursor γ-tocopherol, which remained completely unchanged. High UV seems to have some effect on this lipophilic antioxidant, but without changes in membrane activity or structure. Probably the final decrease of about 20% in α-tocopherol was balanced by other antioxidants in the plastid.

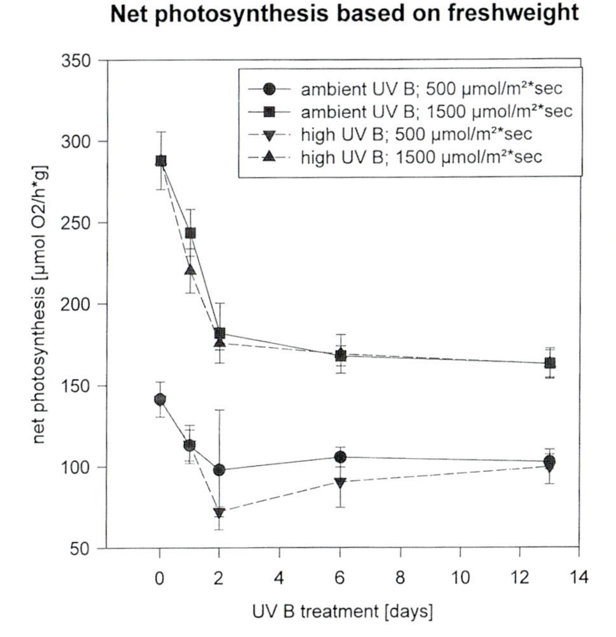


**Fig. 4.12** Net photosynthesis of *Deschampsia antarctica* under climate simulation with field conditions (control) and enhanced UV-B. Two different light intensities (PAR) during measurement of oxygen development were applied

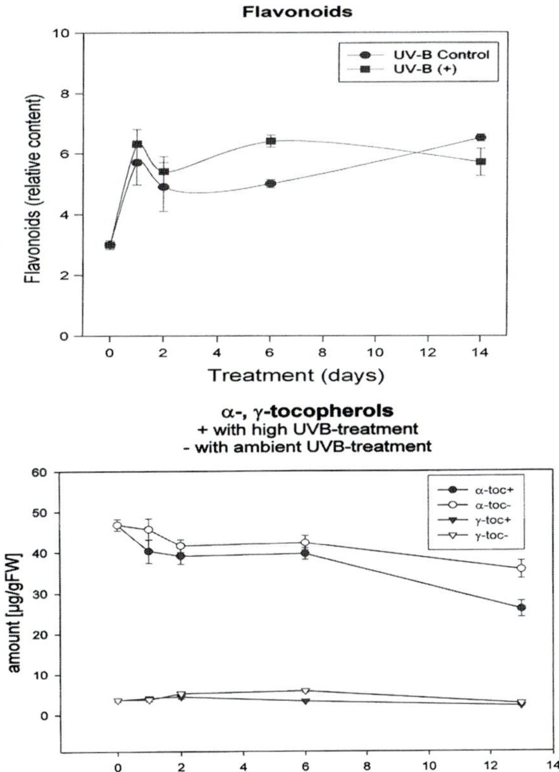


**Fig. 4.13** (**a**) *Deschampsia antarctica*: changes in the relative content of total flavonoids (see Fig. 4.4) during the simulation experiment. (**b**) Changes in the content of the two main tocopherols during simulation. FW: fresh weight

As expected, the pool of soluble flavonoids increased about threefold from the beginning of the experiment in both treatments, with the high UV-B leading (Fig. 4.13b). At the end of the experiment, the differences between control and high UV-B seem to disappear.

The simulation experiment shows that DA has a strong genetic competence to cope with UV irradiation, even if this exposure is not present for a long time (over a year). An onset of UV irradiation in the experiment, perhaps comparable with snow melt in the field, will immediately increase flavonoids, and protection of photosynthetic metabolism and cell structures is guaranteed. This also holds for UV-B irradiation much higher than the plant would experience due to antarctic ozone holes. The stability can be explained by the original occurrence of DA (and in all probability also for the other antarctic phanerogam, *Colobanthus quitensis*), because they grow in altitudes over 3,000 m a.s.l. in the High Andes (Casanova-Katny et al. 2006; Bravo et al. 2007). An alpine location absolutely requires best UV protection.

## Conclusion

It can be expected that a possible increase in UV-B by stratospheric ozone holes will not threaten alpine vegetation in the future, which supports a similar, generalized statement for plants made by Xu and Sullivan (2010). Our experimental results are supported by an observation of Shi et al. (2004), which suggests that mountain plants in 3,200 m a.s. l. in the Himalayas did not reduce photosynthetic activity under additional UV irradiation in the field. In addition to a perfect set of flavonoids and precursors, the radicals formed by such a surplus of irradiation can effectively be removed by several antioxidative mechanisms (Mackerness and Thomas 1999; Brosché and Strid 2003; Smirnoff 2005).

Ozone will remain a dangerous pollutant for most species in alpine vegetation, especially from

1,500 m a.s.l and higher, because it persists in the air longer than at lower altitudes (Herman et al. 1998; Lütz et al. 1998; Smidt 1993, 1998). Alpine plants will sometimes need more time to cope with both stressors than a high altitude vegetation period may allow, especially if senescence starts earlier because of pollutant activity. The results presented here are supported by observations made by Thron (1996), who studied different high alpine plants from the less polluted Central Alps, taken from between 2,400 m and 3,000 m a.s.l.. Under ozone fumigation in the large plant climate chambers described above he found yellowing of leaf tips of *Carex curvula* and yellowing plus undulating leaf surface in *Leontodon helveticus*, together with reductions in photosynthesis.

Future simulation studies on ozone effects should be conducted with arctic plants growing in a mostly unpolluted atmosphere and lower UV-B irradiation; several related species also occur in the Alps, and arctic species could be planted in alpine plots to follow different or common adaptation strategies. This will help to understand the broad spectrum of defence reactions in alpine / polar plants in general.

**Acknowledgements** This work is mainly the result of two research projects to C.L. by the Bavarian Ministry for Environment and Health; we are grateful for this support. The following scientists and technicians have highly contributed to this work over a period of more than 6 years: I. Darchinger, A. Haniss, Dipl. Biol. K. Heinz, Dr. G. Lehner, P. Kary, R. Kuhnke, Dipl. Phys. A. Kuhnt, Dr. B. Wallner.

## References

Alberdi M, Bravo LA, Gutiérrez A, Gidekel M, Corcuera LJ (2002) Ecophysiology of Antarctic vascular plants. Physiol Plantarum 115:479–486

Albert KR, Mikkelsen TN, Ro-Poulsen H (2008) Ambient UV-B radiation decreases photosynthesis in high arctic *Vaccinium uliginosum*. Physiol Plantarum 133:199–210

Albert A, Sareedenchal V, Heller W, Seidlitz HK, Zidorn C (2009) Temperature is the key to altitudinal variation of phenolics in *Arnica montana* L. Cv. ARBO. Oecologia 160:1–8

Amils R, Ellis-Evans C, Hinghofer-Szalkay H (eds) (2007) Life in extreme environments. Springer, Dordrecht

Ballaré CL, Caldwell MM, Flint SD, Robinson SA, Bornman JF (2011) Effects of solar ultraviolet radiation on terrestrial ecosystems. Patterns, mechanisms, and interaction with climate change. Photochem Photobiol Sci 10:226–241

Bargagli R (2008) Antarctic ecosystems. Environmental contamination, climate change, and human impact, vol 175, Ecolog. Studies. Springer, Berlin

Barnes JD, Pfirrmann T, Steiner K, Lütz C, Busch U, Küchenhoff H, Payer HD (1995) Effects of elevated $CO_2$, $O_3$ and K deficiency on Norway spruce (Picea abies): seasonal changes in photosynthesis and non-structural carbohydrate content. Plant Cell Environ 18:1345–1357

Björn LO, Callaghan TV, Gehrke C, Gwynn-Jones D, Lee JA, Johanson U, Sonesson M, Buck ND (1999) Effects of ozone depletion and increased ultraviolet-B radiation on northern vegetation. Polar Res 18(2):331–337

Blank LW, Lütz C (1990) Tree exposure experiment in closed chambers. Environ Pollut 64(Special issue):189–399

Blumthaler M (2007) Factors, trends and scenarios of UV radiation in arctic-alpine environments. In: Oerbaeck JB, Kallenborn R, Tombre I, Hegseth EN, Falk-Petersen S, Hoel AH (eds) Arctic Alpine ecosystems and people in a changing environment. Springer, Berlin

Blumthaler M, Ambach W (1990) Indications of increasing solar ultraviolet-B radiation flux in alpine regions. Science 248:206–208

Blumthaler M, Ambach W, Ellinger R (1996) UV-Bestrahlung von horizontalen und vertikalen Flächen im Hochgebirge. Wetter Leben 1:25–31

Blumthaler M, Ambach A, Ellinger R (1997) Increase in solar UV radiation with altitude. J Photochem Photobiol B 39:130–134

Bornman JF, Reuber S, Cen L, Weissenböck G (1997) Ultraviolet as a stress factor and the role of protective pigments. In: Lumsden P (ed) Plants and UV-B: responses to environmental change. Cambridge University Press, Cambridge, pp 157–168

Boxler-Baldoma C, Lütz C, Heumann HG, Siefermann-Harms D (2006) Structural changes in the vascular bundles of light-exposed and shaded spruce needles suffering from Mg deficiency and ozone pollution. J Plant Physiol 163:195–205

Bravo LA, Ulloa N, Zuñiga GE, Casanova A, Corcuera LJ, Alberdi M (2001) Cold resistance in Antarctic angiosperms. Physiol Plantarum 111:55–65

Bravo LA, Saavedra-Mella FA, Vera F, Guerra A, Cavieres LA, Ivanov AG, Huner NPA, Corcuera LJ (2007) Effect of cold acclimation on the photosynthetic performance of two ecotypes of *Colobanthus quitensis* (Kunth.) Bartl. J Exp Bot 58(13):3581–3590

Brosché M, Strid A (2003) Molecular events following perception of ultraviolet-B radiation by plants. Physiol Plantar 117:1–10

Caldwell MM, Robberecht R, Nowak R (1982) Differential photosynthetic inhibition by ultraviolet radiation in species from the arctic-alpine life zone. Arctic Alpine Res 14:195–202

Caldwell MM, Bornman JF, Ballaré CL, Flint SD, Kulandaivelu G (2007) Terrestrial ecosystems, increased solar ultraviolet radiation, and interactions with other climate change factors. Photochem Photobiol Sci 6:252–266

Casanova-Katny MA, Bravo LA, Molina-Montenegro M, Corcuera LJ, Cavieres LA (2006) Photosynthetic performance of *Colobanthus quitensis* (Kunth) Bartl.

(Caryophyllaceae) in a high-elevation site of the Andes of central Chile. Rev Chil Hist Nat 79:41–53

Crawford RMM (1997) Habitat fragility as an aid to long-term survival in arctic vegetation. In: Woodin SJ, Marquiss M (eds) Ecology of Arctic environments. Blackwell, Oxford, pp 113–136, ISBN 0-632-04218-4

Crawford RMM (2008) Plants at the margin. Ecological limits and climate change. Cambridge University Press, Cambridge

Day TA, Ruhland CT, Grobe CW, Xiong F (1999) Growth and reproduction of Antarctic vascular plants in response to warming and UV radiation reductions in the field. Oecologia 119:24–35

Döhring T, Köfferlein M, Thiel S, Seidlitz HK (1996) Spectral shaping of artificial UV-B irradiation for vegetation stress research. J Plant Physiol 148:115–119

Fiscus E, Booker L (1995) Is increased UV-B a threat to crop photosynthesis and productivity? Photosynth Res 43:81–92

Flint SD, Caldwell MM (2003) A biological weighting function for ozone depletion research with higher plants. Physiol Plantar 117:137–144

Frederick JE (1990) Trends in atmospheric ozone and ultraviolet radiation: mechanisms and observations for the Northern Hemisphere. Photochem Photobiol 51:757–763

Fujibe T, Saji H, Arakawa K, Yabe N, Takeuchi Y, Yamamoto K (2004) A methyl-viologen-resistant mutant of Arabidopsis, which is allelic to ozone-sensitive rcd1, is tolerant to supplemental ultraviolet-B irradiation. Plant Physiol 134:275–285

Götz M, Albert A, Stich S, Heller W, Scherb H, Krins A, Langebartels C, Seidlitz HK, Ernst D (2010) PAR modulation of the UV-dependent levels of flavonoid metabolites in *Arabidopsis thaliana* (L.) HEYNH. leaf rosettes: cumulative effects after a whole vegetative growth period. Protoplasma 243:95–103

Gould KS, Mckelvie J, Markham K (2002) Do anthocyanins function as antioxidants in leaves? Imaging of $H_2O_2$ in red and green leaves after mechanical injury. Plant Cell Environ 25:1261–1269

Griffith OW (1980) Determination of glutathione and glutathione disulphide using glutathione reductase and 2-vinylpyridine. Anal Biochem 106:207–212

Herman F, Lütz C, Smidt S (1998) Pollution related stress factors to forest ecosystems - Synopsis. Environ Sci Pollut Res 1(Special issue No):2–15

Hessen DO (2002) UV radiation and Arctic ecosystems, vol 153, Ecological Studies. Springer, Berlin Heidelberg

Huiskes AHL, Gieskes WWC, Rozema J, Schorno RML, van der Vies SM, Wolff WJ (eds) (2003) Antarctic biology in a global context. Backhuys Publishers, Leiden, ISBN 90-5782-079-X

Hutzler P, Fischbach R, Heller W, Jungblut T, Reuber S, Schmitz R, Veit M, Weissenböck G, Schnitzler J (1998) Tissue localization of phenolic compounds in plants by confocal laser scanning microscopy. J Exp Bot 49: 953–965

Johnson GA, Day TA (2002) Enhancement of photosynthesis in Sorghum bicolor by ultraviolet radiation. Physiol Plantar 116:554–562

Kappen L (1983) Anpassungen von Pflanzen an kalte Extremstandorte. Ber Dtsch Bot Ges 96:87–101

Kirchhoff VWJH, Echer E (2001) Erythema UV-B exposure near the Antarctic Peninsula and comparison with an equatorial site. J Photochem Photobiol B 60:102–107

Körner C (2003) Alpine plant life, 2nd edn. Springer, Berlin

Körner C, Larcher W (1988) Plant life in cold climates. Symp Soc Exp Biol 42:25–57

Kotzabasis K, Christakis-Hampsas MD, Roubelakis-Angelakis KA (1993) A narrow-bore HPLC method for the identification and quantitation of free, conjugated and bound polyamines Anal. Biochem 214:484–489

Kotzabasis K, Strasser B, Navakoudis E, Senger H, Dörnemann D (1999) The regulatory role of polyamines in structure and functioning of the photosynthetic apparatus during photoadaptation. J Photochem Photobiol B Biol 50:45–52

Kramer GF, Wang CY (1989) Effects of chilling and temperature preconditioning on the activity of polyamine biosynthetic enzymes in zucchini. J Plant Physiol 136:115–119

Kramer GF, Norman HA, Krizek DT, Mirecki RM (1991) Influence of UV-B radiation on polyamines, lipid peroxidation and membrane lipids in cucumber. Phytochemistry 30:2101–2108

Krupa SV, Kickert RN (1989) The greenhouse effect: impact of ultraviolet-B (UV-B) radiation, carbon dioxide ($CO_2$) and ozone ($O_3$) on vegetation. Environ Pollut 61:263–293

Langebartels C, Ernst D, Heller W, Lütz C, Payer HD, Sandermann H (1997) Ozone responses of trees: results from controlled chamber exposures at the GSF Phytotron. In: Sandermann H, Wellburn A, Heath R (eds) Forest decline and ozone: a comparison of controlled chamber and field experiments, vol 127, Ecological Studies. Springer Verlag, Berlin, pp 163–200

Larcher W (2001) Ökophysiologie der Pflanzen, 6th edn. Ulmer, Stuttgart

Larcher W, Wagner J (2009) High mountain bioclimate: temperatures near the ground recorded from the timberline to the nival zone in the Central Alps. Contrib Nat Hist 12:857–874

Larcher W, Kainmüller C, Wagner J (2010) Survival types of high mountain plants under extreme temperatures. Flora 205:3–18

Lütz C (1987) Cytology of high alpine plants II. Microbody activity in leaves of *Ranunculus glacialis*. Cytologia 52:679–686

Lütz C (1996) Avoidance of photoinhibition and examples of photodestruction in high alpine *Eriophorum*. J Plant Physiol 148:120–128

Lütz C (2001) Prüfung der möglichen Gefährdung von Hochgebirgspflanzen durch erhöhtes Ozon und durch UV-B Strahlung. 56 p. Report Bavarian Ministry Environment and Health, Munich, Germany

Lütz C (2004) Veränderung der UV-Resistenz von kalkalpinen Hochgebirgspflanzen durch Ozon. 51 p. Report Bavarian Ministry Environment and Health, Munich, Germany

Lütz C (2010) Cell physiology of plants growing in cold environments. Protoplasma 244:53–73, Review

Lütz C, Engel L (2007) Changes in chloroplast ultrastructure in some high alpine plants: adaptation to metabolic demands and climate? Protoplasma 231:183–192

Lütz C, Moser W (1977) Beiträge zur Cytologie hochalpiner Pflanzen. I. Untersuchungen zur Ultrastruktur von *Ranunculus glacialis L.* Flora 166:21–34

Lütz C, Kuhnke-Thoss R, Thiel S (1998) Natural and anthropogenic influences on photosynthesis in trees of alpine forests. Environ Sci Pollut Res 1:88–95

Lütz C, Navakoudis E, Seidlitz HK, Kotzabasis K (2005) Simulated solar irradiation with enhanced UV-B adjust plastid- and thylakoid-associated polyamine changes for UV-B protection. Biochim Biophys Acta 1710:24–33

Lütz C, Blassnigg M, Remias D. (2008) Different flavonoid patterns in *Deschampsia antarctica* and *Colobanthus quitensis* from the maritime Antarctic. In: Wiencke Ch, et al. (eds.) Reports on Polar and Marine Research, Bremerhaven, 571:192–199

Mackerness SAH, Thomas B (1999) Effects of UV-B radiation: gene expression and signal transduction pathways. In: Smallwood MF, Calvert CM, Bowles DJ (eds) Plant Responses to Environmental Stress, BIOS Scientific Publishers, Oxford 17–24

Marchant HJ (1997) Impacts of ozone depletion on Antarctic organisms. In: Battaglia B, Valenica J, Walton DWH (eds) Antarctic communities: species, structure, and survival. Cambridge University Press, Cambridge, pp 367–374

McKenzie RL, Aucamp PJ, Bais AF, Björn LO, Ilyas M, Madronich S (2011) Ozone depletion and climate change: impacts on UV radiation. Photochem Photobiol Sci 10:182–198

Nagy L, Grabherr G (2009) The biology of Alpine habitats. University Press, Oxford

Navakoudis E, Lütz C, Langebartels C, Lütz-Meindl U, Kotzabasis K (2003) Ozone impact on the photosynthetic apparatus and the protective role of polyamines. Biochim Biophys Acta – General Subjects 1621:160–169

Nybakken L, Bilger W, Johanson U, Björn LO, Zielke M, Solheim B (2004) Epidermal UV-screening in vascular plants from Svalbard (Norwegian Arctic). Polar Biol 27:383–390

Oerbaek J, Tombre I, Kallenborn R (2004) Challenges in Arctic–Alpine environmental research. Arctic Antarctic Alpine Res 36:281–283

Oppeneiger C (2008) Einfluss von klimatischen Faktoren auf den Primär- und Sekundärstoffwechsel von Dryas octopetala L. Ph. D. thesis, University of Innsbruck

Payer HD, Blodow P, Köfferlein M, Lippert M, Schmolke W, Seckmeyer G, Seidlitz H, Strube D, Thiel S (1993) Controlled environmental chambers for experimental studies on plant responses to $CO_2$ and interactions with pollutants. In: Schulze ED, Mooney HA (eds) Design and execution of experiments on $CO_2$ enrichment. Commission European Communities, Brussels, pp 127–145

Rozema J, van de Staaij J, Björn LO, Caldwell M (1997) UV-B as an environmental factor in plant life: stress and regulation. Tree 12:22–28

Rozema J, Boelen P, Blokker P (2005) Depletion of stratospheric ozone over the Antarctic and Arctic: responses of plants of polar terrestrial ecosystems to enhanced UV-B, an overview. Environ Pollut 137:428–442

Ruhland CT, Day TA (2000) Effects of ultraviolet-B radiation on leaf elongation, production and phenylpropanoid concentrations of *Deschampsia antarctica* and *Colobanthus quitensis* in Antarctica. Physiol Plantarum 109:244–251

Schiffgens A, Lütz C (1992) Ultrastructure of mesophyll cell chloroplasts of spruce needles exposed to $O_3$, $SO_2$, $NO_2$ alone and in combination. Environ Exp Bot 32:243–254

Schnitzler JP, Jungblut TP, Heller W, Köfferlein M, Hutzler P, Heinzmann U, Schmelzer E, Ernst D, Langebartels C, Sandermann H Jr (1996) Tissue localization of UV-B-screening pigments and of chalcone synthase mRNA in needles of Scots pine seedlings. New Phytol 132:247–258

Senser M, Kloos M, Lütz C (1990) Influence of soil substrate and ozone plus mist on the pigment content and composition of needles from young spruce trees. Environ Pollut 64:295–312

Shi S, Zhu W, Zhou D, Han F, Zhao X, Tang Y (2004) Photosynthesis of Saussurea superba and Gentiana straminea is not reduced after long-term enhancement of UV-B radiation. Env Explt Bot 51:75–83

Siefermann-Harms D, Payer HD, Schramel P, Lütz C (2004) The effect of ozone on the yellowing process of magnesium-deficient clonal Norway spruce grown under defined conditions. J Plant Physiol 162:195–206

Smidt S (1993) Die Ozonsituation in alpinen Tälern Österreichs. Centralbl Ges Forstwesen 110:205–220

Smidt S (1998) Risk assessment of air pollutants for forested areas in Austria, Bavaria and Switzerland. Environ Sci Pollut Res 1:25–31

Smidt S, Englisch M (1998) Die Belastung von österreichischen Wäldern mit Luftverunreinigungen. Centralbl. f.d. gesamt. Forstwesen 115(4):229–248

Smirnoff N (2005) Antioxidants and reactive oxygen species in plants. Blackwell Publishing, Oxford

Strasser RJ, Srivastava A, Govindjee (1995) Polyphasic chlorophyll *a* fluorescence transient in plants and cyanobacteria. Photochem Photobiol 61:32–42

Stratmann J (2003) Ultraviolet-B radiation co-opts defense signaling pathways. Trends Plant Sci 8:526–533

Streb P, Feierabend J, Bligny R (1997) Resistance of photoinhibition of photosystem II and catalase and antioxidative protection in high mountain plants. Plant Cell Environ 20:1030–1040

Tevini M, Teramura AH (1989) UV-B effects on terrestrial plants. Photochem Photobiol 50:479–487

Thiel S, Döhring T, Köfferlein M, Kosak A, Martin P, Seidlitz HK (1996) A phytotron to plant stress research: how far can artificial lighting compare to natural sunlight? J Plant Physiol 148:456–463

Thron C (1996) Auswirkungen globaler Umweltveränderungen auf Photosynthese und Ultrastruktur von alpinen Hochgebirgspflanzen.- Ph. D. thesis, University München

Wielgolaski FE, Karlsen SR (2007) Some views on plants in polar and alpine regions. Rev Environ Sci Biotechnol 6:33–45

Wiencke Ch, Ferreyra GA, Abele D, Marenssi S (eds) (2008) Reports on Polar and Marine Research, Vol. 571, AWI, Bremerhaven, ISSN 1618–3193

Wildi B, Lütz C (1996) Antioxidant composition of selected high alpine plant species from different altitudes. Plant Cell Environ 19:138–146

Xiong FS, Ruhland CT, Day TA (1999) Photosynthetic temperature response of the Antarctic vascular plants *Colobanthus quitensis* and *Deschampsia antarctica*. Physiol Plantarum 106:276–286

Xu Ch, Sullivan J (2010) Reviewing the technical designs for experiments with ultraviolet-B radiation and impact on photosynthesis, DNA and secondary metabolism. J Integr Plant Biol 52:377–387

# 5 Cell Organelle Structure and Function in Alpine and Polar Plants are Influenced by Growth Conditions and Climate

Cornelius Lütz, Paul Bergweiler, Lavinia Di Piazza, and Andreas Holzinger

## Abbreviations

| | |
|---|---|
| ARP | Anti-radical power |
| AWI | Alfred Wegener institute for polar research |
| CP | Chloroplast protrusion |
| Fv/Fm | Photosynthetic optimum quantum yield |
| LM | Light microscopy |
| PAR | Photosynthetic active radiation |
| PS II | Photo system II |
| SEM | Scanning electron microscopy |
| TEM | Transmission electron microscopy |

## 5.1 Introduction

The alpine environment requires high flexibility in morphology, anatomy, cell structures and physiology for all biological life forms that grow and propagate there. These demands also characterize most of the polar environments.

Alpine and polar plants have been studied in several ecophysiological and physiological aspects, especially to describe the light- or temperature adaptations of their unique growth environments (Amils et al. 2007; Billings 1974; Crawford 2008; Körner 2003; Larcher 2001; Larcher et al. 2010; Lütz 2010). In comparison to lowland plants or crops, many biochemical processes were found changed, such as intensified metabolic activities or higher antioxidant amounts as a result of the pressure of the local climate and growth conditions – and the short time span for physiological processes, growth and propagation. Cell membranes as places of vital electron transport chains, of biosyntheses and controlling metabolite exchanges between compartments should also show signs of special adaptations. But cellular functions can only be understood when the physiological and structural (mostly membrane) properties of the cells/tissues are known and interpreted in a common view. A combination of modern TEM/SEM analyses, advanced light microscopy (e.g. confocal) and cell metabolic studies therefore provides best chances to approach an understanding of high alpine and polar plant life on a cellular level.

When investigating higher plants from the Arctic, a detailed description of occurring species and plant associations serves as a good basis (e.g. Möller et al. 2001). Other arctic field studies combine systematics, ecology or geobotany (Eurola 1968; Crawford and Balfour 1983; Hadac 1989; Heide 2005; Wielgolaski and Karlsen 2007). Some ecophysiological pieces of research addressed the cold adaptation of plants (Kappen 1983; Wüthrich et al. 1999; Robinson et al. 1999). The fragility of arctic plant covers and examples of their survival strategies have been described by Crawford (1997, 2008) and Oerbaeck et al. (2007). Several of the investigated species are also found in the High Alps, for which numerous

C. Lütz (✉) • L. Di Piazza • A. Holzinger
Institute of Botany, University of Innsbruck, Innsbruck, Austria
e-mail: Cornelius.luetz@uibk.ac.at

P. Bergweiler
German Aerospace Center Project Management Agency, Bonn, Germany

C. Lütz (ed.), *Plants in Alpine Regions*,
DOI 10.1007/978-3-7091-0136-0_5, © Springer-Verlag/Wien 2012

publications offer a great deal of physiological and ultrastructural data.

A short comparison of ecophysiological data collected for alpine vs. arctic plants is presented in Körner (2003) and in Crawford (2008).

Plant biologists working with higher plant stress- or ecophysiology have only limited sources of plants in the Antarctic. In coastal regions, mostly around and on the Antarctic Peninsula and surrounding islands, only two vascular plants have survived for thousands of years: *Deschampsia antarctica* (Gramineae) and *Colobanthus quitensis* (Caryophyllaceae) (Lewis Smith 2003; Mosyakin et al. 2007). Both plant species can also be found in the High Andes, while other high mountain plants have not been able to establish continuous growth in the maritime Antarctic. The two antarctic species obviously reach their limits of distribution when air humidity becomes too low and water access is too limited because of frozen soil, which happens around the polar circle and southwards. In comparison to the Arctic, the latitude of the distribution limit is much lower here, which is determined by the extension of the cold Antarctic continent. Overviews on stress physiology of biota from Antarctica mostly include marine or microbial systems and human impacts on the environment; plants are respected, but limited (Bargagli 2008; Beyer and Bölter 2002; Huiskes et al. 2003). Several plant physiological studies will be mentioned in Paragraphs 3 and 4.

It is a challenge for modern plant (stress-) physiology or ecophysiology to study the cellular and functional properties of plants from both polar regions and to compare them with the considerably well documented plant adaptations in the European Alps. Physiological and cellular studies on high mountain plants from other mountains of the world are limited, yet some sources can be found in Akhalkatsi and Wagner (1997), Crawford (2008), Körner (2003), Schulze et al. (2005), Nagy and Grabherr (2009).

Alpine plant stress research has clearly demonstrated that physiological or cell biological observations made with plants taken from the field have to be correlated with the ambient micro- and macroclimate as well as with soil and general weather conditions. This is addressed in Chaps. 1–3 in this book, or by Larcher and Wagner (2009), Körner (2003), Nagy and Grabherr (2009).

All these environmental influences exaggerate short- or long-term answers or adaptations in the plant body. This makes it more complex and difficult to compare physiological or cell structural results, especially with low land plants or lab experiments. The human experience that high mountain or polar expeditions can be stressful has often influenced the interpretation of data collected for plants from extreme environments as indications of "stress". However, even the growth form of such plants is mostly a stress avoidance mechanism (Crawford 2008; Körner 2003; Larcher 2001), and the physiological adaptation often does not underpin real stress. Therefore, the term "stress" found in interpretations of metabolic data must be used carefully and in the sense of the stress definitions by Larcher (1987) and Lichtenthaler (1996), as discussed in Lütz (2010).

This chapter preferentially describes ultrastructural and cell physiological aspects of extreme environment adaptations mainly in alpine plants, because they are much easier to study than polar plants – from a logistical point of view and due to a well documented history of plant research. Plants from both polar regions are described and compared under cellular and physiological aspects, with a focus on high Arctic species, which allow for greater flexibility in research objects than the two higher plants in Antarctica. We will not go into any detail with scenarios of future effects of global change; merely the influences of short-wave solar irradiation on plants from extreme environments are discussed and compared by Lütz and Seidlitz in Chap. 4 of this book.

## 5.2 Cell Organelle Structural Adaptations in Alpine and Polar Plants

### 5.2.1 European Alpine Plants

For many decades the variability of plastid shapes has been described in the literature. In developing tissues, the plastids have often not yet reached their final differentiation and appear amoeboid with great plasticity of the envelope membranes. Newcomb (1967) shows variable shapes of protein storing plastids in bean root tips, and mitochondria located in pockets of young plastids have been described by Devidé and Ljubešić (1989) for onion bulbs. The enormous flexibility of the envelopes was shown in a 3-D reconstruction of leukoplasts in pine secretory cells forming protrusion-like outgrowths even larger than observed in alpine plant chloroplasts (see below) (Charon et al. 1987). *Funaria* haustorium cells contain long, thin plastid protrusions (Wiencke and Schulz 1975, 1977). The observed variability of plastid outer membranes seems to be similar to the de-differentiating young chloroplasts to leukoplasts in tissue cultures (Sjolund and Weier 1971). Spencer and Wildman (1962) describe a "mobile jacket" surrounding the chloroplasts, which became visible by LM.

Permanent thylakoid-free regions of the active chloroplast were described for the first time in high alpine plants by Lütz and Moser (1977) and Lütz (1987), earlier referred to as "proliferations", but later as "chloroplast protrusions" (CP). Most herbaceous alpine and polar plants develop CP during photosynthetic activity (Holzinger et al. 2007b; Buchner et al. 2007a; Lütz and Engel 2007). A list of over 30 plant species studied by TEM for CP occurrence is given in Lütz (2010). Their ultrastructural appearance together with the respective plant species growing in the Alps or the High Arctic is given in Figs. 5.1, 5.2, 5.5, 5.9 and 5.10.

Plastids can be artificially induced to form similar protrusion-like structures in different ways, as reported for most of the alpine/polar plants: after tobacco mosaic virus infection (Shalla 1964); under osmotic stress after isolation (Spencer and Unt 1965); manganese deficiency (Vesk et al. 1965); water stress in wheat (Freeman and Duysen 1975), high temperature (32°C) stress in rye (Schäfers and Feierabend 1976); exposure of *Funaria* protonemata to lead (4 µM $PbCl_2$, Krzesłowska and Woźny 2002).

Some reports suggest that plastid structural changes can be seen as an adaptation or stress effect to low temperatures. Musser et al. (1984) decreased the temperature from 25°C to 10°C in soybean and described long protrusion-like plastid outgrowths, more similar to stromules. These TEM observations, however, were made after extremely extended glutaraldehyde fixation (48 h!). However, perturbations in chlorophyll fluorescence because of strong pigment decomposition indicate artificial membrane formation. Similarly, under low temperature plus high light, photooxidation may induce some protrusion-like plastid formations in pea and in cucumber plants combined with strong decreases in $CO_2$ fixation (Wise and Naylor 1987). This is a typical response of non-acclimated systems, different from alpine plants. More comparable with alpine and polar plants are observations that lowland plants undergoing cold acclimation do not develop cell organelle structural changes, but increase metabolic activities (Ciamporova and Trginova 1999). However, an important difference to high alpine plants comes from their way of using longer vegetation periods for their metabolism. Kratsch and Wise (2000) compared chilling resistant vs. chilling sensitive species: only the latter group showed several different injuries in the membrane system of the cells.

In response to the environmental parameters, it has been proven that temperature is a critical factor in CP formation (Buchner et al. 2007a, b), and light intensity contributes only marginally. To investigate the effect of temperature on chloroplasts, a special temperature-controlled chamber for the light- and confocal laser scanning microscope was constructed (LM-TCC; Buchner et al. 2007b). During observation, it controls object temperature in a range of −10°C to + 95°C with an accuracy of ±0.1°C in the stationary phase, therefore preventing uncontrolled overheating of the

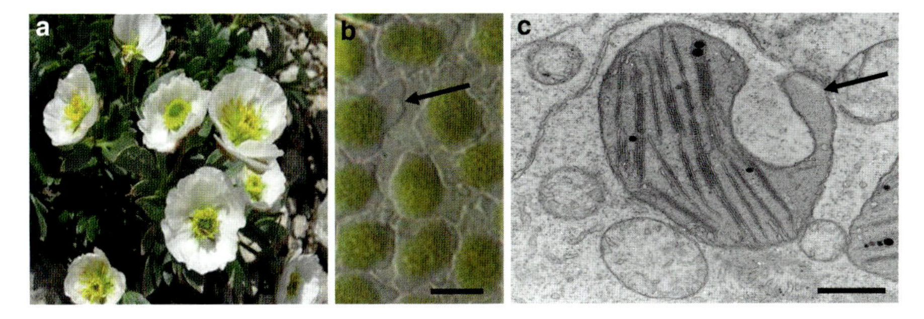


**Fig. 5.1** (**a**) *Ranunculus glacialis* (flowering plants). (**b**) DIC image of leaf mesophyll cell showing chloroplasts with protrusions (*arrow*). (**c**) TEM image of chloroplast with protrusion (*arrow*). DIC images were generated by a Zeiss Axiovert 200 light microscope. *Bars*: (**b**) 5 µm, (**c**) 1 µm

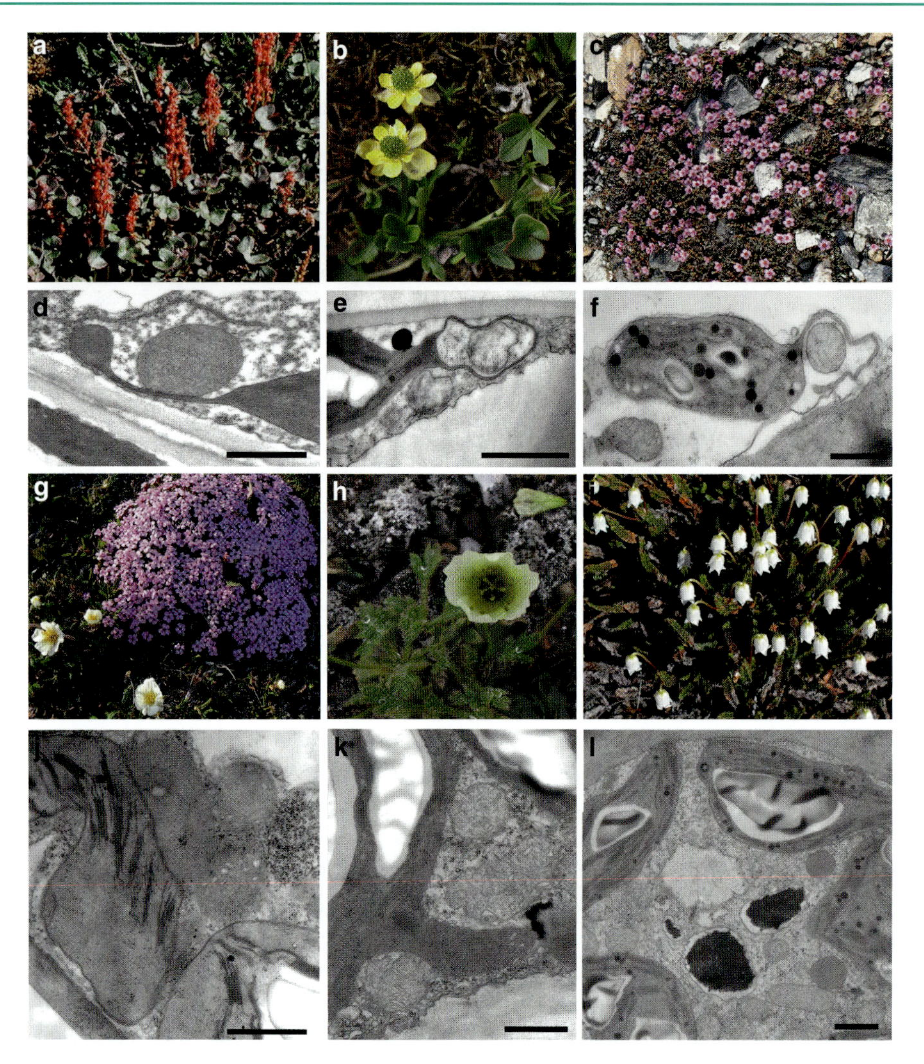

**Fig. 5.2** Flowering plants (**a–c**, **g–i**) and corresponding details of leaf cell ultrastructure (**d–f**, **j–l**) of plants from the arctic Ny Alesund. (**a**, **d**) *Oxyria digyna*; (**b**, **e**) *Ranunculus pygmaeus*; (**c**, **f**) *Saxifraga oppositifolia*; (**g**, **j**) *Silene acaulis*; (**h**, **k**) *Papaver dahlianum*; (**i**, **l**) *Cassiope tetragona*. For all images of the ultrastructure: samples were fixed at the field research site in glutaraldehyde, buffered by sodium cacodylate, postfixed in $OsO_4$ in the same buffer. After dehydration specimens were embedded in epoxy resin. Ultrathin sections were examined under a Zeiss 902 or Libra 120 Transmission electron microscope. *Bars*: 1 μm

samples during microscopy. Using this technique, we were able to follow the formation and shaping of CP such as in *Ranunculus glacialis* and compare this with the long-known TEM appearance (Figs. 5.1 and 5.2, Buchner et al. 2007a). The observed structures are clearly distinct from the thin tubular emergences of the chloroplast termed stromules by Köhler et al. 1997 (for a review, see Kwok and Hanson 2004a).

These stroma-filled, highly dynamic structures may underpin a specific advantage for plants under climatic pressure in that they facilitate metabolite exchange and chloroplast-organelle or chloroplast–chloroplast interaction, as will be explained in the following sections. When considering all of these observations, it is not surprising that plastids are capable of changing the surface and stroma volume according to the physiological demands either of a developmental program or induced by secondary factors. Many alpine and polar plants growing under harsh climate conditions in a limited time of productivity use these properties to ameliorate their energy metabolism.

#### 5.2.1.1 Comparison with the Model Plant *Arabidopsis thaliana*

Stable transformants of *Arabidopsis thaliana*, where the stroma is brightly stained by GFP, gave details of the structural aspects of stromules and chloroplast protrusions and their relation to different temperature scenarios (Holzinger et al. 2007b). Although *A. thaliana* is not an alpine plant, the findings are important in order to understand stromule and CP dynamics. In leaf mesophyll cells of *A. thaliana* plants, stromules with a diameter of about 400–600 nm and a length of up to 20 μm were predominantly observed in cells with spaces between the chloroplasts. They appeared extremely dynamic, occasionally branched or polymorphic. The occurrence of stromules in *A. thaliana* has also been investigated against the background of *arc* mutation (accumulation and replication of chloroplast, Holzinger et al. 2008).

With the above mentioned LM-TCC, a temperature-dependent appearance of chloroplast protrusions was found in *A. thaliana* mesophyll cells (Holzinger et al. 2007b). These structures had a considerably smaller length to diameter ratio than typical stromules and reached lengths of 3–5 μm. At 5–15°C (low temperatures), almost no chloroplast protrusions were observed, but they developed with increasing temperatures. At 35–45°C (high temperatures), numerous chloroplast protrusions with a beak-like appearance extended from a single chloroplast. One can assume that the temperature threshold for CP formation in alpine plants is lower because of the prevailing environmental temperature. Studies on chloroplasts of the cold-adapted plant *R. glacialis* showed a tendency to form stroma-filled protrusions, depending on the exposure temperature in the LM-TCC (Buchner et al. 2007b). The relative number of chloroplasts with protrusions decreased at 5°C when compared to the number at 25°C. This effect was reversible.

The occurrence of chloroplast protrusions may be interpreted as an adaptation to environmental strain, such as cold plus high light values, and the physiological demand for coping with a short vegetation period (Lütz 2010; Lütz and Engel 2007). This is of particular interest, as these plants are generally regarded as being adapted to cold temperatures.

#### 5.2.1.2 Function of the Cytoskeleton in Chloroplast Protrusion Formation: Inhibitor Studies with Alpine *Oxyria digyna*

Most alpine species are characterized by chloroplast structures like CPs, but important questions about their formation remain still open. Are microtubules or microfilaments involved in the development of CPs?

To address how these structures are generated, first detailed investigations of microtubules and actin filaments were undertaken in *O. digyna* (Holzinger et al. 2007a). The aim of this study was to elucidate if chloroplast protrusions are directly dependent on the activity of cytoskeleton components. Leaves from *O. digyna* collected in the Arctic at Svalbard (79°N) and in the Austrian Alps (47°N) were compared at cellular and ultrastructural levels. *O. digyna* plants collected in Svalbard had significantly thicker leaves than the samples collected in the Austrian Alps. This difference was generated by increased thickness of the palisade and spongy mesophyll layers in the arctic plants, while the size of the epidermal cells did not significantly differ in the two habitats.

Arctic-alpine as well as cultivated samples contained CP, 2–5 μm broad and up to 5 μm long. They were positioned in the cells in close spatial contact with other organelles including mitochondria and microbodies (Lütz and Moser 1977; Lütz and Engel 2007), but membrane fusion was never observed. Mitochondria were also present in invaginations of the chloroplasts. A dense network of cortical MTs was found in the mesophyll cells (Fig. 5.3, in addition for *Papaver dahlianum*). However, no direct interactions between MTs and chloroplasts were observed, and disruption of the MT arrays with the anti-MT agent oryzalin at 5–10 μM did not alter the appearance or dynamics of chloroplast protrusions (Holzinger et al. 2007a). These observations suggested that, in contrast to studies on stromule formation in *Nicotiana* (Kwok and Hanson 2004b), MTs were not involved in the formation and morphology of chloroplast protrusions in *Oxyria digyna*.

To address a possible role of actin, the microfibril (MF)-disrupting drug latrunculin B (5–10 μM for 2 h) arrested cytoplasmic streaming and altered the cytoplasmic integrity of mesophyll cells in *O. digyna*.

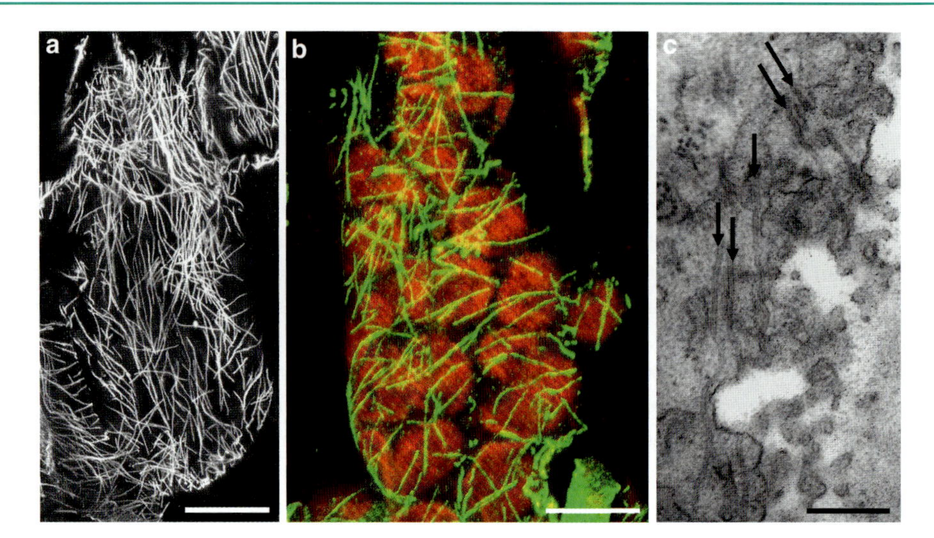


**Fig. 5.3** (**a**, **b**) Visualization of microtubules in alpine *Oxyria digyna*; visualized by immunofluorescence with anti-microtubule antibody, viewed under a Zeiss confocal microscope with 505–550 nm bandpass filter for the *green channel* and 560 nm longpass filter for the *red channel*. (**c**) TEM image of a mesophyll cell cortical section of high Arctic *Papaver dahlianum* showing numerous microtubules (*arrows*). *Bars*: (**a**) 20 μm, (**b**) 10 μm, (**c**) 200 nm

However, at the ultrastructural level, stroma-containing, thylakoid-free areas (CP) were still visible, mostly at the concave sides of the chloroplasts.

### 5.2.1.3 Effects of Ethephon

Ethephon is a pesticide often used in agriculture to make fruits and vegetables mature faster, as it regulates the plant growth by releasing ethylene (Worthing 1983; Thomson 1992). Ethephon changes the disassembly of proteins and causes reorientation of microtubules. There are several studies about the effects of herbicides or ethylene on chloroplasts. Shimokawa et al. (1978), Stoynova et al. (1997), Pechová et al. (2003) as well as Paramonova et al. (2004) demonstrate the occurrence of "finger-like protuberances", "protrusions" and "pockets filled with cytoplasm". This means, the shaping of the plastid and envelope structural adaptations were not hindered or even induced in these (non-alpine) samples.

A different approach to follow possible connections of microtubules and CP formation was taken with the application of ethephon to some alpine plants under field conditions. The following species were studied in their natural habitats in alpine regions: *Cerastium uniflorum*, *Homogyne alpina*, *O. digyna*, *Poa alpina*, *R. glacialis*. Three different concentrations of ethephon were applied as intense spraying on the leaves (0.8 mM, 4 mM, 40 mM), while controls were sprayed with water only. On the 5th and on the 14th day, the plants treated with 40 mM ethephon solution were sampled. Leaf sections were viewed with a Zeiss Axiovert 200 microscope. For each sample, ten cells were chosen randomly and the number of chloroplasts counted under DIC conditions. Then the occurrence of CPs in these cells was recorded. In parallel, leaf samples were fixed for TEM.

Both observations, made under the light microscope and the TEM, confirmed the presence of CPs in all five species (Fig. 5.4). *R. glacialis* showed the highest percentage (up to 70%) of CPs, while *P. alpina* showed the lowest (approximately 10%). The experiment was repeated in intervals throughout the summer, and it was noticed that the number of CPs was higher in early summer then in late summer. Their formation can therefore be modulated by seasonal climatic conditions. Investigation of the ultrastructure showed the typical close association between CPs and other organelles, like mitochondria (Fig. 5.4), even under ethephon treatment.

The observations made in the above cited five plant species support the results of the previous inhibitor

**Fig. 5.4** Ethephon treatments of alpine plants. (**a**) *Cerastium uniflorum* control plants. (**b**) Two weeks after 40 mM ethephone treatment in the field. (**c–e**) Ultrastructure of 40 mM ethephone treated plants: (**c**) *C. uniflorum*, (**d**) *H. alpina*, (**e**) *R. glacialis*. *Bars*: 1 μm

studies. It was observed how ethephon damaged the plants (yellow–brown leaves); this is the proof that ethephon was absorbed by the plants. In the case of *O. digyna*, newly formed leaves under higher herbicide treatment were smaller, but still contained plastids with many CP. In green leaves of all treated plants and the control (water spray), the chloroplasts did not cease CP formation. This indicates that microtubules are not involved in their formation.

Finally, we would like to draw attention to a different system of stress adaptation/avoidance as has been found in plants inhabiting tropical high mountain regions, such as at Mt. Kenya or the Andes (Schulze et al. 2005) in altitudes of about 4,000 m a.s.l.. "Summer every day but winter every night" characterizes these climates. Freezing is avoided by delayed supercooling of the plant body, until sunrise provides higher temperatures. Results of photosynthesis and internal plant temperature measurements as well as light microscopy of the leaves have been reported by Beck (1994), but ultrastructural studies are still missing to our knowledge. It would be a challenging task to try leaf sample fixation for TEM studies at the growth sites, comparing possible effects of different in vivo temperatures on cell ultrastructure.

### 5.2.2 Cell Ultrastructure of Plants from the High Arctic and the Maritime Antarctic

#### 5.2.2.1 Examples from High Arctic Plants

The arctic international research site around Ny Ålesund in North–West Spitsbergen (79°N, Norway) enables a plant stress physiologist or ecophysiologist to perform a broad range of studies and lab experiments close to the growth sites. A variety of climatic and environmental parameters are available for this research settlement (e.g. Wiencke 2004). The arctic environment is characterized by lower irradiation in comparison to the European Alps, but a 24 h daytime. Additionally, rather low temperatures in summer (0°C to ca. 10°C) and air humidity (about 30%) prevail, according to Elberling (2007).

Even UV light does not seem to stress plants in the High Arctic. Detailed studies on vascular plants from Svalbard did not support an impact of increased UV irradiation, because sufficient UV-screening mechanisms are already present in the plants (Caldwell et al. 1998; Nybakken et al. 2004). UV-B, applied additionally in the field for years (experimental setup: C. Gehrke and L.O. Björn, in: Caldwell et al. 1995), did not drastically change the ecosystem. Apart from some single, but minor effects, mainly litter decomposition and biomass changes in mosses have been described as being influenced. Similarly, Phoenix et al. (2000, 2001) described that this heathland vegetation showed some growth responses under increased UV-B, but that these effects disappeared if interactions with the increased summer precipitation were taken into consideration.

An even stronger impact on arctic biota during the next decades may come from the increase in temperature caused by global warming and increases in $CO_2$ and methane; an overview relating to Svalbard can be found in Oerbaek et al. (2004) and Elberling (2007). The possible effects of global warming on arctic ecosystems are discussed by Wüthrich et al. (1999) and Robinson et al. (1997) for Svalbard and Abisko. The latter authors clearly point out that ecophysiological studies in the arctic should consider the different climate and soil conditions prevailing in high arctic (Ny-Alesund) and low arctic (Abisko) regions (for the latter site many experiments are published as "arctic" in a general meaning). Arctic ecosystems in particular have a considerable efflux in $CO_2$ output from the soil (for Svalbard: Thannheiser et al. 1998; Wüthrich et al. 1999; Lloyd 2001), even under the snow, which can be used by small plants such as mosses or several angiosperms, such as *Saxifraga* or *D. octopetala* species. Such important studies on soil–atmosphere connections have been made in different places in the Arctic, including Svalbard (Laurila et al. 2001; Möller et al. 2001). Robberecht and Junttila (1992) studied the cold tolerance mechanisms of cushion plants from Ny Alesund and Tromsoe. Barsig et al. (1998) present one of the rare cytological studies on arctic plants, in this case mosses.

A further good basis for comparisons with alpine plants are the investigations of *D. octopetala* and *B. vivipara* responding to simulated global change carried out by Wookey et al. (1995), and Robinson et al. (1997). A short comparative summary of ecophysiological data collected for alpine vs. arctic plants is given in Körner (2003). However, little remains known about the physiology and ultrastructure of plants from high latitudes.

A comparison of the climate adaptation mechanisms and their amplitude in high arctic and high alpine plants on a cellular level is of scientific interest *per se*: knowledge about the usually occurring range of adaptations to these "extreme" environments is the basis for discussing possible effects of man-made stress on vegetation, like UV-B irradiation or high $CO_2$.

In this part, we describe some ultrastructural features of higher plants growing in the Kongsfjord area near Ny-Ålesund (Figs. 5.2 and 5.5). Specimens of *Cerastium arcticum*, *Saxifraga oppositifolia*, *P. dahlianum*, *O. digyna*, *D. octopetala*, *Silene acaulis*, *Ranunculus pygmaeus*, *Bistorta vivipara* were collected during several field campaigns (2002–2008) for comparison with alpine species. From the investigated species, seven species showed marked chloroplast protrusions (compare also the list of plants in Lütz 2010, and Fig. 5.2), *R. pygmaeus* shows CPs, occasionally ring-like structures are observed, together with high amounts of starch grains. In *S. acaulis* large and broad chloroplast protrusions have been observed, and the cellular structures are similar to the alpine species. *O. digyna* developed numerous CP, like in plants from alpine origins, frequently seen in association with other organelles like mitochondria. *S. oppositifolia* again

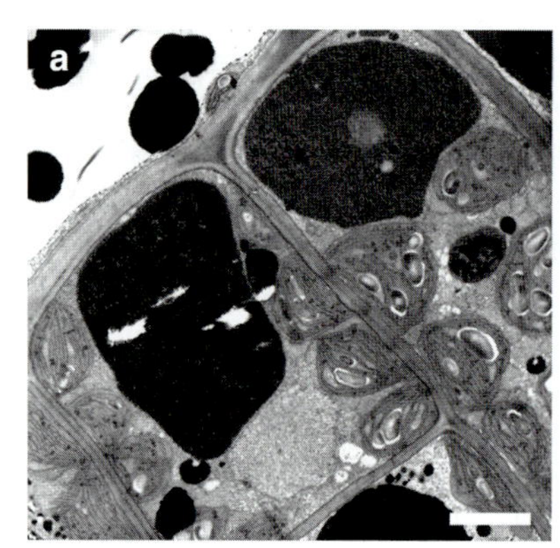


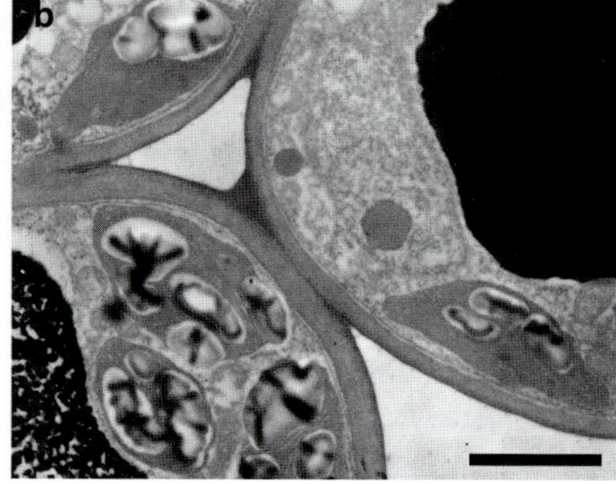


**Fig. 5.5** (**a**) Details of the leaf ultrastructure of *Dryas octopetala* collected in Ny Alesund, Spitsbergen. (**b**) Plants collected at 2,200 m a.s.l. in the Northern Limestone Alps. *Bars*: 2.5 μm

possesses ring-shaped chloroplast protrusions. In *C. arcticum* the chloroplast protrusions appear plain and contain small membranous structures, apparently not thylakoids. When sectioned transversally, those membranes are especially obvious. In this species, again close spatial contact with mitochondria is observed. In *P. dahlianum* the amount of microbodies appears to be higher than in the other species. This species is the highest climbing angiosperm in the Svalbard archipelago (up to 1,200 m a.s.l, Rønning 1996). It would be interesting to investigate the special climate adaptation of this species. In *B. vivipara* both ring-shaped and plain CP have been observed, mostly in close proximity to mitochondria. Among the investigated species *D. octopetala* is the only one not containing chloroplast protrusions. In *Dryas*, especially vacuoles appear heavily stained, as in the alpine species from 2,200 m altitude with very similar cell structures (Fig. 5.5). This observation is of interest because *D. octopetala* leaves remain green during winter and as a permanent defence screen have large amounts of anti-radical power, more than herbs with annual leaves (see below, and Lütz 2010; Oppeneiger 2008). This holds also for *C. tetragona* from Svalbard. Remarkable stability shows the thylakoid and grana formation between those growth sites, despite the fact that at 79°N plants are exposed to 24 h of light vs. the diurnal conditions in the Alps – with much higher average light intensities. All in all there are no clear differences in ultrastructure between plants from the High Arctic, harvested at about sea level, and European alpine plants taken from 2,200 to 3,200 m a.s.l..

#### 5.2.2.2 Cell Structures of Antarctic Phanerogams

Only in some parts of the maritime Antarctic two higher plants survived after isolation of this continent from South America about 30 million years ago (Huiskes et al. 2003; Parnikoza et al. 2007), when living conditions became harder and harder: *Deschampsia antarctica* (Gramineae) and *Colobanthus quitensis* (Caryophyllaceae). Their physiology is described better (see next paragraph) than their cytology. An introduction into ecophysiology of both plants is given by Alberdi et al. (2002), with a characterization of their environment by Beyer and Bölter (2002), Bergstrom et al. (2006) and by Wiencke et al. (2008).

The first ultrastructural studies (TEM) on *D. antarctica* were performed by Gielwanowska and Szczuka (2005), Piotrowicz-Cieślak et al. (2005) and Giełwanowska et al. (2005), and recently for this plant and for *C. quitensis* by Lütz (2010). The general feature of leaf ultrastructure for both plants (Fig. 5.6)

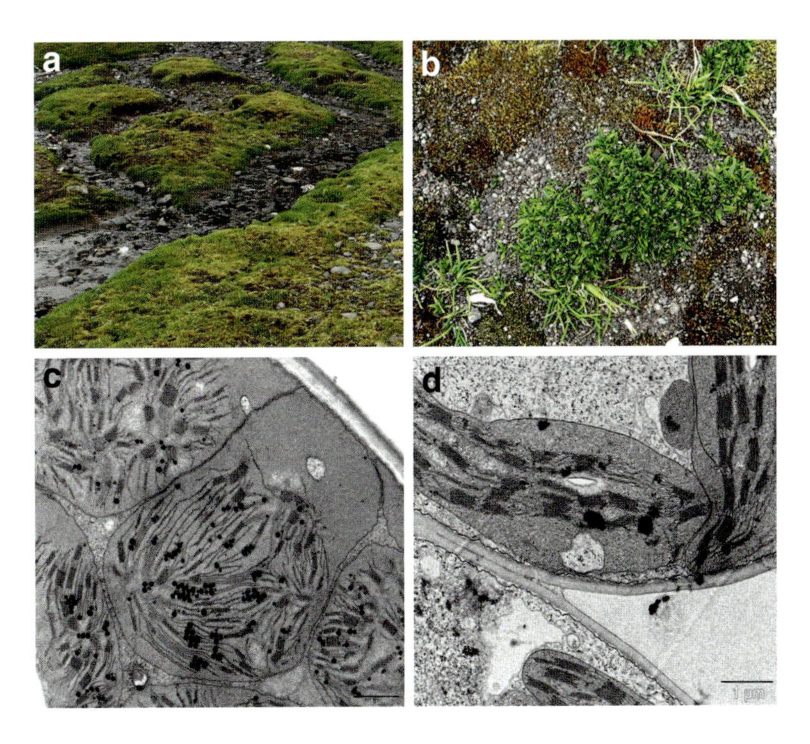


**Fig. 5.6** (**a**) Growth site of *Deschampsia antarctica* and *Colobanthus quitensis* near melting water in the coastal tundra of King George Island, maritime Antarctic. (**b**) Colonisation of tundra soil by both plants. (**c**) Ultrastructure of a leaf cell of *Deschampsia*. (**d**) Ultrastructure of leaf cells of *Colobanthus*. *Bars*: 1 μm

points to an active metabolism in the mesophyll cells, but screening of many tissue areas did not indicate any specific visible ultrastructural adaptation to the environment of the maritime Antarctic. The only exception is that plastids in both plants form CP in considerable amounts, as has been described in the previous paragraph or, in summary, by Lütz (2010) for many arctic and high alpine species. Cell organelles and internal cell walls appear very similar to other plants, e.g. to *S. acaulis* (Fig. 5.2), which may be taken instead of *C. quitensis,* as a similar Caryophyllaceae from alpine and arctic sites with a similar cushion growth form.

In *D. antarctica* the occurrence of CP was found to depend on growth conditions: plants stored in an artificial climate chamber environment for several weeks did not form CP in the newly developed leaf tissue (Lütz and Seidlitz, Chap. 4 of this book). This observation corroborates the observation described by Lütz and Engel (2007) with *O. digyna*; this plant did not form CP under moderate temperatures and the longer vegetation period of valley conditions in a botanical garden.

Climatic growth conditions in the maritime Antarctic seem to affect plant metabolism (see below) and cellular ultrastructure in a similar way as do the conditions in the High Alps or the High Arctic.

## 5.3 Chloroplast Functional Adaptations

Plants growing at the margins of life (Crawford 2008) may have developed extraordinary functional adaptations in the cell organelles, especially in the chloroplast. For many years it has been well described that high alpine plants are able to run photosynthesis below zero, individual species at down to −7°C (Körner 2003; Larcher and Wagner 1976; Larcher 1977; Larcher et al. 2010; Moser 1970; Moser et al. 1977). Furthermore, a broad amplitude in light intensity usage has been reported for species such as *R. glacialis* and *C. alpinum* (Bergweiler 1987; Moser et al. 1977). These range from no photoinhibition in full sunlight at 3,180 m altitude to survival for months under a snow cover or clouds, with energy uptake remaining active all the same, as was shown for snow covered *S. alpina*.

Similarly, several arctic species efficiently photosynthesize even at lower average temperatures (Jones and Demmers-Derks 1999) compared to their alpine counterparts, where leaf temperatures in the field can easily reach 30°C or more (Larcher and Wagner 1976, Larcher, Chap. 3 of this book; Neuner and Buchner, Chap. 6 of this book).

Our own assays of photosynthetic oxygen production (polarographic method, as is described in Bergweiler (1987), Lehner and Lütz (2003), Lütz (1996)) was performed in plants taken from the field to the research station or the institute.

Figure 5.7 compares the photosynthetic oxygen production (net photosynthesis) as dependent on light intensity for several high arctic plants and the two antarctic species. It is remarkable that no species became photoinhibited over the period of about 1 h required to complete the whole run from dark adaptation to the highest intensities. The antarctic species are within a similar range as the arctic plants. *D. octopetala* shows less activity, which is not a matter of relation to fresh weight, but remains similar if leaf

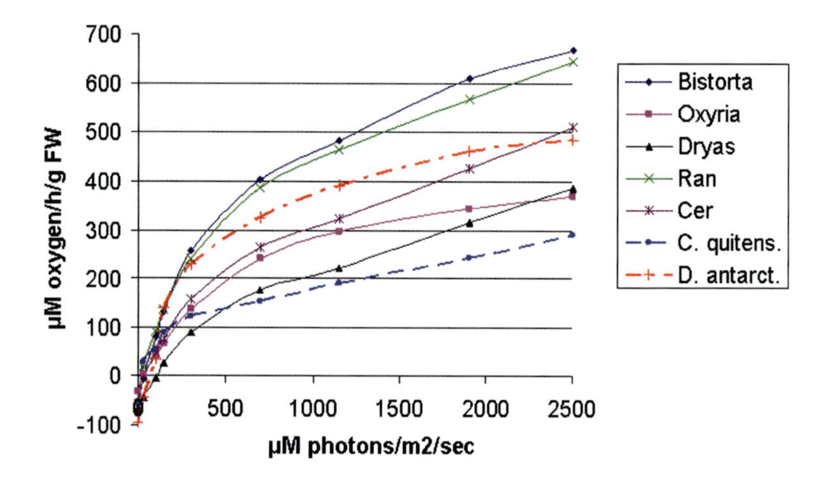


**Fig. 5.7** Light-dependent net photosynthesis per fresh weight of plants from Svalbard (*Bistorta vivipara, Oxyria digyna, Dryas octopetala, Ranunculus pygmäus, Cerastium arcticum*) and from the maritime Antarctic (*Colobanthus quitensis, Deschampsia antarctica*). Measuring temperature: 15°C. Samples were taken directly from field-grown plants

surface is taken as reference. *D. octopetala* is the only shrub in this collection, and it has leaves that remain green during winter. In general, plants with long-living leaves show somewhat slower photosynthetic activities, but can keep up this process much longer than most herbaceous plants (own observations). In contrast to our observations, Jones and Demmers-Derks (1999) measured down-regulation of photosynthesis at higher light intensities in *D. octopetala* and *B. vivipara* sampled near Ny-Alesund, Svalbard. The results, however, were generated only by leaf fluorescence measurements. This downregulation should avoid damage to photosynthesis.

The resistance of activity even under much higher light intensities than usually experienced at their growth sites (approx. 800 μMol photons at 79°N, NW-Svalbard; and approx. 1,000 μMol photons at the maritime Antarctic, 62°S, own measurements and Wiencke et al. (2008)) points to genetically preserved photosynthetic adaptation to withstand light stress. Both antarctic plants, *D. antarctica* and *C. quitensis*, still grow in the High Andes, as ecotypes adapted to strong irradiation (Casanova-Katny et al. 2006), much stronger also in the UV region of the spectrum. The arctic species either grow in the high Alps, in the form of same species, or are closely related to alpine species (like *R. pygmaeus* and *C. arcticum*). The measured activities are in the same range as Moser (1970), Moser et al. (1977) and Bergweiler (1987) reported for several high alpine plants, not inhibited until 2,200 μM photons. Obviously evolution has prepared plants growing in "extreme" environments with a broad plasticity of photosynthetic fitness, which guarantees survival.

The measurements do not provide a difference between the plants growing under permanent light for months (high Arctic) or long-day conditions like on King George Island (maritime Antarctic).

For insight into whether arctic plants keep a diurnal rhythm under continuous daylight in photosynthesis, fluorescence measurements of photosystem II were been performed from 9 a.m. in the morning until about 3 p.m. at "night" (Fig. 5.8a). The recorded Fv/Fm values are all in the positive activity range between 0.7 and 0.8, when determined in the leaves under outdoor conditions. This observation is different from European lowland species, where a noon depression of the values is found under high light values, which recovers during the night. These data point to a perfect acclimation of photosynthesis to the cold environment. The photosynthetic pigments of some alpine and arctic plants have been compared by Lütz and Holzinger (2004). Despite the differences in the light regimes during the vegetation period, the main pigments of the thylakoids show similar amounts, only accumulation of the xanthophyll cycle pigments was slightly higher in the alpine species – an influence of the higher PAR exposure above timberline compared to near sea level growth sites in the High Arctic.

This long-term physiological stability may also be based on a good radical defence power (ARP), because all electron-bound processes, especially in

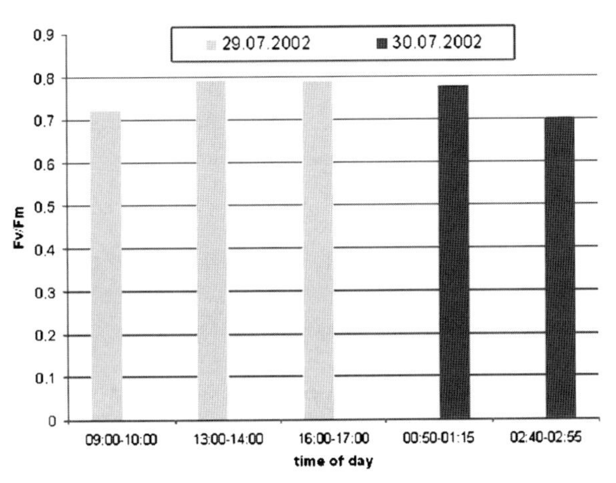


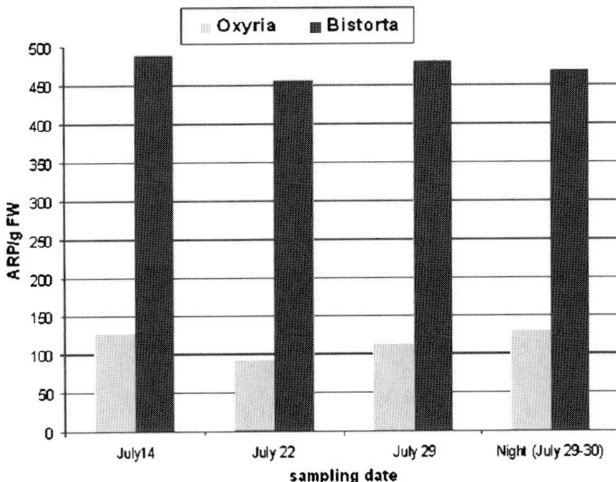


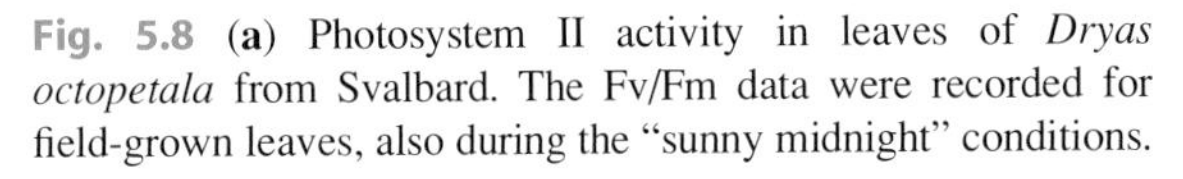

**Fig. 5.8** (**a**) Photosystem II activity in leaves of *Dryas octopetala* from Svalbard. The Fv/Fm data were recorded for field-grown leaves, also during the "sunny midnight" conditions. Means of ten assays per time. (**b**) Status of ARP in *Oxyria digyna* and in *Bistorta vivipara* over 2 weeks and during a polar "night". Daytime samples were taken between 10 a.m. and 12 o'clock

photosynthesis, can form radicals. The main ARP-active compounds in higher plants are flavonoids, ascorbic acid, glutathione, ß-carotene and tocopherols (Wildi and Lütz 1996). A survey of ARP values from plants in polar and alpine regions is given by Lütz (2010). The highest values were measured in plants with overwintering leaves, such as *D. octopetala* or *C. tetragona*, and the lowest in *O. digyna*. A more detailed view of two high arctic plants' ARP, sampled according to the timetable as for the Fv/Fm measurements, is shown in Fig. 5.8b. While *O. digyna* contains drastically less ARP activity compared with *B. vivipara*, there is no remarkable change over a period of 2 weeks, and also none in the polar "night". The small depression at "July 22" is due to rainy weather during sampling, when plants did not require much ARP.

The stability of photosynthesis under conditions of high light and cold values can additionally be explained by the activity of plastid-located alternative oxidase (PTOX) (Streb et al. 2005). *R. glacialis* contains this protein in large amounts. It dissipates excess electrons, if photorespiration is blocked or disturbed. The positive role of photorespiration as a valve in energy metabolism under extreme climate conditions has been discussed in more detail by Lütz and Engel (2007) and by Lütz (2010).

Several experiments addressed the photosynthetic performance of both antarctic phanerogams. Chlorophyll fluorescence as a measure for PS II activity and $CO_2$ fixation experiments were conducted by Xiong et al. (1999) in the field for both plants; they showed a broad range of activity from about $-3°C$ as low temperature compensation point to about 26°C as high temperature compensation point. Other authors studied photosynthesis with plants collected in the maritime Antarctic after transferring them to their home laboratory (Bravo et al. 2007; Edwards and Smith 1988; Montiel et al. 1999; Perez-Torres et al. 2007). These data match measurements for alpine plants (see above) and can be explained by the High Andean origin of the species. The low temperature resistance (Bravo et al. 2001, 2007; Bravo and Griffiths 2005) may indicate well developed cellular protection mechanisms like dehydrin formation (Olave-Concha et al. 2005) together with carbohydrate accumulation (Piotrowicz-Cieslak et al. 2005; Zuniga-Feest et al. 2005). The high light photostability, as shown by Peres-Torres et al. (2004a, b, 2007), can also be explained by the genetic history of the plants, which maintained this means of protection under the much lower light intensities present at about sea level and higher latitude of the growth sites compared to Andean growth sites. Therefore, we can conclude that polar plants with their counterparts in high mountain regions (see Crawford 2008; Körner 2003 and Larcher 2001 for an overview), have developed a broad range of stress adaptation to high light values, as is demonstrated in the comparison shown in Fig. 5.7, and also to short-wave irradiation, as is discussed in Chap. 4 of this book.

## 5.4 Developmental Aspects

Spring time is a critical growth period for herbaceous plants. Young, developing tissues need protection against high PAR and UV irradiation; the young meristematic plant parts contain relatively more water, which, if not adapted, may freeze during frequent cold events.

The development of chloroplasts is retarded until average tissue temperatures allow for enzyme and transport activities without blockage. Otherwise the biophysical processes of light energy being transformed into chemical energy, which are largely temperature-independent, will continuously produce electrons which accumulate and soon exceed the antioxidant capacity of the cells. Bleaching and destruction follows, as has been described for sudden summer snow situations in alpine plants (Lütz 1996). Most alpine plants therefore retard greening until growth conditions allow for more or less undisturbed metabolism (Körner and Larcher 1988), because the last enzyme in chlorophyll synthesis, the chlorophyll synthetase complex, is inhibited at temperatures lower than approx. 10°C (Blank-Huber 1986). Hence, young light-exposed leaves of alpine, but also of spring vegetation at lower altitudes remain yellow-greenish until temperatures rise. Deep yellow leaves of young developing *Taraxacum alpinum* from sites at 2,500 m contain etioplasts when taken out from an approx. 20 cm snow cover, which allows for penetration of sufficient light for greening under normal temperatures (Fig. 5.9 and Bergweiler 1987), but the inner structures are still small prolamellar bodies (as typical for dark grown plants, Lütz 1981, 1986). Near to the snow, and fully sun-exposed, etio-chloroplasts indicate an arrested intermediate stage, with

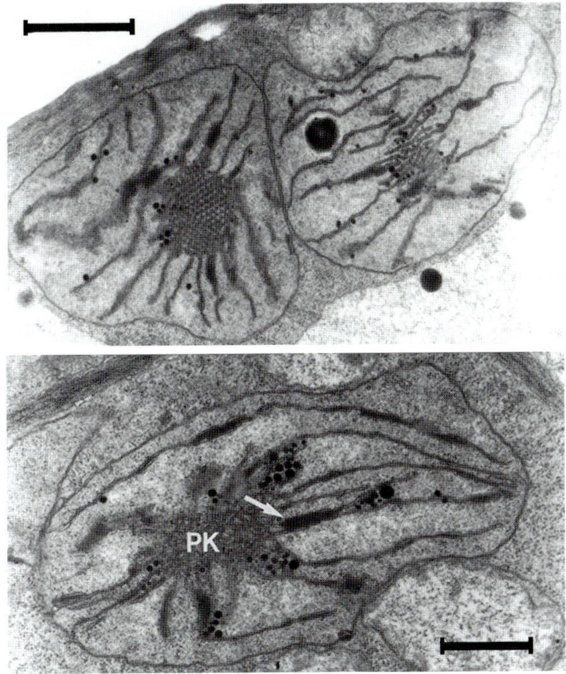


**Fig. 5.9** Ultrastructure of plastids from yellow leaves of *Taraxacum alpinum* (2,500 m altitude) taken directly out of a 20 cm snow cover (*top*), and from yellow-greenish leaves exposed to the sun near a snowfield for min. 2 days (*bottom*). *Arrow*: first small grana formation, *PK* prolamellar body, disintegrating. *Bar*: 1 μm

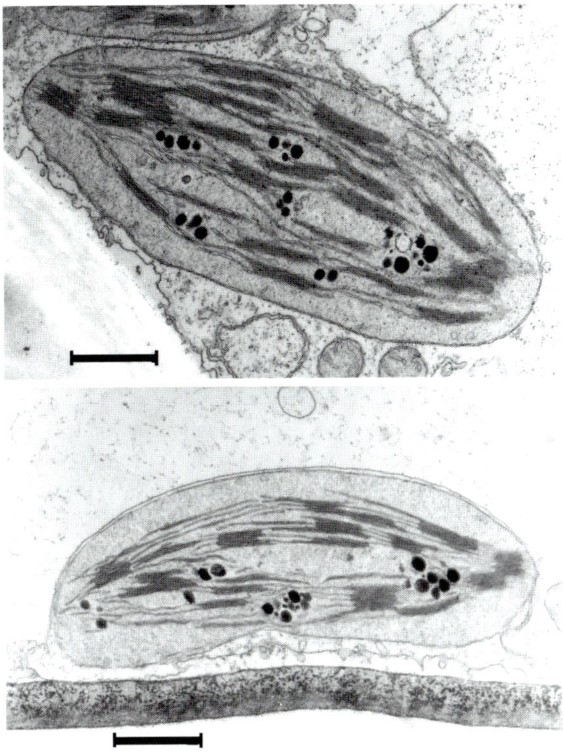

**Fig. 5.10** Chloroplasts from *Soldanella alpina* leaves, which were snow-free for about 2 weeks before sampling (*top*), and taken out of a 30 cm snow cover of the last winter (*bottom*). *Bar*: 1 μm

disintegrating prolamellar bodies as well as first thylakoids, occasionally with small grana (Fig. 5.9).

Lütz (1996) reported a similar development for the alpine *Eriophorum angustifolium*, including pigment assays and photosynthesis comparisons between the yellow, greening and green parts of the leaves. Some herbaceous plants (not shrubs) keep their green leaves over winter, like *S. alpina*. Bergweiler (1987) compared the ultrastructure of green leaves taken out from ca. 30 cm of snow in springtime with those exposed to full sunlight for several days (Fig. 5.10). No obvious difference can be found despite the snow cover being from the last winter.

The light penetration of the snow (data in Lütz (1996)) seems enough for these leaves to maintain full structural competence, the chlorophyll contents are similar (Bergweiler 1987). In the same study, the chlorophyll protein complex composition found in alpine plants was described for the first time (Fig. 5.11a). As expected from TEM, there is no striking difference between the composition of subunits resolved from both photosystems or the light harvesting complex subunits. Merely light harvesting complex I seems to be enriched – an indication of the lower light intensity under snow. This analysis was extended to other high alpine species (light-exposed), and again the samples from *R. glacialis, O. digyna, T. alpinum and P. alpina* (Fig. 5.11b) developed very similar chlorophyll protein complex compositions in the leaves. Isolated thylakoids of *S. alpina* were measured for electron transport rates, and the rates measured at 21°C did not differ markedly. On the other hand, measurements of oxygen development in *S. alpina* intact leaves revealed a four times higher activity at 20°C in the sun-exposed samples vs. the samples taken from the snow cover (Bergweiler 1987). This points to a physiological rather than a membrane-structural adaptation in the leaf cells: the complex architecture of a membrane seems to be sufficiently prepared for a range of temperature/light changes.

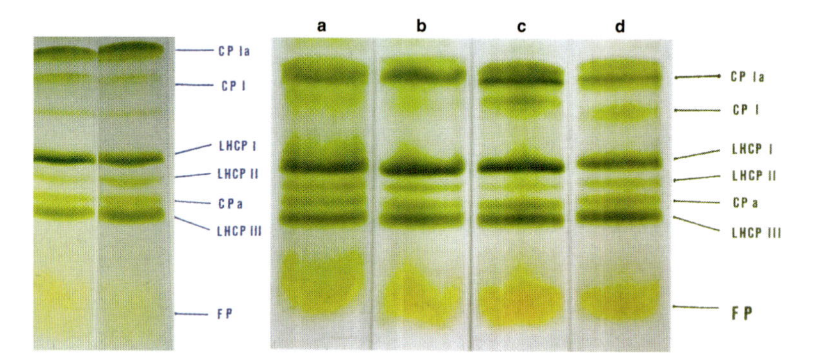


**Fig. 5.11** (**a**) Chlorophyll protein complex separation of leaves from *S. alpina* samples. *Left*: leaves sun-exposed for about 2 weeks, *right*: leaves taken out of 30 cm of old snow cover. (**b**) Chlorophyll protein complex separations of leaves from (**a**) *Ranunculus glacialis,* (**b**) *Oxyria digyna,* (**c**) *Tanacetum alpinum,* (**d**) *Poa alpina. CP I, CP Ia* complexes derived from photosystem I, *LHCP I, II, III* subunits of the light harvesting complex, *CP a* reaction center of photosystem II, *FP* free pigment. Separation methods followed Anderson et al. (1978) and Argyroudi-Akoyunoglou and Akoyunoglou (1979) with a mild detergent mixture according to Sarvari et al. (1984)

Spring aspects of alpine plants, which overwinter with green leaves, have further been discussed by Lehner and Lütz (2003) and Lütz et al. (2005).

Therefore, we can assume that the photosynthetic apparatus is constructed relatively conservatively, in a way that, irrespective of the species, best adaptation is guaranteed by more or less identical principles of membrane construction and metabolic performance.

## 5.5 Concluding Remarks

The large number of different "stressors" in alpine environments has resulted in a broad spectrum of different life forms. There is no "dominant type" of acclimation as a general solution, but a surprisingly large variety of strategies to cope with the environmental pressure and the short vegetation period. This generalization also characterizes the polar higher plants.

In comparison to the well-documented lowland plants of temperate regions, plants from extremely cold environments use a larger span of temperatures, where photosynthesis, respiration and most enzymes work, and adaptation to low as well as to high light intensities. An additional strategy in many alpine/polar species is a result of enlargement of the plastid outer membrane surface for faster and more efficient molecule transport combined with an increase in stroma volume, visible as chloroplast protrusions described herein and in related articles (Lütz 2010). The plastids are often found close to mitochondria and peroxisomes, indicating photorespiration as a valve for energy surplus under light/cold stress conditions. On the other hand, when comparing leaf cell ultrastructures of herbs and trees from lowland, high mountains or polar sites, all other cell organelles and membranes are indistinguishable between the plants. Evolution has obviously constructed mesophyll cells in a way to cope with most environmental pressures without the need for complex reconstruction of cellular architectures when climate conditions change. Cell organelle structures seem to be more conservative than functional processes.

Alpine and polar climates may induce a spectrum of radicals in the cells, but measurements have shown that the plants are well equipped with anti-radical activities. These protective mechanisms work on different physiological levels in the cells and tissues; plants with perennial leaves have by far the highest ARP-activities. Measurements of such plants taken from the snow showed that their high ARP may also protect them against biotic attack during the winter. Winter-green leaves are ready to start their metabolism even under a shallow snow cover: this elegant adaptation leads to an extension of the short vegetation period.

Cell physiology and metabolic activities have been found to be more intense in plants from extreme ecosystems than in those growing under temperate conditions. Therefore, alpine and polar plants can be used as excellent study objects to understand the adaptation to multiple so-called stressors on plants such as high and low (UV-) light intensities, contrasting temperatures, retarded mineral turnover in the soil, mechanical inputs from storm and ice, pathogen attacks under long snow cover etc.. This opportunity should be taken particularly by cell physiology, plant molecular biology and "metabolomics". A long tradition of research in ecophysiology, geobotany, systematics, soil ecology, climate research and in the growing field of plant stress research has laid a useful and indispensable basis for these modern research fields. And one should not forget:

Organisms in alpine and polar climates are not designed to maximize any kind of yield, but to stabilize at least the minimum life functions for continuous survival of a species (Larcher 1987).

**Acknowledgements** The authors thank the Ny-Ålesund International Research and Monitoring Facility for their support. Financial support by the Norsk Polar Institute and the LSF for C. L. and A. H. is kindly acknowledged. We also thank Ch. Wiencke and his group from the AWI for the possibility to be guests at their research stations in Svalbard (High Arctic) and on King George Island (Antarctica). Part of this work was supported by the Austrian Science Fund (FWF) to C.L.

## References

Akhalkatsi M, Wagner J (1997) Comparative embryology of three *Gentianaceae* species from the Central Caucasus and the European Alps. Plant Syst Evol 204:39–48

Alberdi M, Bravo LA, Gutiérrez A, Gidekel M, Corcuera LJ (2002) Ecophysiology of Antarctic vascular plants. Physiol Plant 115:479–486

Amils R, Ellis-Evans C, Hinghofer-Szalkay H (eds) (2007) Life in extreme environments. Springer, Dordrecht

Anderson JM, Waldron JC, Thorne SW (1978) Chlorophyll protein compexes of spinach and barley thylakoids. Spectral characteristics of six complexes resolved by an improved electrophoretic procedure. FEBS Lett 92(2):227–233

Argyroudi-Akoyunoglou JH, Akoyunoglou G (1979) The chlorophyll protein complexes of the thylakoids in greening plastids of *Phaseolus vulgaris*. FEBS Lett 104(1):78–84

Bargagli R (2008) Antarctic ecosystems. Environmental contamination, climate change, and human impact, vol 175, Ecological studies. Springer, Berlin

Barsig M, Gehrke C, Schneider K (1998) Effects of ultraviolet-B radiation on leaf ultrastructure, carbohydrates and pigmentation in the moss *Polytrichum commune* in the subarctic. Bryologist 101:357–365

Beck E (1994) Cold tolerance in tropical alpine plants. In: Rundel PW, Smith AP, Meinzer FC (eds) Tropical alpine environments. Plant form and function. Cambridge University Press, Cambridge, pp 77–110

Bergstrom D, Convey P, Huiskes A (2006) Trends in Antarctic terrestrial and limnetic ecosystems. Springer, Dordrecht

Bergweiler P (1987) Charakterisierung von Bau und Funktion der Photosynthese-Membranen ausgewählter Pflanzen unter den Extrembedingungen des Hochgebirges. Ph.D. thesis, University of Köln

Beyer L, Bölter M (2002) Geoecology of Antarctic ice-free coastal landscapes, vol 154, Ecological studies. Springer, Berlin/Heidelberg

Billings WD (1974) Adaptations and origins of alpine plants. Arct Alp Res 6(2):129–142

Blank-Huber M (1986) Untersuchungen zur Chlorophyll Biosynthese. Solubilisierung und Eigenschaften der Chlorophyll-Synthetase. Ph.D. thesis, University of Munich

Bravo LA, Griffith M (2005) Characterization of antifreeze activity in Antarctic plants. J Exp Bot 56(414):1189–1196

Bravo LA, Ulloa N, Zuñiga GE, Casanova A, Corcuera LJ, Alberdi M (2001) Cold resistance in Antarctic angiosperms. Physiol Plant 111:55–65

Bravo LA, Saavedra-Mella FA, Vera F, Guerra A, Cavieres LA, Ivanov AG, Huner NPA, Corcuera LJ (2007) Effect of cold acclimation on the photosynthetic performance of two ecotypes of *Colobanthus quitensis* (Kunth.) Bartl. J Exp Bot 58(13):3581–3590

Buchner O, Holzinger A, Lütz C (2007a) Effects of temperature and light on the formation of chloroplast protrusions in leaf mesophyll cells of high alpine plants. Plant Cell Environ 30:1347–1356

Buchner O, Lütz C, Holzinger A (2007b) Design and construction of a new temperature-controlled chamber for light and confocal microscopy under monitored conditions: biological applications for plant samples. J Microsc 225 (2):183–191

Caldwell MM, Teramura AH, Tevini M, Bornman JF, Björn LO, Kulandaivelu G (1995) Effects of increased solar ultraviolet radiation on terrestrial plants. Ambio 24:166–173

Caldwell MM, Björn LO, Bornman JF, Flint SD, Kulandaivelu G, Teramura AH, Tevini M (1998) Effects of increased solar ultraviolet radiation on terrestrial ecosystems. J Photochem Photobiol 46(1–3):40–52

Casanova-Katny MA, Bravo LA, Molina-Montenegro M, Corcuera LJ, Cavieres LA (2006) Photosynthetic performance of *Colobanthus quitensis* (Kunth) Bartl. (Caryophyllaceae) in a high-elevation site of the Andes of central Chile. Rev Chil Hist Nat 79:41–53

Charon J, Launay J, Carde J-P (1987) Spatial organization and volume density of leucoplasts in pine secretory cells. Protoplasma 138:45–53

Ciamporova M, Trginova I (1999) Modifications of plant cell ultrastructure accompanying metabolic responses to low temperatures. Biol Bratisl 54(4):349–360

Crawford RMM (1997) Habitat fragility as an aid to long-term survival in arctic vegetation. In: Woodin SJ, Marquiss M (eds) Ecology of Arctic environments. Blackwell, Oxford, pp 113–136. ISBN 0-632-04218-4

Crawford RMM (2008) Plants at the margin. Ecological limits and climate change. Cambridge University Press, Cambridge

Crawford RMM, Balfour J (1983) Female predominant sex ratios and physiological differentiation in arctic willows. J Ecol 71:149–160

Devidé Z, Ljubešić N (1989) Plastid transformation in greening scales of the onion bulb (*Allium cepa, Alliaceae*). Plant Syst Evol 165:85–89

Edwards JA, Lewis Smith RI (1988) Photosynthesis and respiration of *Colobanthus quitensis* and *Deschampsia antarctica* from the maritime Antarctic. Br Antarct Surv Bull 81:43–63

Elberling B (2007) Annual soil $CO_2$ effluxes in the High Arctic: the role of snow thickness and vegetation type. Soil Biol Biochem 39:646–654

Eurola S (1968) Über die Feldheidevegetation in den Gebirgen von Isfjorden und Hornsund in Westspitzbergen. Aquilo. Botanica 7:1–56

Freeman TP, Duysen ME (1975) The effect of imposed water stress on the development and ultrastructure of wheat chloroplasts. Protoplasma 83:131–145

Gielwanowska I, Szczuka E (2005) New ultrastructural features of organelles in leaf cells of *Deschampsia antarctica* Desv. Polar Biol 28:951–955

Giełwanowska I, Szczuka E, Bednara J, Górecki R (2005) Anatomical features and ultrastructure of *Deschampsia antarctica* (Poaceae) leaves from different growing habitats. Ann Bot 96:1109–1119

Hadac E (1989) Notes on plant communities of Spitsbergen. Folia Geobot Phytotax 24:131–169

Heide OM (2005) Ecotypic variation among European arctic and alpine populations of *Oxyria digyna*. Arct Antarct Alp Res 37(2):233–238

Holzinger A, Wasteneys G, Lütz C (2007a) Investigating cytoskeletal function in chloroplast protrusion formation in the arctic-alpine plant *Oxyria digyna*. Plant Biol 9:400–410

Holzinger A, Buchner O, Lütz C, Hanson MR (2007b) Temperature-sensitive formation of chloroplast protrusions and stromules in mesophyll cells of *Arabidopsis thaliana*. Protoplasma 230:23–30

Holzinger A, Kwok EY, Hanson MR (2008) Effects of *arc3*, *arc5* and *arc6* mutations on plastid morphology and stromule formation in green and nongreen tissues of *Arabidopsis thaliana*. Photochem Photobiol 84:1324–1335

Huiskes AHL, Gieskes WWC, Rozema J, Schorno RML, van der Vies SM, Wolff WJ (2003) Antarctic biology in a global context. Backhuys, Leiden. ISBN 90-5782-079-X

Jones HG, Demmers-Derks HHWM (1999) Photoinhibition as a factor in altitudinal for latitudinal limits of species. Phyton 39(4):91–98

Kappen L (1983) Anpassungen von Pflanzen an kalte Extremstandorte. Ber Deutsch Bot Ges 96:87–101

Köhler RH, Cao J, Zipfel W, Webb WW, Hanson MR (1997) Exchange of protein molecules through connections between higher plant plastids. Science 276:2039–2042

Körner C (2003) Alpine plant life, 2nd edn. Springer, Berlin

Körner C, Larcher W (1988) Plant life in cold climates. Symp Soc Exp Biol 42:25–57

Kratsch HA, Wise RR (2000) The ultrastructure of chilling stress. Plant Cell Environ 23:337–350

Krzesłowska M, Woźny A (2002) Why chloroplasts in apical cell of *Funaria hygrometrica* protonemata treated with lead are distributed in different way than in control. Biol Plant 45 (1):99–104

Kwok EY, Hanson MR (2004a) Stromules and the dynamic nature of plastid morphology. J Microsc 214:124–137

Kwok EY, Hanson MR (2004b) In vivo analysis of interactions between GFP-labeled microfilaments and plastid stromules. BMC Plant Biol 10:2

Larcher W (1977) Ergebnisse des IBP-Projektes "Zwergstrauchheide Patscherkofel"Sitzungsber. Österr. Akad. Wiss. Abt I, Vol 186. Springer Berlin

Larcher W (1987) Streß bei Pflanzen. Naturwissenschaften 74:158–167

Larcher W (2001) Ökophysiologie der Pflanzen, 6th edn. Ulmer, Stuttgart

Larcher W, Wagner J (1976) Temperaturgrenzen der $CO_2$-Aufnahme und Temperaturresistenz der Blätter von Gebirgspflanzen im vegetationsaktiven Zustand. Oecol Plant 11:361–374

Larcher W, Wagner J (2009) High mountain bioclimate: temperatures near the ground recorded from the timberline to the nival zone in the Central Alps. Contrib Nat Hist 12:857–874

Larcher W, Kainmüller C, Wagner J (2010) Survival types of high mountain plants under extreme temperatures. Flora 205:3–18

Laurila T, Soegaard H, Lloyd CR, Aurela M, Tuovinen JP, Nordstroem C (2001) Seasonal variations of net $CO_2$ exchange in European Arctic ecosystems. Theor Appl Climatol 70:183–201

Lehner G, Lütz C (2003) Photosynthetic functions of cembran pines and dwarf pines during winter at timberline as regulated by different temperatures, snowcover and light. J Plant Physiol 160:153–166

Lewis Smith R, Lewis Smith RI (2003) The enigma of *Colobanthus quitensis* and *Deschampsia antarctica* in Antarctica. In: Huiskes AHL, Gieskes WWC, Rozema J, Schorno RML, van der Vies SM, Wolff WJ (eds) Antarctic biology in a global context. Backhuys, Leiden. ISBN 90-5782-079-X

Lichtenthaler HK (1996) Vegetation stress: an introduction to the stress concept in plants. J Plant Physiol 148:4–14

Lloyd CR (2001) On the physical controls of the carbon dioxide balance at a high arctic site in Svalbard. Theor Appl Climatol 70:167–182

Lütz C (1981) On the significance of prolamellar bodies in membrane development of etioplasts. Protoplasma 108:99–115

Lütz C (1986) Prolamellar bodies. Review article in: "photosynthetic membranes," section lipids. In: Arntzen C, Staehelin A (eds) Encyclopedia of plant physiology, vol 19. Springer, Berlin, pp 683–692

Lütz C (1987) Cytology of high alpine plants II. Microbody activity in leaves of *Ranunculus glacialis* L. Cytologia 52:679–686

Lütz C (1996) Avoidance of photoinhibition and examples of photodestruction in high alpine *Eriophorum*. J Plant Physiol 148:120–128

Lütz C (2010) Cell physiology of plants growing in cold environments. Protoplasma 244:53–73 (Review)

Lütz C, Engel L (2007) Changes in chloroplast ultrastructure in some high alpine plants: adaptation to metabolic demands and climate? Protoplasma 231:183–192

Lütz C, Holzinger A (2004) A comparative analysis of photosynthetic pigments and tocopherol of some arctic-alpine plants from the Kongsfjord area, Spitzbergen, Norway. In: Wiencke Ch (ed) Reports on polar research, vol 492. AWI, Bremerhaven, pp 114–122, 1618-3193

Lütz C, Moser W (1977) Beiträge zur Cytologie hochalpiner Pflanzen. I. Untersuchungen zur Ultrastruktur von Ranunculus glacialis L. Flora 166:21–34
Lütz C, Schönauer E, Neuner G (2005) Physiological adaptation before and after snow melt in green overwintering leaves of some alpine plants. Phyton 45:139–156
Möller I, Wüthrich Ch, Thannheiser D (2001) Changes of plant community patterns, phytomass and carbon balance in a high arctic tundra ecosystem under a climate of increasing cloudiness. Biomonitoring 35:225–242
Montiel P, Smith A, Keiller D (1999) Photosynthetic responses of selected Antarctic plants to solar radiation in the southern maritime Antarctic. Polar Res 18(2):229–235
Moser W (1970) Ökophysiologische Unersuchungen an Nivalpflanzen. Mittl Ostalp-din. Ges f Vegetkde 11:121–134
Moser W, Brzoska W, Zachhuber K, Larcher W (1977) Ergebnisse des IBP-Projekts "Hoher Nebelkogel 3184 m". Sitzungsber Österr Akad Wiss. Math-naturw Kl Abt 1 186:387–419
Mosyakin SL, Bezusko LG, Mosyakin AS (2007) Origins of native vascular plants of Antarctica: comments from a historical phytogeography viewpoint. Cytol Genet 41(5):308–316
Musser RL, Thomas SA, Wise RR, Peeler TC, Naylor AW (1984) Chloroplast ultrastructure, chlorophyll fluorescence, and pigment composition in chilling-stressed soybeans. Plant Physiol 74:749–754
Nagy L, Grabherr G (2009) The biology of Alpine habitats. Oxford University Press, Oxford
Newcomb EH (1967) Fine structure of protein-storing plastids in bean root tips. J Cell Biol 33:143–163
Nybakken L, Bilger W, Johanson U, Björn LO, Zielke M, Solheim B (2004) Epidermal UV-screening in vascular plants from Svalbard (Norwegian Arctic). Polar Biol 27:383–390
Oerbaeck JB, Kallenborn R, Tombre I, Hegseth EN, Falk-Petersen S, Hoel AH (eds) (2007) Arctic Alpine ecosystems and people in a changing environment. Springer, Berlin
Oerbaek J, Tombre I, Kallenborn R (2004) Challenges in Arctic–Alpine environmental research. Arct Antarct Alp Res 36:281–283
Olave-Concha N, Bravo LA, Ruiz-Lara S, Corcuera LJ (2005) Differential accumulation of dehydrin-like proteins by abiotic stresses in *Deschampsia antarctica* Desv. Polar Biol 28:506–513
Oppeneiger C (2008) Einfluss von klimatischen Faktoren auf den Primär- und Sekundärstoffwechsel von Dryas octopetala L. Ph.D. thesis, University of Innsbruck
Paramonova NV, Shevyakova NI, VlV K (2004) Ultrastructure of chloroplasts and their storage inclusions in the primary leaves of *Mesembryanthemum crystallinum* affected by putrescine and NaCl. Russ J Plant Physiol 51(1):86–96
Parnikoza IY, Maidanuk DN, Kozeretska IA (2007) Are *Deschampsia antarcica* Desv. and *Colobanthus quitensis* (Kunth) Bartl. migratory relicts? Cytol Genet 41:226–229
Pechová R, Kutík J, Holá D, Kocová M, Haisel D, Vicánková A (2003) The ultrastructure of chloroplasts, content of photosynthetic pigments and photochemical activity of maize (*Zea mays* L.) as influenced by different concentrations of the herbicide amitrole. Photosynthetica 41(1): 127–136
Pérez-Torres E, García A, Dinamarca J, Alberdi M, Gutiérrez A, Gidekel M, Ivanov AG, Hüner NPA, Corcuera LJ, Bravo LA (2004a) The role of photochemical quenching and antioxidants in photoprotection of *Deschampsia antarctica*. Funct Plant Biol 31(7):731–741
Pérez-Torres E, Dinamarca J, Bravo LA, Corcuera LJ (2004b) Responses of *Colobanthus quitensis* (Kunth) Bartl. to high light and low temperature. Polar Biol 27:183–189
Pérez-Torres E, Bravo LA, Corcuera LJ, Johson GN (2007) Is electron transport to oxygen an important mechanism in photoprotection? Contrasting responses from Antarctic vascular plants. Physiol Plant 130:185–194
Phoenix GK, Gwynn-Jones D, Lee JA, Callaghan TV (2000) The impacts of UV-B radiation on the regeneration of a subarctic heath community. Plant Ecol 146:67–75
Phoenix GK, Gwynn-Jones D, Callaghan TV, Sleep D, Lee JA (2001) Effects of global change on a sub-arctic heath: effects of enhanced UV-B radiation and increased summer precipitation. J Ecol 89:256–267
Piotrowicz-Cieślak AI, Gielwanowska I, Bochenek A, Loro P, Górecki RJ (2005) Carbohydrates in *Colobanthus quitensis* and *Deschampsia antarctica*. Acta Soc Bot Pol 74(3):209–217
Robberecht R, Junttila O (1992) The freezing response of an arctic cushion plant, *Saxifraga caespitose* L.: acclimation, freezing tolerance and ice nucleation. Ann Bot 70:129–135
Robinson CH, Michelsen A, Lee JA, Whitehead SJ, Callaghan TV, Press MC, Jonasson S (1997) Elevated atmospheric $CO_2$ affects decomposition of *Festuca vivipara* litter and roots in experiments simulating environmental change in two contrasting arctic ecosystems. Glob Change Biol 3:37–49
Robinson CH, Kirkham JB, Littlewood R (1999) Decomposition of root mixtures from high arctic plants: a microcosm study. Soil Biol Biochem 31:1101–1108
Rønning OI (1996) Svalbards flora, 3rd edn. Norsk Polarinstitutt, Oslo. ISBN 82-7666-101-7
Sarvari E, Nyitrai P, Gyöve K (1984) Chlorophyll protein derivative of the peripheral light-harvesting antenna of photosystem I. Photobiophy 8:229–237
Schäfers H-A, Feierabend J (1976) Ultrastructural differentiation of plastids and other organelles in rye leaves with a high-temperature-induced deficiency of plastid ribosomes. Cytobiol/Eur J Cell Biol 14:75–90
Schulze ED, Beck E, Müller-Hohenstein K (eds) (2005) Plant ecology. Springer, Berlin
Shalla TA (1964) Assembly and aggregation of tobacco mosaic virus in tomato leaflets. J Cell Biol 21:253–264
Shimokawa K, Sakanoshita A, Horiba K (1978) Ethylene-induced changes of chloroplast structure in Satsuma mandarin (Citrus unshiu Marc.). Plant Cell Physiol 19:229–236
Sjolund RD, Weier TE (1971) An ultrastructural study of chloroplast structure and dedifferentiation in tissue cultures of *Streptanthus tortosus* (Cruciferae). Am J Bot 58:172–181
Spencer D, Unt H (1965) Biochemical and structural correlations in isolated spinach chloroplasts under isotonic and hypotonic conditions. Aust J Biol Sci 18:197–210
Spencer D, Wildman SG (1962) Observations on the structure of grana-containing chloroplasts and a proposed model of chloroplast structure. Aust J Biol Sci 15:599–610
Stoynova E, Petrov P, Semerdjieva S (1997) Some effects of chlorsulfuron on the ultrastructure of root and leaf cells in pea plants. J Plant Growth Regul 16:1–5

Streb P, Josse E-M, Gallouët E, Baptist F, Kuntz M, Cornic G (2005) Evidence for alternative electron sinks to photosynthetic carbon assimilation in the high mountain species *Ranunculus glacialis*. Plant Cell Environ 28:1123–1135

Thannheiser D, Möller I, Wüthrich Ch (1998) A case study of the vegetation, the carbon budget and possible consequences of climatic changes in western Spitsbergen. Verh Ges Ökologie 28:475–484

Thomson W (1992) Agricultural chemicals. Book II: herbicides. Thomson, Fresno

Vesk M, Mercer FV, Possingham JV (1965) Observations on the origin of chloroplasts and mitochondria in the leaf cells of higher plants. Aust J Bot 13:161–169

Wielgolaski FE, Karlsen SR (2007) Some views on plants in polar and alpine regions. Rev Environ Sci Biotechnol 6:33–45

Wiencke Ch (2004) Reports on polar and marine research, vol 492. AWI, Bremerhaven, 1618-3193

Wiencke C, Schulz D (1975) Sporophyte development of *Funaria hygrometrica* Sibith. I. Structural data of water and nutrient uptake in the haustorium. Protoplasma 86:107–117

Wiencke C, Schulz D (1977) The development of transfer cells in the haustorium of the *Funaria hygrometrica* sporophyte. Bryophytorum Bibliotheca 13:147–167

Wiencke Ch, Ferreyra GA, Abele D, Marenssi S (2008) Reports on polar and marine research, vol 571. AWI, Bremerhaven, 1618-3193

Wildi B, Lütz C (1996) Antioxidant composition of selected high alpine plant species from different altitudes. Plant Cell Environ 19:138–146

Wise RR, Naylor AW (1987) Chilling-enhanced photooxidation. The peroxidative destruction of lipids during chilling injury to photosynthesis and ultrastructure. Plant Physiol 83:272–277

Wookey PA, Robinson CH, Parsons AM, Welker JM, Press MC, Callaghan TV, Lee JA (1995) Environmental constraints on the growth, photosynthesis and reproductive development of *Dryas octopetala* at a high arctic polar semi-desert, Svalbard. Oecologia 102:478–489

Worthing CR (1983) The Pesticide manual: a world compendium, 7th edn. The British Crop Protection Council, Champaign, USA

Wüthrich CH, Möller I, Thannheiser D (1999) $CO_2$ fluxes in different plant communities of a high-arctic tundra watershed (Western Spitsbergen). J Vegetat Sci 10:413–420

Xiong FS, Ruhland CT, Day TA (1999) Photosynthetic temperature response of the Antarctic vascular plants *Colobanthus quitensis* and *Deschampsia antarctica*. Physiol Plant 106: 276–286

Zuñiga-Feest A, Ort DR, Gutiérrez A, Gidekel M, Bravo LA, Corcuera LJ (2005) Light regulation of sucrose-phosphate synthase activity in the freezing-tolerant grass *Deschampsia antarctica*. Photosynth Res 83:75–86

# Dynamics of Tissue Heat Tolerance and Thermotolerance of PS II in Alpine Plants

# 6

Gilbert Neuner and Othmar Buchner

## 6.1 Heat Strain in Alpine Environments

At first sight heat may not be expected to be an environmental constraint of significant importance in alpine environments, as low atmospheric temperatures are among the well-known common features of the alpine macroclimate (see Körner 2003). Although atmospheric temperatures are low, alpine plants – due to their small, prostrate growth form – often grow very close to the soil surface and can be surrounded by bare soil, causing a decoupling from ambient air temperature. In addition, the decoupling effect is promoted by an appropriate protection from cooling winds, a favourable slope, and exposure to the usually increased solar irradiation at high altitudes. However, the major determinant of plant temperature in alpine habitats is plant stature (see Salisbury and Spomer 1964; Körner and Cochrane 1983; Körner and Larcher 1988). This effect of plant stature on daily leaf temperature courses is shown in Fig. 6.1a for four species with contrasting growth forms growing close together at an alpine site (2,200 m). Infrared imaging reveals the small scale scattering of temperatures at alpine sites where temperature differences of more than 30°C can occur within centimetres (Fig. 6.1b).

Compared to plants, the bare soil surface is exposed to even more extreme temperatures. It is usually the coldest during the night but exposed to considerable daytime heating. South-facing bare, dark, raw humus was found to heat to temperatures of up to 80°C (Turner 1958). Soil temperatures drop rapidly with soil depth. A comparison of maximum leaf temperatures of six alpine plant species of diverse growth forms determined at three alpine sites (1,950, 2,200 and 2,600 m) throughout three successive growing periods (1998, 1999, 2000) reveals the frequency of heat events (Fig. 6.2). Across species, daily maximum leaf temperatures exceed 30°C quite frequently (13%; Fig. 6.3). 30°C is the temperature threshold for the onset of heat hardening in *Silene acaulis* (Neuner et al. 2000; >32°C in higher plants: Alexandrov 1977). 42°C is the lowest heat killing temperature of the most susceptible alpine plant species (Pisek and Kemnitzer 1968; Larcher and Wagner 1976; Smillie and Nott 1979; Gauslaa 1984) which is still surpassed on about 1% of days.

Overheating and heat stress appeared to be particularly pronounced in compact cushion plants, with cushions being 15–24.5°C warmer than the air on clear days (Salisbury and Spomer 1964; Körner and De Moraes 1979; Gauslaa 1984; Larcher and Wagner 2009). Under drought conditions, *S. paniculata* leaves were up to 36°C warmer than the air (Neuner et al. 1999). Although heating above ambient temperature in a cold climate can be advantageous for carbon uptake, growth, and reproductive processes, the heat-trapping stature may occasionally expose plants to temperatures that cause heat damage. Severe heat damage on an individual cushion of *M. recurva* growing at 2,600 m on Mt. Glungezer was observed after a cold spell and snow cover in mid-summer when leaf temperatures had reached 57°C during the middle of the day within 4 days (Fig. 6.4, Buchner and Neuner 2003). Heat damage to alpine plants in their natural habitat appears to be a real

G. Neuner (✉) • O. Buchner
Institute of Botany, University of Innsbruck, Innsbruck, Austria
e-mail: Gilbert.neuner@uibk.ac.at

C. Lütz (ed.), *Plants in Alpine Regions*,
DOI 10.1007/978-3-7091-0136-0_6, © Springer-Verlag/Wien 2012

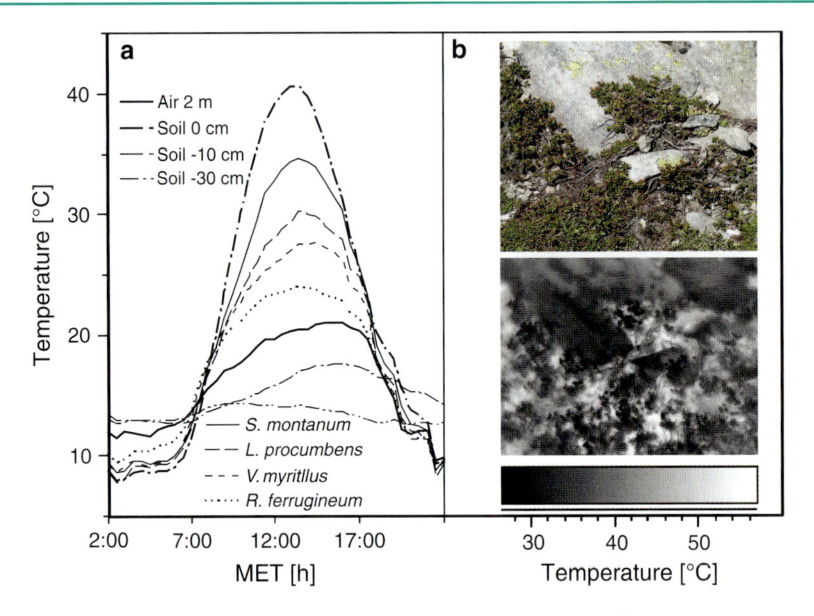


**Fig. 6.1** (**a**) Comparison of the diurnal course of temperatures recorded in the air, soil and on leaves of plants growing in close proximity, at 2,200 m a.s.l. close to the summit of Mt. Patscherkofel on a clear summer day (26.6.2009). (**b**) Digital and infrared image of an alpine plant community on a clear summer day (26.6.2009) close to the summit of Mt. Patscherkofel at 2,200 m a.s.l. The horizontal grey scale bar indicates the range of temperatures recorded with the infrared camera. Within the measurement range the lowest temperature was 26.6°C and the highest was 56.6°C

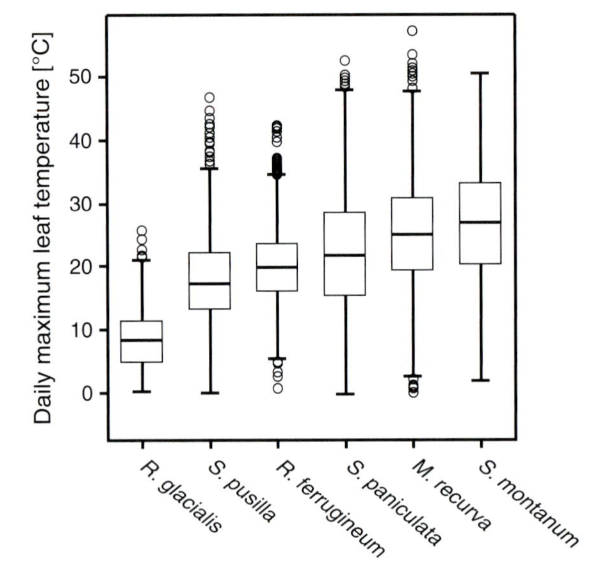


**Fig. 6.2** Daily maximum leaf temperatures measured on leaves of six alpine plant species (*Minuratia recurva*, *Ranunculus galcialis*, *Rhododendron ferrugineum*, *Saxifraga paniculata*, *Sempervivum montanum*, *Soldanella pusilla*; 15 sensors per species) with diverse growth forms determined throughout three successive summer (June–July–August) growing periods (1998, 1999, 2000), except for *R. glacialis* which was measured only at 2,600 m, at three alpine sites (1,950, 2,200 and 2,600 m). The box plots show the median (line inside the box) and the 25 and 75 percentiles. The whiskers are Tukey's hinges. Values more than 1.5 interquartile range's (IQR's) but less than 3 IQR's from the end of the box are labelled as outliers (O)

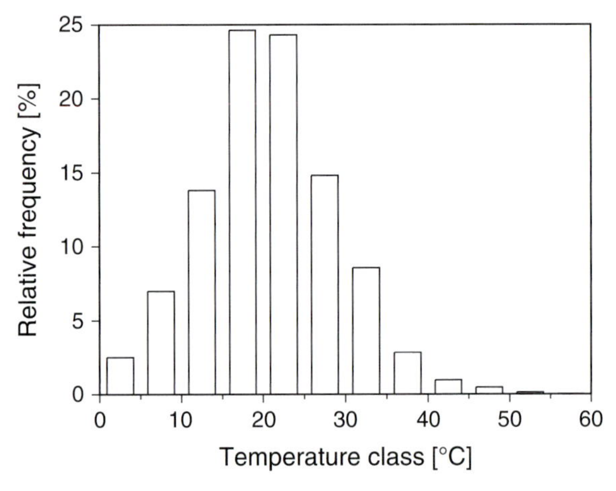


**Fig. 6.3** Relative frequency of daily maximum leaf temperatures measured on leaves of six alpine plant species (for details see legend to Fig. 6.2)

possibility if the sky is clear, the soil is dry and there is little wind (Larcher and Wagner 1976; Gauslaa 1984; Neuner et al. 1999; Körner 2003).

The maximum heat plants have to cope with on earth varies little between plant biomes – however, tropical, arctic, and alpine environments differ largely in the annual and daily duration of exposure to such extremes (see Körner 2003). Plants in alpine environments are exposed to heat for only a short period of time around midday. Daily leaf temperature amplitudes are known to

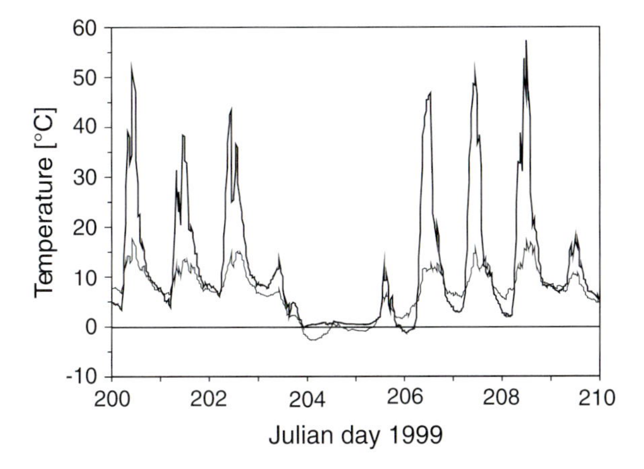


**Fig. 6.4** Temperature course measured on leaves of a *M. recurva* cushion (*thick solid line*) at an alpine site close to the Mt. Glungezer hut (2,600 m) in July 1999 compared to air temperature (*thin solid line*)

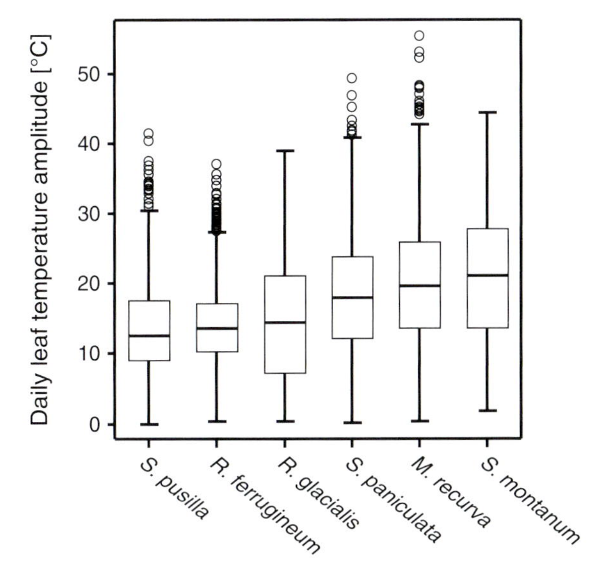


**Fig. 6.5** Daily leaf temperature amplitudes recorded on leaves of six alpine plant species (for details see legend to Fig. 6.2). The box plots show the median (line inside the box) and the 25 and 75 percentiles. The whiskers are Tukey's hinges. Values more than 1.5 interquartile range's (IQR's) but less than 3 IQR's from the end of the box are labelled as outliers (O)

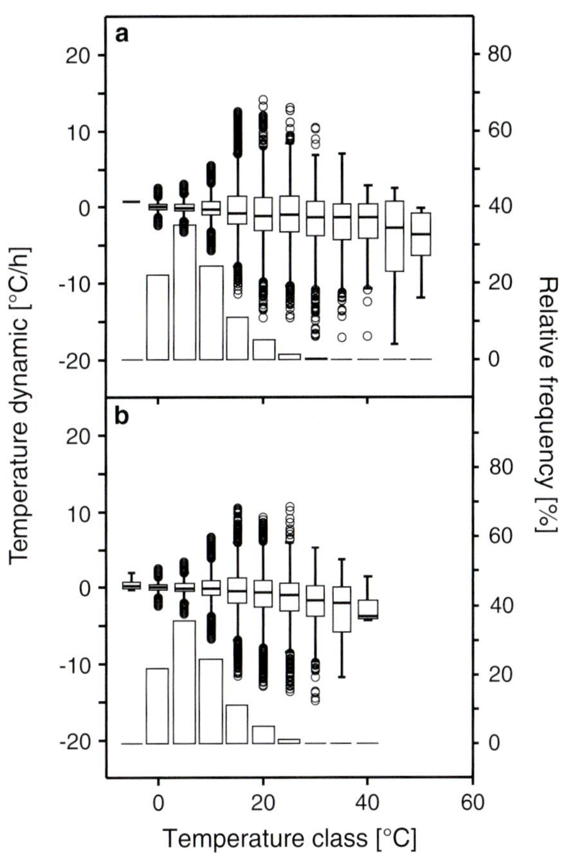


**Fig. 6.6** Rate of temperature change (°C/h) for a given starting temperature and relative frequency of temperature classes measured on leaves of (**a**) *M. recurva* and (**b**) *R. ferrugineum* at three alpine sites (1,950, 2,200 and 2,600 m) throughout three successive summer (June–October) growing periods (1998, 1999, 2000). The box plots show the median (line inside the box) and the 25 and 75 percentiles. The whiskers are Tukey's hinges. Values more than 1.5 interquartile range's (IQR's) but less than 3 IQR's from the end of the box are labelled as outliers (O)

reach up to 50°C (see Körner and Larcher 1988). Leaf temperature data obtained throughout three successive growing periods on three sites and for various plant species occasionally revealed even higher temperature amplitudes and provided information about the frequency of such events (Fig. 6.5). The data also revealed growth form specific differences. The highest temperature amplitude of 55.4°C (1998–2000) was recorded on leaves of the cushion plant *M. recurva*. Mean daily leaf temperature amplitudes ranged from 13.9°C (*R. ferrugineum*) to 21.6°C (*S. montanum*). High leaf temperature amplitudes were observed throughout the whole summer season (data not shown).

High daily leaf temperature amplitudes imply exposure to fast temperature changes. The velocity of the temperature change depends largely on the starting temperature (Fig. 6.6). The highest heating rates were usually found at low leaf temperatures and successively lower rates with increasing leaf temperature. The highest cooling rates were linked to high leaf temperatures. In the subzero temperature range, cooling rates averaged at −0.3°C/h and did not exceed −2°C/h.

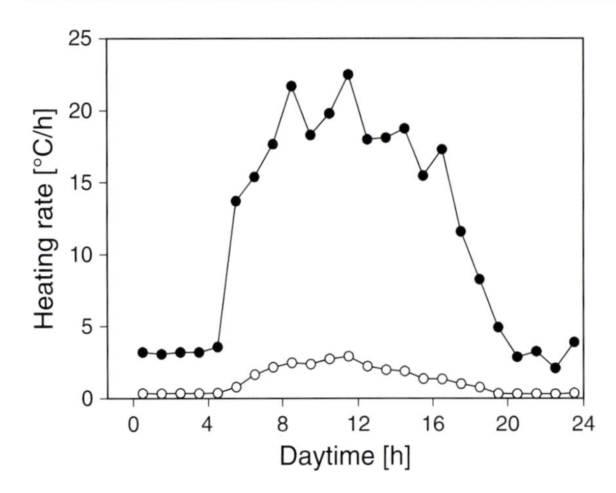


**Fig. 6.7** Diurnal course of mean (○) and maximum (●) heating rates (°C/h) measured on leaves of six alpine plant species (for details see legend to Fig. 6.2)

Heating rates were generally highest (mean >2°C/h) during morning hours from 7:00 till 12:00 and across species, highest mean heating rates of 2.9°C/h were observed to occur between 11:00 and 12:00 (Fig. 6.7). between 7:00 and 17:00 maximum heating rates could exceed 17°C/h. The highest recorded heating rate was as high as 22.5°C/h on a cushion of *M. recurva* (1,100). For alpine plants the heating rates occurring from morning till midday are of particular interest. In the cushion plant *M. recurva* leaf temperatures increased at hourly rates >4°C/h from morning (700) till midday (1,400) with a frequency of 24.5%, but only occasionally (3.6%) at rates higher than 10°C/h. During the daytime mean leaf heating rates were lowest in *Ranunculus glacialis* (1.3°C/h) and *Soldanella alpina* (1.9°C/h), followed by the woody shrub *R. ferrugineum* (2.1°C/h), and highest in cushion plants (2.7–3.8°C/h). These mean rates can occasionally be significantly exceeded in all species. This occasional rapid heating is peculiar to the alpine microclimate and poses a significant challenge to plant physiology. As a result, a sufficiently rapid heat hardening capacity must be considered particularly important in alpine habitats.

## 6.2 Tissue Heat Tolerance

### 6.2.1 Inter-Specific Variation

Some alpine plants are particularly susceptible to high temperatures as they are unable to survive temperatures above 43°C, others such as *Sempervivum arachnoideum* tolerate temperatures as high as 64°C (Fig. 6.8). Variability between species is still high in plants growing in close proximity at the same growing site. This is shown for species growing at an altitude of 2,600 m close to the Mt. Glungezer hut (grey bars in Fig. 6.8). With respect to growth form, succulent plants and grasses exhibit the highest heat tolerances, followed by cushion plants. Herbs and woody plants show the lowest mean heat tolerances.

Heat tolerance of alpine plants has been determined by various authors, however, with one exception (Buchner and Neuner 2003), they represent only snap-shots of a particular environmental situation. A comprehensive comparison of species requires many more measurements under various environmental conditions (Buchner and Neuner 2003). Additionally, most of the heat tolerance data are not fully comparable due to differing measurement procedures and high variability of natural preconditions before sampling. To be able to assess the potential heat threat threshold temperature, a realistic comparison would need to include the heat hardening capacity of a species and its maximum heat tolerance.

### 6.2.2 Within-Species Variation

#### 6.2.2.1 Heat Hardening Capacity

Sufficient data on heat tolerance to draw a clear picture of the intra-specific heat tolerance variability are available for some species only. Variability of heat tolerance might reflect differences between provenances as found, for instance, between arctic and alpine populations of *Oxyria digyna* (Mooney and Billings 1961), but generally seems to be the result of short-term heat hardening, i.e. a reversible increase of high temperature thresholds for damage to plant cells, which is a well-known phenomenon (Alexandrov 1977; Kappen and Lösch 1984). Early investigations on heat hardening by exposure of randomly collected leaf samples to sublethal temperatures for 24 h suggested species-specific differences in heat hardening capacities from 1°C to 5°C (Gauslaa 1984). Higher heat hardening capacities of up to 8°C were measured in leaves of alpine *Sempervivum* species, with daily rates ranging between 2°C/d and 4°C/d (Larcher et al. 2010). For alpine *S. acaulis* cushions the range of heat tolerance thresholds under summer field conditions was much greater (9°C, 45.5–54.5°C; Neuner et al.

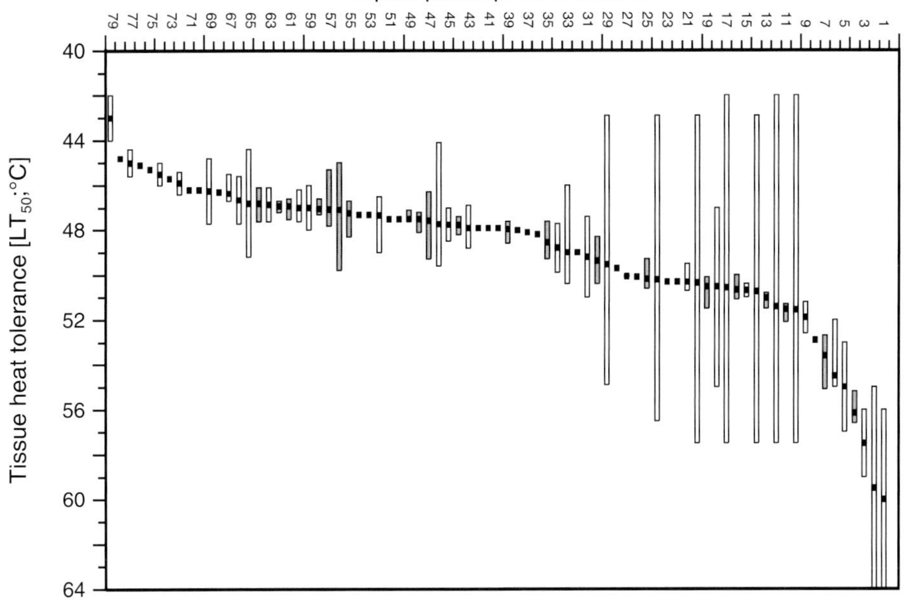


**Fig. 6.8** Tissue heat tolerance (mean, minimum and maximum $LT_{50}$) of leaves from alpine and high alpine plants species. *Grey bars* indicate 25 species that were tested at 2,600 m on Mt. Glungezer during midday hours of 27 July 2000 (Buchner and Neuner 2001). **1** *Sempervivum tectorum*[7], **2** *Sempervivum arachnoideum*[7], **3** *Carex firma*[8], **4** *Poa alpina ssp. vivipara*[3], **5** *Soldanella alpina*[7], **6** *Sempervivum monatnum*[7], **7** *Carex curvula*[3], **8** *Arctostaphylos uva-ursi*[5,8], **9** *Juniperus communis*[5,6], **10** *Saxifraga paniculata* [1,3,4,5,6,12,13,14], **11** *Linaria alpina*[3], **12** *Ranunculus glacialis*[4,5,6,8,9,11], **13** *Potentilla crantzii*[3], **14** *Silene acaulis*[2,4,5,6,7], **15** *Campanula rotundifolia*[5,6], **16** *Phyteuma hemisphaericum*[3], **17** *Loiseleuria procumbens*[4,5,6,8,12,16], **18** *Saxifraga oppositifolia*[5,7,8], **19** *Saxifraga moschata*[3], **20** *Soldanella pusilla*[4,8,12], **21** *Hieraceum alpinum*[5], **22** *Cotoneaster integerimus*[5], **23** *Lychnis alpina*[5], **24** *Minuartia recurva*[4], **25** *Gentiana bavarica*[3], **26** *Galium mollugo*[5], **27** *Sedum alpestre*[3], **28** *Draba incana*[5], **29** *Rhododendron ferrugineum*[4,12,13,15], **30** *Leontodon pyrenaicus*[3], **31** *Saxifraga caespitosa*[5,6], **32** *Saxifraga cotyledon*[5], **33** *Primula minima*[8,12], **34** *Viola biflora*[5,6], **35** *Achillea moschata*[3], **36** *Draba alpina*[5], **37** *Draba dovrensis*[5], **38** *Draba oxycarpa*[5], **39** *Ligusticum muttelina*[3], **40** *Veronica alpina*[5], **41** *Papaver radicatum ssp. ovatilobum*[5], **42** *Astragalus alpinus*[5], **43** *Picea abies*[13], **44** *Veronica bellidioides*[3], **45** *Veronica fruticans*[5,6], **46** *Arabis alpina*[5,6], **47** *Pedicularis asplenifolia*[3], **48** *Euphrasia minima*[3], **49** *Leucanthemopsis alpina*[3], **50** *Gentiana nivalis*[5], **51** *Gentiana campestris ssp. campestris*[5], **52** *Salix herbacea*[2,5,6], **53** *Saxifraga nivalis*[5], **54** *Astragalus frigidus*[5], **55** *Homogyne alpina*[3], **56** *Dryas octopetala*[5,6,16], **57** *Trifolium thalli*[3], **58** *Salix retusa*[3], **59** *Carex sempervirens*[16], **60** *Pinus cembra*[13], **61** *Doronicum clusii*[3], **62** *Cerastium uniflorum*[3], **63** *Salix reticulata*[5,6], **64** *Geum reptans*[3,11], **65** *Arctostaphylos alpinus*[5,6], **66** *Saxifraga cernua*[5,6], **67** *Oxyria digyna*[3,6,8,10,11], **68** *Cirsium helenioides*[5], **69** *Saussurea alpina*[5,6], **70** *Ranunculus pygmaeus*[5,6], **71** *Pedicularis oederi*[5], **72** *Cerastium cerastoides*[5,6], **73** *Alchemilla alpina*[5], **74** *Achillea millefolium*[16], **75** *Chrysosplenium alternifolium*[5], **76** *Pinguicula vulgaris*[5], **77** *Bartsia alpina*[5,6], **78** *Saxifrga adscendens*[5], **79** *Senecio carniolicus*[16]. (Heat tolerance data have been compiled from [1] Biebl and Maier (1969), [2] Biebl (1968), [3] Buchner und Neuner (2001), [4] Buchner und Neuner (2003), [5] Gauslaa (1984), [6] Kjelvik (1976), [7] Larcher et al. (2010), [8] Larcher and Wagner (1976), [9] Larcher et al. (1989), [10] Mooney and Billings (1961), [11] Moser (1965), [12] Neuner (unpublished), [13] Neuner und Pramsohler (2006), [14] Sapper (1935), [15] Schwarz (1970), [16] Wildner-Eccher (1988))

2000). Recent field data obtained for seven species throughout two successive growing periods revealed a generally high total heat hardening capacity of 6–11.7°C, and within the course of a single day, heat hardening was found to be 4.8–9.5°C (Buchner and Neuner 2003). Heat tolerance of individual species differed significantly between investigation sites and years for *S. paniculata* (Fig. 6.9), stressing that comparisons across species using data from various investigations may not be expedient, as the actual heat tolerance level is strongly influenced by environmental preconditioning.

Seasonal changes in heat tolerance have been reported to occur (Larcher 1984) although they are generally difficult to assess, as the diurnal amplitude is usually similar to the seasonal amplitude – nevertheless predawn values suggest some changes with decreases (*R. ferrugineum*, *Pinus cembra*) or increases

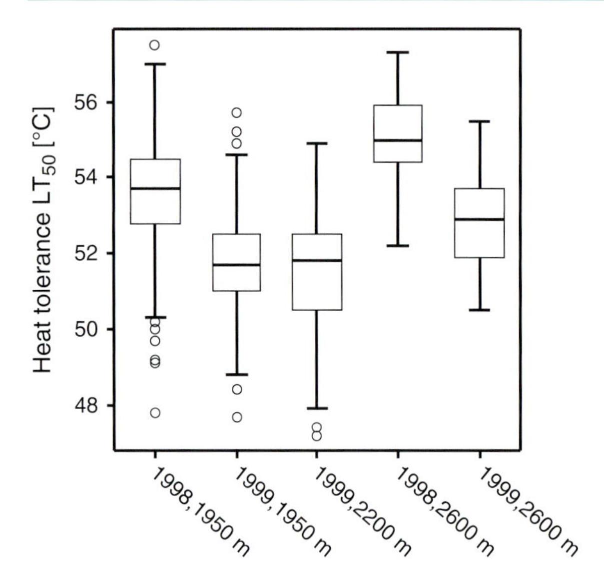


**Fig. 6.9** Annual mean heat tolerance ($LT_{50}$) of leaves of *S. paniculata* monitored throughout two successive summer (June–July–August) growing periods (1998, 1999) at three alpine sites (1,950, 2,200 and 2,600 m). The box plots show the median (line inside the box) and the 25 and 75 percentiles. The whiskers are Tukey's hinges. Values more than 1.5 interquartile range's (IQR's) but less than 3 IQR's from the end of the box are labelled as outliers (O)

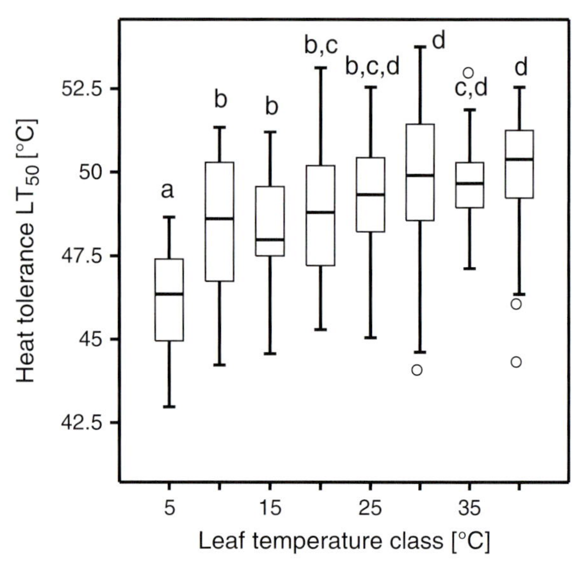


**Fig. 6.10** Heat tolerance ($LT_{50}$) of leaves of *L. procumbens* at the given sampling temperature prior to the determination of heat tolerance. Measurements were conducted in June and July 2000, and data were collected during 20 independent in situ heat treatment experiments where leaf temperatures were successively increased by +3°C/h from morning to afternoon. The box plots show the median (line inside the box) and the 25 and 75 percentiles. The whiskers are Tukey's hinges. Values more than 1.5 interquartile range's (IQR's) but less than 3 IQR's from the end of the box are labelled as outliers (O)

(*S. paniculata*) from winter to summer or even no change as in *Picea abies* (Neuner and Pramsohler unpublished).

### 6.2.2.2 Heat Hardening Velocity, Thresholds and Driving Forces

Heat hardening of the leaf tissue can occur within hours or even minutes when critical high temperature thresholds are surpassed (Alexandrov 1977). In nature diurnal heat tolerance changes greater than ±1.5°C occurred on 18% of summer days at a velocity of 0.4–2.2°C/h (Buchner and Neuner 2003). In *S. acaulis* changes in heat tolerance of the leaf tissue were found to occur frequently and at high diurnal rates of up to 4.7°C/d with leaf heating rates of up to 10°C/h at leaf temperatures higher than 30°C (Neuner et al. 2000; Buchner and Neuner 2001). Threshold temperatures for the onset of heat hardening in most higher plants were temperatures exceeding 32°C (Alexandrov 1977). In *S. acaulis* cushions, heat tolerance started to increase at temperatures above 30°C in response to a controlled in situ heat treatment with infrared lamps under moderate irradiation (Neuner et al. 2000). To test the effect of temperature on heat hardening under field conditions, *L. procumbens* plants were exposed in situ to controlled heating treatments (3°C/h) with infrared lamps. The results of 20 field heating experiments are contrary to the laboratory experiments, as no threshold temperature for heat hardening was found (Fig. 6.10). Instead, heat hardening appeared to be complete by 30°C. Under alpine field conditions at leaf temperatures higher than 30°C, naturally occurring maximum heating rates in cushion plants ranged from 11.6°C/h to 14.7°C/h and in herbs from 1.9°C/h to 6.4°C/h (Table 6.1). Critical high temperature thresholds for initial heat damage could be reached within 1 h of surpassing 30°C. Therefore, a heat hardening velocity of 0.4–2.2°C/h (Neuner et al. 2000) could be too slow to ensure survival at high temperatures. In nature, mean heating rates above 30°C are lower than at lower temperatures ranging from 1.1°C/h in *R. glacialis* to 3.9°C/h in *S. montanum*.

Little is known about the driving forces of heat hardening in nature, as most investigations have been carried out under controlled environmental conditions and mostly at low irradiation. In a field study the

**Table 6.1** Mean heating rates and maximum heating rates (°C/h) measured between 8:00 and 14:00 on leaves of six alpine plant species: *M. recurva*, *R. glacialis*, *R. ferrugineum*, *S. paniculata*, *S. montanum* and *S. pusilla* using 15 thermocouple sensors per species

| Species | Heating rate (°C/h) | Maximum heating rate (°C/h) | N |
|---|---|---|---|
| *S. pusilla* | 2.6 ± 1.6 | 6.4 | 5074 |
| *S. paniculata* | 2.5 ± 2.1 | 12.1 | 9365 |
| *R. ferrugineum* | 2.4 ± 2.3 | 14.6 | 6852 |
| *S. montanum* | 3.9 ± 2.8 | 11.6 | 600 |
| *R. glacialis* | 1.1 ± 0.6 | 1.9 | 2398 |
| *M. recurva* | 1.9 ± 1.7 | 14.7 | 5027 |

Data were determined at three alpine sites (1,950, 2,200 and 2,600 m) throughout three successive summer (June–July–August) growing periods (1998, 1999, 2000)

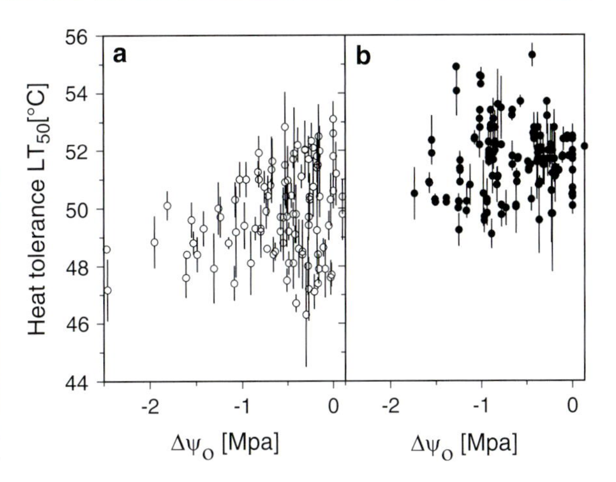


**Fig. 6.11** Effect of naturally occurring drought stress indicated by decreasing $\Delta\psi_o$ values on heat tolerance [$LT_{50}$; °C] of (**a**) *S. pusilla* and (**b**) *S. paniculata* (acc. to Buchner and Neuner (2003))

significant differences in heat tolerance of *S. paniculata* observed between investigation sites and years could be explained by differences in the leaf temperature climate (see Fig. 6.9). For example, in 1998 leaf temperatures were significantly higher than in 1999 and the Glungezer site was the warmest, which resulted in increased mean heat tolerance levels in both cases. Another example of the effect of temperature on heat tolerance was obtained by moderate long-term heating of *L. procumbens* canopies in the field. Canopy leaf temperatures were increased by +3°C in reference to a control plot using large infrared lamps throughout a whole summer growing period. This moderate warming treatment resulted in a significant increase of heat tolerance in heated *L. procumbens* plants by +0.6°C (Buchner and Neuner 2003). Similar effects of leaf temperature on $LT_{50}$ are reported for *Potentilla gracilis*, a widespread herb of subalpine meadows (Loik and Harte 1996).

In some plants water deficiency can also increase heat tolerance (Alexandrov 1977). Drought stress is known to lead to an increase in heat tolerance at low leaf temperatures, particularly in xerophytic plants (Bannister 1970; Falkova 1973; Alexandrov 1977; Larcher et al. 2010). By contrast, mesophytic species often do not respond to decreasing water content with an increase in heat tolerance (Zavadskaya and Denko 1968). While in an alpine field study, high heat tolerances were absent in most species under drought stress (Fig. 6.11a), in the xerophytic species *S. paniculata*, maximum heat tolerance values were found under drought conditions (Fig. 6.11b). This appears to be ecologically significant for a species that grows preferentially on dry sites where a decrease in osmotic water potential occurs with high leaf temperatures and strong irradiation (Neuner et al. 1999).

Irradiation is a major factor for diurnal leaf temperature increase in alpine environments, and leaf temperatures critical for survival will only occur when plants are exposed to full solar irradiation. Hence strong irradiation in the morning hours could act as an early indicator of potential heat stress during the middle of the day. However, until now there has been no direct evidence for this, except that on days with clear skies high heat tolerance levels were observed in the morning when leaf temperatures were still low (Neuner and Eder unpublished).

While heat hardening is known to proceed rapidly, heat dehardening is much less well understood but seems to occur within days during cold weather conditions (Alexandrov 1977). In mosses, heat dehardening took 2–3 days (Meyer and Santarius 1998).

### 6.2.2.3 Developmental Changes

While heat tolerance of adult alpine plant species has been studied to some extent, we still lack information on heat tolerance of seeds, seedlings and juvenile stages. These developmental stages must be considered particularly vulnerable to heat stress, as seed storage and establishment of seedlings takes place in the soil, and the soil surface experiences the most extreme temperatures (see above). The heat tolerance of dry seeds, imbibed seeds and seedlings of some alpine plant species was investigated by Wildner-Eccher (1988): *A. millefolium*, *C. sempervirens*, *D. octopetala*, *L. procumbens* and *S. incanus* ssp. *carniolicus*. In this study significant differences in heat tolerance between seeds, juvenile stages and adults were noticed. Seedlings were most heat-susceptible and died after exposure to temperatures of between 42°C and 48°C. Imbibed seeds were damaged at between 50°C and 57°C. Dry seeds were killed between 59°C and 74°C, i.e. they would not survive at sites investigated by Turner (1958) with temperatures around 80°C at the soil surface. Heat tolerances of dry seeds, imbibed seeds, germinated seeds, seedlings and juveniles were compared to heat tolerances of adults in pioneer species in the glacier foreland, as heat damage was considered to be a possible threat given its contribution to high seedling mortalities on the recently deglaciated moraines (Marcante et al. 2008; Erschbamer et al. 2008). Under laboratory growth conditions, imbibed seeds of *S. acaulis* became heat-damaged at 54.3°C, heat tolerance in germinating seeds was reduced to 48.2°C and seedlings tolerated only 41.6°C (Table 6.2; Marcante et al. unpublished), which is distinctly less than that of leaf tissues from adult plants (mean 50.3°C, 45.5°C–54.9°C; Buchner and Neuner 2003). The ecological significance of the lower heat tolerance of seedlings is evident, as their tolerance lies within the naturally occurring soil surface temperature range, so naturally occurring temperatures could potentially kill them (Marcante et al. unpublished). Still, preliminary results suggest that seedlings have a heat hardening potential, as seedlings taken from the field exhibited a significantly higher heat tolerance (5.6°C higher) than those grown in Petri-dishes in growth chambers (Marcante et al. unpublished).

**Table 6.2** Mean heat tolerance $LT_{50}$ (°C) of adult versus juvenile (seedlings, imbibed seeds, germinating seeds) *S. acaulis* plants grown either in the field (F) or in an environmental growth chamber (G)

| Developmental stage | | Heat tolerance $LT_{50} \pm SD$ (°C) | References |
|---|---|---|---|
| Adult | F | $50.7 \pm 2.4^{a}$ | Buchner and Neuner 2003 |
| Seedling | F | $47.2 \pm 0.7^{b}$ | Marcante et al. unpublished |
| Seedling | G | $41.6 \pm 2.1^{c}$ | Marcante et al. unpublished |
| Germinating seeds | G | $48.3 \pm 2.7^{b}$ | Marcante et al. unpublished |
| Imbibed seeds | G | $54.2 \pm 1.1^{d}$ | Marcante et al. unpublished |

Different letters indicate significant differences between means at $P > 0.01$

#### 6.2.2.4 Vegetative Versus Reproductive Tissues

The heat tolerance of the reproductive processes may influence the distribution of alpine plants, as sexual reproduction is essential for them to maintain their presence and to colonize new sites. However, until now all knowledge concerning the heat tolerance of alpine plant species originates nearly exclusively from vegetative tissues, except for the most heat tolerant *Sempervivum* species (Larcher and Wagner 1983; Larcher et al. 2010). Leaves of *S. montanum* have been shown to tolerate temperatures of up to 57°C, while the flowers are significantly more heat-susceptible. While the gynoeceum was heat-damaged at 52°C, the stamens were already injured at temperatures above 48°C (Larcher et al. 2010). The heat tolerance ($LT_{50}$) of reproductive organs of High-Andean plants in Central Chile ranged from 47.8°C in *Chuquiraga oppositifolia* to 51.7°C in *Cerastium arvense* and was up to 2°C lower than that of the vegetative tissues of the same species (Table 6.3; Ladinig et al. 2009). A broad survey on the heat tolerance of different developmental stages of reproductive tissues in European alpine plants corroborated this trend of lower heat tolerance in reproductive tissues than in vegetative ones. The stigmas and ovules were found to be particularly heat-susceptible parts, and the flower bud sprouting stage was usually more susceptible in most of the tested species.

## 6.3 Thermotolerance of PS II

Overheating in alpine habitats is usually coupled with high photon flux densities (Körner and Larcher 1988). In terms of photosynthesis, the most susceptible site for both heat damage and functional disturbances induced by strong irradiation is associated with the same component of the photosynthetic system located in the thylakoid membranes, most probably photosystem II (PS II). Using chlorophyll fluorescence measurements, the response of PS II to increasing

**Table 6.3** Mean heat tolerance $LT_{50}$ (°C) of leaves versus reproductive organs of high Andean plants measured in Faralleones, Chile

| Species | Heat tolerance $LT_{50} \pm SD$ (°C) Leaves | Heat tolerance $LT_{50} \pm SD$ (°C) Reproductive organs |
|---|---|---|
| *Adesmia sp.* | 49.5 ± 0.4 | 50.0 ± 0.3 |
| *Cerastium arvense* | 52.9 ± 0.2* | 51.6 ± 0.2* |
| *Chugiraga oppositifolia* | 47.9 ± 0.4 | 47.2 ± 1.6 |
| *Loasa caespitosa* | 51.4 ± 0.8 | 50.3 ± 0.3 |
| *Oxalis compacta* | 50.2 ± 0.8* | 47.9 ± 0.2* |
| *Oxalis squamata* | 49.0 ± 0.5* | 47.0 ± 0.4* |
| *Senecio sp.* | 50.2 ± 1.2 | 48.0 ± 2.4 |

Asterisks indicate significant differences between means at $P > 0.01$ ($N = 5$)

temperatures in darkness can be easily measured by the so-called $F_0$/T-curves (Schreiber and Berry 1977; Bilger et al. 1984; Bilger et al. 1987; Neuner and Pramsohler 2006). $F_0$/T-curves yield a critical temperature ($T_c$) and a lethal temperature ($T_p$) for PS II. Thermotolerance of PS II is then calculated from $T_c$ and $T_p$ as: $(T_c + T_p)/2$ (Fig. 6.12). Thermotolerance of PS II has been shown to correspond to some extent with leaf tissue heat tolerance (Bilger et al. 1984), but this may not hold true in all cases (Neuner and Pramsohler 2006).

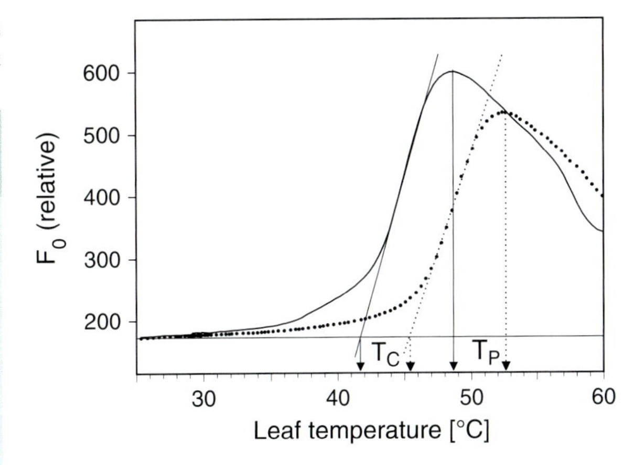


**Fig. 6.12** $F_0$/T-curves determined on leaves of *S. paniculata* during progressive heating yield a critical temperature ($T_c$) and a lethal temperature ($T_p$) for PS II. Preconditioning of plants (*dotted line*) with increased temperatures (5°C higher than control plants) for 2 weeks caused an increase in PS II thermotolerance in *S. paniculata* by +3.4°C

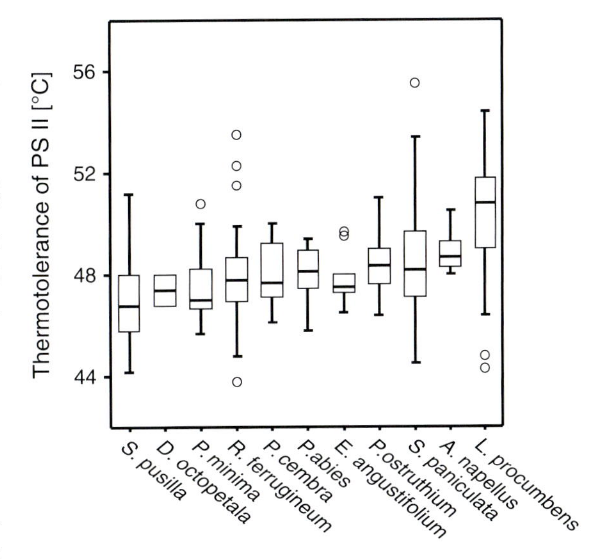


**Fig. 6.13** Variability of thermotolerance of PS II determined on various alpine plants species (mean N = 30). The box plots show the median (line inside the box) and the 25 and 75 percentiles. The whiskers are Tukey's hinges. Values more than 1.5 interquartile range's (IQR's) but less than 3 IQR's from the end of the box are labelled as outliers (O)

### 6.3.1 Intra-Specific Variation

PS II thermotolerance for alpine plants has been studied to a lesser degree than tissue heat tolerance, but sufficient data are available to draw comparisons for eleven alpine species including two timberline conifers (Fig. 6.13). The thermotolerance of PS II of these alpine species is in the range reported for various lowland plant species (Bilger et al. 1984; Havaux 1993; Valladares and Pearcy 1997). Mean values of PS II thermotolerance vary little between species, ranging from 47.1°C (*S. pusilla*) to 50.5°C (*S. paniculata*), but in most other species they are close to 48.4°C. For the nival species *R. glacialis*, a significantly lower thermotolerance value of 42.7°C has been reported (Larcher et al. 1997).

The mean critical temperature threshold for functional disturbances of PS II varies between 43.9°C and 47.4°C. In cushion plants such leaf temperatures occurred on 1.9% of summer days at alpine sites. Thermotolerance of PS II, however, does not appear to appropriately describe tissue heat tolerance, as these parameters may deviate by as much as 3.3°C (Fig. 6.14). A single physiological function, such as PS II efficiency, may not sufficiently describe the overall heat tolerance of cells and tissues.

### 6.3.2 Within-Species Variation

#### 6.3.2.1 Heat Hardening of Thermotolerance of PS II

The daily time course of thermotolerance determined under alpine field conditions for leaves of *R. ferrugineum* is shown in Fig. 6.15. Adjustments of PS II thermotolerance were observed to occur frequently and rapidly in *L. procumbens*, *S. paniculata* and *S. pusilla* throughout alpine summers (Braun et al. 2002). The thermotolerance of PS II can vary significantly for a single species over a summer (up to 9.6°C), and in *L. procumbens* a high diurnal variability of up to 4.8°C was found to occur (Braun et al. 2002).

Seasonal variations in thermotolerance of PS II are difficult to assess, as during winter, due to the naturally occurring depression of the potential efficiency of PS II (low $F_V/F_M$-value), the $F_0$-T curve technique cannot successfully be applied, since the curve progression is changed significantly by reduced $F_V/F_M$-values, making the determination of $T_c$ difficult (Neuner and Pramsohler 2006).

#### 6.3.2.2 Driving Forces of Heat Hardening of Thermotolerance of PS II

Under alpine field conditions elevated leaf temperatures increased the degree of heat hardening of PS II thermotolerance (Braun et al. 2002). Two weeks of preconditioning with increased temperatures (5°C higher than control plants) caused an increase in PS II thermotolerance in *S. paniculata* by +3.4°C (see Fig. 6.12). This leaf temperature effect under natural conditions corroborates earlier findings that long-term and short-term elevated leaf temperatures are able to increase the thermotolerance of PS II (Weis and Berry 1988; Havaux 1993; Huxman et al. 1998; Larcher et al. 1990; Froux et al. 2004). Similarly, slight changes in leaf temperature yielded a 5°C increase in PS II thermotolerance in potato leaves. This increase occurred as quickly (5°C/h; Havaux 1993) as found for *L. procumbens*. Rapid adjustments in PS II thermotolerance may hence not be peculiar to alpine plant species. However, as discussed earlier (Havaux 1993; Valladares and Pearcy 1997), dynamic as opposed to intrinsic thermotolerance is probably ecologically more significant, particularly for alpine plant species.

In nature high leaf temperatures co-occur with high irradiation intensities. We therefore cannot exclude effects of irradiation intensity (Schreiber and Berry

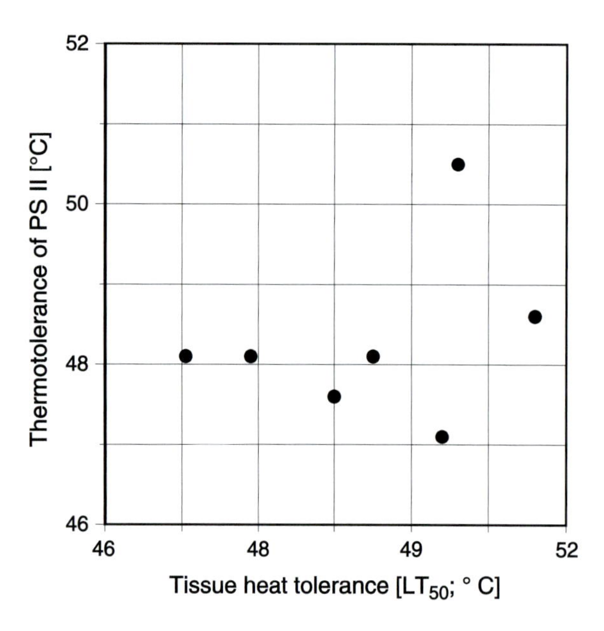


**Fig. 6.14** Thermotolerance of PS II in relation to tissue heat tolerance correlated for seven alpine plants species

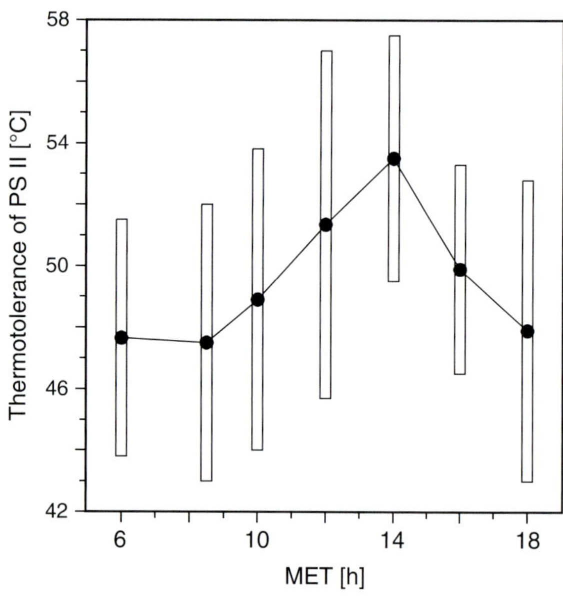


**Fig. 6.15** Diurnal change of PS II thermotolerance determined on leaves of *R. ferrugineum* on 1 June 1994 (mean values ±SD)

1977; Weis 1982) and quality (Havaux and Strasser 1992) that may also affect PS II thermotolerance.

During periods with slight water stress, a high variation in PS II thermotolerance and the highest overall values were recorded for leaves of *R. ferrugineum* and *S. pusilla*, for *L. procumbens* no distinct effect of the osmotic leaf water potential on PS II thermotolerance was observed during the growing period (Braun et al. 2002). Thermostability of PS II is known to be affected by drought stress (Seemann et al. 1986; Havaux 1992; Epron 1997; Valladares and Pearcy 1997; Hamerlynck et al. 2000; Froux et al. 2004) or osmotic stress, such as high salinity (Larcher et al. 1990). In several species dehydration (Seemann et al. 1986; Havaux 1992; Epron 1997; Lu and Zhang 1998) increased the thermotolerance of PS II. However, drought-preconditioning can increase the thermotolerance of PS II over a long period of up to 1 month (Valladares and Pearcy 1997), even when drought stress is removed (Ladjal et al. 2000). These observations may explain the difficulties in correlating thermotolerance of PS II with actual osmotic leaf water potentials.

Thermostability of PS II can also be affected by elevated $CO_2$ (Huxman et al. 1998; Taub et al. 2000).

## 6.4 Future Prospects

Adaptations to heat stress will become increasingly important as climate changes. The mean global surface temperature increased by +0.6 ±0.2°C during the twentieth century and is predicted to increase by a further +1.4°C to +5.8°C by 2100 (IPCC 2001). On a global scale 11 of the 12 years from 1995 to 2006 have been the warmest since 1850 (IPCC 2007). The frequency and severity of 'heat-waves' in temperate zones is predicted to increase further in the future (Semenov 2007; Semenov and Halford 2009). Climate change at high elevation sites is characterized by a high degree of complexity and uncertainty, and it is likely that the impacts of future, accelerated climate change will be proportionally more perceptible at high elevations (Beniston et al. 1997). In the Alps atmospheric temperature increase was found to be more than twice (Böhm et al. 2001) that of the global data. This has been most apparent for the summer periods since 1970 (Casty et al. 2005). A reconstructed temperature course from 755 to 2004 revealed that in the European Alps the summer of 2003 was the warmest in 1,250 years (Büntgen et al. 2006). Furthermore, in alpine glacier forefields the impacts of climate change are clearly accelerating (Cannone et al. 2008). High mountain vegetation is considered to be particularly vulnerable to climate change (Pauli et al. 2003), as in alpine environments abiotic factors, especially climate and soil chemistry, overrule biotic ones (Cannone et al. 2007).

In the first instance the predicted temperature increase will induce acclimations. In the long run, when the acclimation potential of the species is fully utilized, the changing climate will affect the presence or absence of certain plant species at certain sites and lead to alterations in the composition of plant communities and their abundance (Körner 1994; Grabherr et al. 1994). Changes in plant phenology (Menzel and Fabian 1999; Bertin 2008), growth and reproduction (Kudernatsch et al. 2008), plant migration and upward shifts (see Körner 1994; Holzinger et al. 2008) of vegetation belts will be the natural consequences and can already be monitored (Grabherr et al. 1994; Walther et al. 2005). There is a risk of drastic area losses, extinction, disintegration of the current vegetation patterns and impacts on the stability of high mountain ecosystems (Pauli et al. 2003).

All of these facts emphasize the urgency and the importance of developing a comprehensive understanding of the maximum heat tolerance or heat hardening capacity of plants and of developing techniques and field testing methods to assess these important plant functional traits. However, gathering data for a comprehensive understanding of species-specific heat tolerance has been very time-consuming until now. There is a need for new in situ measuring techniques that avoid abscising plant parts. Determination of the de facto and maximum heat tolerance of plants at certain growing sites must also include mechanisms of recuperation and repair.

In addition, the methods should allow for measuring heat tolerance under fully natural environmental conditions. This is a crucial point, as the methods employed for tissue heat tolerance and PS II thermotolerance are applied to samples kept in darkness. However, this does not match the natural situation and is in fact quite unrealistic, as overheating in alpine habitats is usually coupled with high photon flux densities (Körner and Larcher 1988). Physiologically the responses to moderate heat and strong irradiation appear to be closely related (Ludlow 1987; Havaux and Tardy 1996), and the need for co-adaptations to heat and high irradiation stress in alpine plants must hence be considered to be quite evident. However, very little is known about the combined effects of these stresses on alpine plants.

**Acknowledgements** We wish to thank Prof. Larcher for helpful suggestions on the manuscript and for access to the data of the PhD thesis of Maria Wildner-Eccher.

## References

Alexandrov VY (1977) Cells, molecules and temperature, vol 21, Ecological studies. Springer, New York

Bannister P (1970) The annual course of drought and heat resistance in heath plants from an oceanic environment. Flora 159:105–193

Beniston M, Diaz HF, Bradley RS (1997) Climate change at high elevation sites: an overview. Climatic Change 36:233–251

Bertin RI (2008) Plant phenology and distribution in relation to recent climate change. J Torrey Bot Soc 135:126–146

Biebl R (1968) Über Wärmehaushalt und Temperaturresistenz arktischer Pflanzen in Westgrönland. Flora 157:327–354

Biebl R, Maier R (1969) Tageslänge und Temperaturresistenz. Österr Bot 117:176–194

Bilger H-W, Schreiber U, Lange OL (1984) Determination of leaf heat resistance: comparative investigation of chlorophyll fluorescence changes and tissue necrosis methods. Oecologia 63:256–262

Bilger W, Schreiber U, Lange OL (1987) Chlorophyll fluorescence as an indicator of heat induced limitation of photosynthesis in *Arbutus unedo* L. In: Tenhunen JD, Catarino FM, Lange OL, Oechl WC (eds) Plant responses to stress. Springer, Berlin, pp 391–399

Böhm R, Auer I, Brunetti M, Maugeri M, Nanni T, Schöner W (2001) Regional temperature variability in the European Alps: 1760–1998 from homogenized instrumental time series. Int J Climatol 21:1779–1801

Braun V, Buchner O, Neuner G (2002) Thermotolerance of photosystem 2 of three alpine plant species under field conditions. Photosynthetica 40(4):587–595

Buchner O, Neuner G (2001) Determination of heat tolerance: a new equipment for field measurements. J Appl Bot 75 (3–4):130–137

Buchner O, Neuner G (2003) Variability of heat tolerance in alpine plant species measured at different altitudes. Arct Antarct Alp Res 35(4):411–420

Büntgen U, Frank DC, Niedergelt D, Esper J (2006) Summer temperature variations in the European Alps, A.D. 755–2004. J Climate 19(21):5606–5623

Cannone N, Sgorbati S, Guglielmin M (2007) Unexpected impacts of climate change on alpine vegetation. Front Ecol Environ 5(7):360–364

Cannone N, Diolaiuti G, Guglielmin M, Smiraglia C (2008) Accelerating climate change impacts on alpine glacier forefield ecosystems in the European Alps. Ecol Appl 18 (3):637–648

Casty C, Wanner H, Luterbacher J, Esper J, Böhm R (2005) Temperature and precipitation variability in the European Alps since 1500. Int J Climatol 15:1855–1880

Epron D (1997) Effects of drought on photosynthesis and on the thermotolerance of photosystem II in seedlings of cedar (*Cedrus atlantica* and *C. libani*). J Exp Bot 48 (10):1835–1841

Erschbamer B, Niederfriniger-Schlag R, Winkler E (2008) Colonization processes on a central Alpine glacier foreland. J Veg Sci 19(6):855–862

Falkova TV (1973) Seasonal changes of thermoresistance of higher plants cells under the conditions of the Mediterranean type subtropics. Bot Zeit 58:1424–1438

Froux F, Ducrey M, Epron D, Dreyer E (2004) Seasonal variations and acclimation potential of the thermostability of photochemistry in four Mediterranean conifers. Ann For Sci 61:235–241

Gauslaa Y (1984) Heat resistance and energy budget in different Scandinavian plants. Hol Ecol 7:1–78

Grabherr G, Gottfried M, Pauli H (1994) Climate effects on mountain plants. Nature 369:448

Hamerlynck EP, Huxman TE, Loik ME, Smith SD (2000) Effects of extreme high temperature, drought and elevated $CO_2$ on photosynthesis of the Mojave desert evergreen shrub, *Larrea tridentata*. Plant Ecol 148(2):183–193

Havaux M (1992) Stress tolerance of photosystem II in vivo. Antagonistic effects of water, heat, and photoinhibition stresses. Plant Physiol 100:424–432

Havaux M (1993) Rapid photosynthetic adaptation to heat stress triggered in potato leaves by moderately elevated temperatures. Plant Cell Environ 16:461–467

Havaux M, Strasser RJ (1992) Dynamics of electron transfer within and between PS II reaction center complexes indicated by the light-saturation curve of in vivo variable chlorophyll emission. Photosynth Res 31:149–156

Havaux M, Tardy F (1996) Temperature-dependent adjustment of the thermal stability of photosystem II in vivo: possible involvement of xanthophyll-cycle pigments. Planta 198:324–333

Holzinger B, Hülber K, Camenisch M, Grabherr G (2008) Changes in plant species richness over the last century in the eastern Swiss Alps: elevational gradient, bedrock effects and migration rates. Plant Ecol 195:179–196

Huxman TE, Hamerlynck EP, Loik ME, Smith SD (1998) Gas exchange and chlorophyll fluorescence responses of three south-western *Yucca* species to elevated $CO_2$ and high temperature. Plant Cell Environ 21(12):1275–1283

IPCC Intergovernmental Panel on Climate Change (2001) Climate change 2001: synthesis report – summary for policymakers. Cambridge University Press, UK

IPCC Intergovernmental Panel on Climate Change (2007) Climate Change 2007. Synthesis Report. In: Pachauri RK, Reisinger A (eds) Contribution of Working Groups I, II and III to the Fourth Assessment Report of the Intergovernmental Panel on Climate Change. IPCC

Kappen L, Lösch R (1984) Diurnal patterns of heat tolerance in relation to CAM. Z Pflanzenphysiol 114:87–96

Kjelvik S (1976) Varmeresistens og varmeveksling for noen planter, vesentlig fra Hardangervidda. Blyttia 34:211–226

Körner C (1994) Impact of atmospheric changes on high mountain vegetation. In: Beniston M (ed) Mountain environments in changing climates. Routledge, London, New York, pp 155–166

Körner C (2003) Alpine plant life. Functional plant ecology of high mountain ecosystems. Springer, Berlin

Körner C, Cochrane P (1983) Influence of plant physiognomy on leaf temperature on clear midsummer days in the snowy mountains, south-eastern Australia. Acta Oecol 4 (18):117–124

Körner C, De Moraes JAPV (1979) Water potential and diffusion resistance in alpine cushion plants on clear summerdays. Acta Oecol 14(2):109–120

Körner C, Larcher W (1988) Plant life in cold climates. In: Long S, Woodward FI (eds) Plants and temperature. Comp Biol Ltd, Cambridge, pp 25–57

Kudernatsch T, Fischer A, Bernhardt-Römermann M, Abs C (2008) Short-term effects of temperature enhancement on growth and reproduction of alpine grassland species. Basic Appl Ecol 9:263–274

Ladinig U, Hacker J, Neuner G, Steinacher G, Cavieres L, Wagner J (2009) Temperature resistance of vegetative and reproductive tissues of high-Andes plants in Central Chile. Plant abiotic stress tolerance, Vienna, 8. 11.2.2009

Ladjal M, Epron D, Ducrey M (2000) Effects of drought preconditioning on thermotolerance of photosystem and susceptibility of photosynthesis to heat in cedar seedlings. Tree Physiol 20:1235–1241

Larcher W (1984) Ökologie der Pflanzen auf physiologischer Grundlage. Eugen Ulmer, Stuttgart, Berlin

Larcher W, Wagner J (1976) Temperaturgrenzen der $CO_2$-Aufnahme und Temperaturresistenz der Blätter von Gebirgspflanzen im vegetativen Zustand. Acta Oecol 11 (4):361–374

Larcher W, Wagner J (1983) Ökologischer Zeigerwert und physiologische Konstitution von *Sempervivum montanum*. Verh GfÖ 11:253–364

Larcher W, Wagner J (2009) High mountain bioclimate: temperatures near the ground recorded from the timberline to the nival zone in the Central Alps. Contrib Nat Hist 12:857–874

Larcher W, Holzner M, Pichler J (1989) Temperaturresistenz inneralpiner Trockenrasen. Flora 183:115–131

Larcher W, Wagner J, Thammathaworn A (1990) Effects of superimposed temperature stress on *in vivo* chlorophyll fluorescence of *Vigna unguiculata* under saline stress. J Plant Physiol 136:92–102

Larcher W, Wagner J, Lütz C (1997) The effect of heat on photosynthesis, dark respiration and cellular ultrastructure of the arctic-alpine psychrophyte *Ranunculus glacialis*. Photosynth 34(2):219–232

Larcher W, Kainmüller C, Wagner J (2010) Survival types of high mountain plants under extreme temperatures. Flora 205 (1):3–18

Loik ME, Harte J (1996) High-temperature tolerance of *Artemisia tridentata* and *Potentilla gracilis* under a climate change manipulation. Oecologia 108:224–231

Lu CM, Zhang JH (1998) Thermostability of photosystem II is increased in salt-stressed *sorghum*. Aust J Plant Physiol 25 (3):317–324

Ludlow MM (1987) Light stress at high temperature. In: Kyle DJ, Osmond CB, Arntzen CJ (eds) Photoinhibition. Elsevier, Amsterdam, pp 89–109

Marcante S, Winkler E, Erschbamer B (2008) Population dynamics in a glacier foreland: do alpine species fit into demographic successional models? Ber Nat med Verein Innsbruck 18:22

Menzel A, Fabian P (1999) Growing season extended in Europe. Nature 397:659

Meyer H, Santarius K (1998) Short-term thermal acclimation and heat tolerance of gametophytes of mosses. Oecologia 115:1–8

Mooney HA, Billings WD (1961) Comparative physiological ecology of arctic and alpine populations of *Oxyria digyna*. Ecol Monogr 31:1–29

Moser W (1965) Temperatur- und Lichtabhängigkeit der Photosynthese sowie Frost- und Hitzeresistenz der Blätter von drei Hochgebirgspflanzen (*Ranunculcus glacialis*, *Geum reptans*, *Oxyria digyna*). PhD thesis, University of Innsbruck

Neuner G, Pramsohler M (2006) Freezing and high temperature thresholds of photosystem 2 compared to ice nucleation, frost and heat damage in evergreen subalpine plants. Physiol Plant 126:196–204

Neuner G, Braun V, Buchner O, Taschler D (1999) Leaf rosette closure in the alpine rock species *Saxifraga paniculata* Mill.: significance for survival of drought and heat under high irradiation. Plant Cell Environ 22:1539–1548

Neuner G, Buchner O, Braun V (2000) Short-term changes in heat tolerance in the alpine cushion plant *Silene acaulis ssp. excapa* [All.] J. Braun at different altitudes. Plant Biol 2:677–683

Pauli H, Gottfried M, Grabherr G (2003) Effects of climate change on the alpine and nival vegetation of the Alps. J Mount Ecol 7:9–12

Pisek A, Kemnitzer R (1968) Der Einfluß von Frost auf die Photosynthese der Weißtanne (*Abies alba* MILL.). Flora 157:314–326

Salisbury FB, Spomer GG (1964) Leaf temperatures of alpine plants in the field. Planta 60:497–505

Sapper I (1935) Versuche zur Hitzeresistenz der Pflanzen. Planta 23:518–556

Schreiber U, Berry JA (1977) Heat-induced changes of chlorophyll fluorescence in intact leaves correlated with damage of the photosynthetic apparatus. Planta 136:233–238

Schwarz W (1970) Der Einfluß der Photoperiode auf das Austreiben, die Frosthärte und die Hitzeresistenz von Zirben und Alpenrosen. Flora 159:258–285

Seemann JR, Downton WJS, Berry JA (1986) Temperature and leaf osmotic potential as factors in the acclimation of photosynthesis to high temperature in desert plants. Plant Physiol 80:926–930

Semenov MA (2007) Development of high resolution UKCIPO2-based climate change scenarios in the UK. Agri For Meteo 144:127–138

Semenov MA, Halford NG (2009) Identifying target traits and molecular mechanisms for wheat breeding under a changing climate. J Exp Bot 60:2791–2804

Smillie RM, Nott R (1979) Heat injury in leaves of alpine, temperate and tropical plants. Aust J Plant Physiol 6:135–141

Taub DR, Seemann JR, Coleman JS (2000) Growth in elevated $CO_2$ protects photosynthesis against high-temperature damage. Plant Cell Environ 23(6):649–656

Turner H (1958) Maximaltemperaturen oberflächennaher Bodenschichten an der alpinen Waldgrenze. Wald Leben 10:1–12

Valladares F, Pearcy RW (1997) Interactions between water stress, sun-shade acclimation, heat tolerance and photoinhibition in the sclerophyll *Hetereomeles arbutifolia*. Plant Cell Environ 20:25–36

Walther GR, Beißner S, Burga CA (2005) Trends in the upward shift of alpine plants. J Veg Sci 16:541–548

Weis E (1982) Influence of metal cations and pH on the heat sensitivity of photosynthetic oxygen evolution and chlorophyll fluorescence in spinach chloroplasts. Planta 154:41–47

Weis E, Berry JA (1988) Plants and high temperature stress. In: Long SP, Woodward FI (eds) Plants and temperature. Comp Biol Ltd, Cambridge, pp 329–346

Wildner-Eccher MT (1988) Keimungsverhalten von Gebirgspflanzen und Temperaturresistenz der Samen und Keimpflanzen. PhD thesis, University of Innsbruck

Zavadskaya IG, Denko EI (1968) The effect of insufficient water supply on the stability of leaf cells of certain plants of the Pamirs. Bot Zeit 53:795–805

# 7 Photosynthesis and Antioxidative Protection in Alpine Herbs

Peter Streb and Gabriel Cornic

## Abbreviations

| | |
|---|---|
| APx | Ascorbate peroxidase |
| $C_a$ | Concentration of $CO_2$ outside the leaf |
| $C_c$ | Concentration of $CO_2$ at the chloroplast level |
| $C_i$ | Concentration of $CO_2$ inside the leaf |
| Car | Carotenoid |
| Chl | Chlorophyll |
| $g_{CO2}$ | Conductance for $CO_2$ |
| LHC | Light harvesting complex |
| m.a.s.l. | Meter above sea level |
| NPQ, qN | Non-photochemical fluorescence quenching (for details see Maxwell and Johnson 2000) |
| PFD | Photon flux density |
| PS | Photosystem |
| PTOX | Plastid terminal oxidase |
| qP | Photochemical fluorescence quenching |
| ROS | Reactive oxygen species |
| $S_{CO2/O2}$ | Specificity factor of Rubisco for $CO_2$ relative to $O_2$ |
| SOD | Superoxide dismutase |

P. Streb (✉) • G. Cornic
Laboratoire Ecologie Systématique et Evolution, University of Paris-Sud, Orsay, France

CNRS, Orsay, France

AgroParisTech, Paris, France
e-mail: peter.streb@u-psud.fr

Plant Species Often Mentioned

*Colobanthus quitensis: C. quitensis*
*Geum reptans (montanum): G. reptans (montanum)*
*Homogyne alpina: H. alpina*
*Polygonum cuspidatum: P. cuspidatum*
*Ranuculus glacialis: R. glacialis*
*Soldanella alpina (pusilla): S. alpina (pusilla)*

## 7.1 Introduction

Alpine environments are found all over the world, from the south over the tropics to the north. Alpine herbs are defined here as higher plant species growing above the tree line up to and within the persisting snow line (nival life zone). The altitude of their occurrence varies strongly from around sea level in the far north and south to elevations above 4,000–5,000 m.a.s.l. in Africa and the Himalaya (Körner 2003). In the European Alps the alpine life zone starts at approximately 2,000 m elevation, depending on local microclimatic conditions. The mean annual air temperature in the Alps at this elevation is approximately 0°C (Friend and Woodward 1990) and on average the vegetation period is limited to 5 months per year.

At higher elevation, mean air temperatures are considerably lower and vegetation periods are shorter (Körner 2003). Plants surviving under these extreme conditions are acclimated to complete their life cycle within a short vegetation period and to accumulate sufficient reserves for a long-lasting winter period. Moser et al. (1977) described *R. glacialis* plants which were capable of surviving 1 year completely covered by snow.

C. Lütz (ed.), *Plants in Alpine Regions*,
DOI 10.1007/978-3-7091-0136-0_7, © Springer-Verlag/Wien 2012

The air pressure declines with altitude, and the concomitant lower partial pressure of $CO_2$ has often been considered as a significant restriction to photosynthesis. Other environmental factors, such as precipitation and wind speed, depend strongly on local climatic conditions and may vary from dry to wet conditions as well as from areas exposed to strong wind to sheltered areas (Körner 2003).

The high biodiversity of alpine environments (4% of total plant biodiversity) as compared to the land surface covered by alpine environments (3% of total land surface) (Körner 2003) suggests that alpine plants grow in a rather favourable environment supporting reproduction and biodiversity by efficient photosynthetic activity during the short growing season. However, several alpine plant species grow outside the meadows with high biodiversity either at the limit of higher plant distribution or in particular environmental niches. The present contribution is focused on such plants that are regularly exposed to extreme and rapidly changing climatic stress events, namely *R. glacialis* and *G. reptans* from highest elevations in the Alps and *S. alpina, G. montanum* and *H. alpina* from snow valleys.

## 7.2 Microclimatic Conditions for Selected Plant Species

Plants growing at the highest altitude limit of plant distribution in the Alps experience mainly low air temperature at moderately to high PFD (Moser et al. 1977). During the vegetation period, mean photon flux densities at high altitude are often comparable to the PFDs in the valleys (Körner 2003). However, peaks of high light intensity may occur at low or moderately high temperatures. Moser et al. (1977) described photosynthetic carbon assimilation of *R. glacialis* leaves at variable climatic conditions *in situ*, such as low temperatures (0–10°C) and higher temperatures (up to 30°C) at high as well as low PFDs. Surprisingly, efficient photosynthetic activity was measured under all these contrasting conditions. According to Diemer and Körner (1996) *R. glacialis* plants need more than 30 days of carbon exchange to amortize the carbon investment for leaves. This means, that photosynthetic carbon assimilation should not be affected by extreme climate events in order to guarantee survival during prolonged short vegetation periods.

Some other plants species, like *S. alpina*, *H. alpina* or *G. montanum*, grow at lower altitudes in snowbeds. These plants survive the long winter period with green leaves. During snowmelt, green leaves are covered by cold melting water and are irradiated by high photon flux densities. Figure 7.1 shows variations of leaf temperature and PFD at the leaf level of an *S. alpina* plant during snowmelt. As long as the leaf is covered by snow, the PFD is low (mean values between 50 μmol $m^{-2}$ $s^{-1}$ and 100 μmol $m^{-2}$ $s^{-1}$) but rises markedly during snowmelt to maximum values of more than 2,000 μmol $m^{-2}$ $s^{-1}$. Maximum PFDs

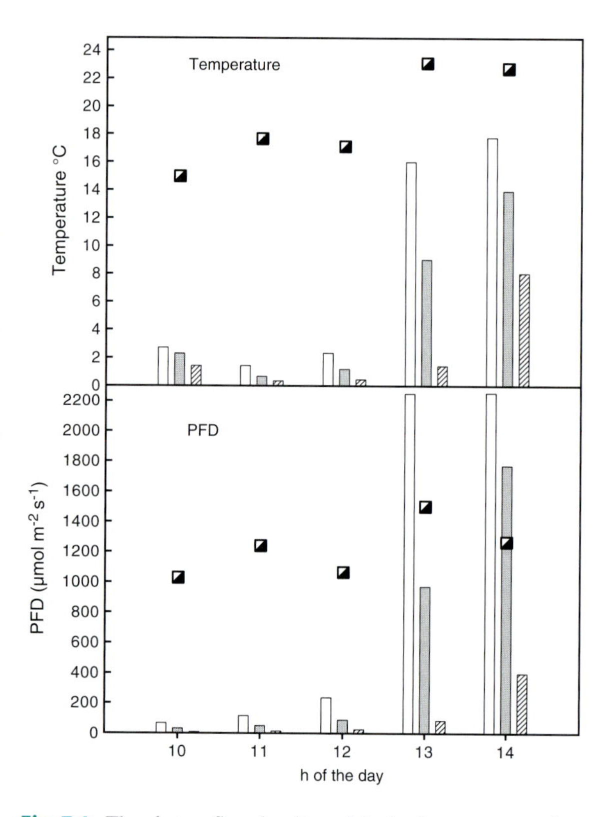


**Fig. 7.1** The photon-flux-density and the leaf temperature of an *S. alpina* leaf during the snowmelt. To fix the thermocouple at the lower side of an *S. alpina* leaf and the light sensor directly beneath the leaf, the snow was removed. Subsequently the leaf and the light sensor were covered by the same amount of snow (approximately 5 cm high). The leaf temperature and the PFD at the leaf level were recorded every 30 s with a Campbell datalogger. Every hour the maximum, the minimum and the average temperature and PFD were calculated, and the results are displayed in the figure. *Grey*: average leaf temperature and PFD, *white*: maximum leaf temperature and PFD, *stripes*: minimum leaf temperature and PFD. For comparison, the average PFD outside the snowfield of a horizontally orientated light sensor and the average air temperature were measured (*semi-filled squares*) (Laureau and Streb, unpublished)

measured in the vicinity of this plant were higher than 3,000 μmol $m^{-2}$ $s^{-1}$ (Streb, unpublished). This strong increase in PFD is accompanied by an increase in the leaf temperature by 15°C. While snowbed plants grow solitarily during and for a short time after the snowmelt, they are overgrown during the later season. The PFD at the leaf level is then much lower because plants are shaded, but leaf temperatures may exceed 30°C. In autumn before snowfall, snowbed plants may again become isolated and exposed.

Körner (2003); Kleier and Rundel (2009); Larcher et al. (2010) described even more pronounced temperature extremes in alpine plants ranging from values below 0°C to more than 40°C in full sunlight and varying between leaves and roots during the same daytime. The fact that temperature as well as PFD extremes are both very high, but mean PFDs in the Alps are rather moderate (Körner 2003; Streb, unpublished) raises the question of whether alpine plants are acclimated to high PFD and temperature or to low PFD and temperature and how these plants tolerate either one of the two extremes.

## 7.3 Acclimation to Light and Temperature

### 7.3.1 General Features of Light and Temperature Acclimation

Plants mostly acclimate to their respective growing light (Anderson and Osmond 1987; Bailey et al. 2001; Walters 2005) and temperature (Huner et al. 1993, 1998). Some characteristics of high-light-grown plants are: (1) thicker leaves, (2) a higher ratio of Chl a/b and of β-carotene/xanthophyll accompanied by a lower ratio of LHC/PS, (3) a lower ratio of Chl/Car which results mainly from higher contents of xanthophylls and is interpreted as a higher capacity for non-radiative dissipation of excitation energy and antioxidative protection by carotenoids, (4) a higher photosynthetic capacity and the requirement of higher PFDs for light compensation and light saturation of photosynthesis as compared to plants acclimated to low light (Anderson and Osmond 1987; Demmig-Adams and Adams 1992b; Logan et al. 1996; Demmig-Adams 1998; Bailey et al. 2001; Walters 2005). Similar adjustments in photosynthetic parameters are observed during acclimation to low temperature (Huner et al. 1993, 1998; Streb et al. 1999) although Chl a/b ratios do not increase during cold acclimation when the PFD was kept constant (Huner et al. 1984; Streb 1994; Bravo et al. 2007). However, the above mentioned parameters are not linearly related to the growth PFD (Bailey et al. 2001), and the response of these parameters is plant species specific (Murchie and Horton 1998).

At high PFD, absorption of light might exceed the capacity for carbon assimilation. At low temperatures the enzymatic reactions of carbon assimilation and sugar synthesis slow down, while light absorption is almost temperature-independent (Falk et al. 1996). In both cases, this may lead to a higher reduction state of the plastoquinone pool, which might serve as a signalling cascade inducing acclimation (Huner et al. 1998; Pfannschmidt 2003; Walters 2005). Overall, acclimation to high light and low temperature reduces both the capacity for primary photosynthetic reactions relative to the capacity for carbon assimilation and sugar synthesis, thus enabling the plant to use high PFD more efficiently but limiting photosynthesis at low PFD without changes in quantum yield (Huner et al. 1998; Walters 2005).

### 7.3.2 Light and Temperature Acclimation in Alpine Plants

Alpine plants growing at high elevation have thicker leaves than their lowland counterparts. This was shown by microscopic analysis as well as by measuring the leaf mass area ratio (Körner and Diemer 1987; Friend and Woodward 1990; Diemer and Körner 1996; Körner 2003; Kogami et al. 2001; Shi et al. 2006). Thicker leaves are also induced by lower growth temperatures, irrespectively of light conditions in alpine species (Mächler and Nösberger 1978; Medek et al. 2007).

The ratio of Chl a/b and the content of xanthophyll pigments increase, and the ratio of Chl/Car decreases with altitude in the Alps (Table 7.1), indicating acclimation to high light intensities. Chl a/b ratios of around 4.0 were also measured in 6 Tasmanian alpine shrubs at much lower altitude in autumn and spring (Williams et al. 2003). However, lower Chl a/b ratios were determined in several alpine species growing at altitudes of 2,400–2,700 m in the French Alps, in *P. cuspidatum* leaves from Mt Fuji (2,500 m.a.s.l.),

**Table 7.1** Chlorophyll a/b ratio, chlorophyll a + b/carotenoid ratio and xanthophyll cycle pigments in several alpine species growing at different altitude

| Elevation (m) | Chl a/b | Chl/Car | Xanth µg/g Fw | n species |
|---|---|---|---|---|
| 1,000 | 3.18 ± 0.06 | 7.55 ± 0.27 | 83.3 ± 13.6 | 4 |
| 2,000 | 3.92 ± 0.14 | 6.09 ± 0,57 | 122.9 ± 15.7 | 9 |
| 3,000 | 4.13 ± 0.37 | 5.97 ± 0,55 | 139.3 ± 30.8 | 6 |

The number of different species investigated is shown (Data were taken from Wildi and Lütz (1996) and mean values were calculated)

in an Andes ecotype of *C. quitensis* and in several high mountain plants from the Anatolian mountains (Streb et al. 1997; Kogami et al. 2001; Öncel et al. 2004; Bravo et al. 2007, Streb unpublished). In *P. cuspidatum* leaves, no difference in Chl a/b ratios was found between the highland and the lowland species (Kogami et al. 2001), and in *C. quitensis* no difference in Chl a/b was measured in plants acclimated or not to cold or grown at high or low light intensities (Bravo et al. 2007). Furthermore Chl/Car ratios and contents of xanthophyll cycle pigments vary strongly in alpine species. More than twice as much carotenoids and xanthophyll cycle pigments were observed in leaves of *S. alpina* and *H. alpina* as compared to *R. glacialis* leaves (Streb et al. 1997). Similar significant differences were reported from other species from the Alps, the Tasmanian as well as Anatolian mountains (Wildi and Lütz 1996; Williams et al. 2003; Öncel et al. 2004). While in five species measured by Wildi and Lütz (1996) the xanthophyll cycle pigments increase with altitude, these pigments are independent of altitude in two species, including *R. glacialis*, and even decrease in two other species.

Photosynthetic carbon assimilation under ambient $CO_2$ and optimum temperature in various alpine species is often not saturated at PFD of 2,000 µmol $m^{-2}$ $s^{-1}$, clearly indicating acclimation to high-light conditions (Körner and Diemer 1987; Germino and Smith 2001; Streb et al. 2005; Hacker and Neuner 2006; Kleier and Rundel 2009). The light compensation point in leaves of *S. alpina* and *R. glacialis* was determined to be between 25 µmol $m^{-2}$ $s^{-1}$ and 50 µmol $m^{-2}$ $s^{-1}$ at temperatures between 10°C and 25°C (Streb et al. 2005 and unpublished). In *Saxifraga paniculata* a higher light compensation point was reported for plants collected from an alpine, in particular a snowfree site, as compared to plants from the valley or the greenhouse (Hacker and Neuner 2006). In alpine and Antarctic ecotypes of *C. quitensis* the light compensation point increased due to cold acclimation (Bravo et al. 2007). The increase in the light compensation point normally coincides with an increase in leaf respiration. The photosynthetic capacity of alpine plants varied strongly (Körner 2003) and cannot be correlated to acclimation to high or low light intensities, suggesting that factors other than PFD influence photosynthetic capacity. Alpine plants often show a particular morphological structure (rosettes, cushions, tussocks) which may reduce light absorption as well as increasing temperature in light (Körner 2003).

Overall, several general properties suggest acclimation to high PFD and low temperature in high mountains plants. Nevertheless, exceptions can be found in several microclimates, especially regarding the pigment content and the ratio of Chl a/b and regarding plants with reduced light absorption resulting from their anatomy. It should, however, be noted that pigment contents are not linearly related to light acclimation and may be species-specific. Clearly, more work is necessary to understand light and temperature acclimation in alpine plant species.

## 7.4 Photosynthesis in Alpine Plants

Carbon assimilation studies using the $CO_2$ titration method under atmospheric conditions in alpine herbaceous plant species were performed by Bonnier (1888, 1895, cited in Henrici 1918), Henrici (1918) and Cartellieri (1940) in the French Alps, in Switzerland and Austria. Bonnier (cited in Henrici 1918) found assimilation rates up to twice as high in alpine as compared to valley species measured under the same conditions. Henrici (1918) concluded that photosynthesis in alpine, as compared to lowland plants, is induced at higher PFD, has a lower temperature optimum and is higher at high PFD. Cartellieri (1940) measured assimilation rates during the course of the year and the day. He found the highest assimilation rate in leaves of *R. glacialis* at full sunlight and concluded that this species is especially acclimated to high light intensities. Cartellieri (1940) reported photosynthetic activity also at temperatures lower than 0°C. Mooney and Billings (1961) compared arctic and alpine *Oxyria digyna* species, showing that the alpine variety is saturated at higher PFD than the arctic variety, while the arctic variety has a lower temperature optimum. Detailed photosynthetic studies involving measurements during the course of the year were published by Moser et al. (1977) and performed with a great variety of plant species at high and low altitudes by Körner and Diemer (1987).

### 7.4.1 Light and Temperature Effects on Photosynthesis in Alpine Species

Körner (2003) concluded that photosynthetic productivity of alpine plants is limited by the PFD rather than by low temperature, despite the fact that very high PFDs were measured in alpine environments (Streb, unpublished). Körner (2003) argued that the leaf temperature increases if PFDs are high and therefore plants experience moderate to high temperatures at high PFD. This temperature increase in light can also be seen in Fig. 7.1 for an *S. alpina* leaf during the snowmelt. An even further increase in leaf temperature is observed in typical alpine plant species with growth forms such as cushions, rosettes and tussocks in sunlight (Körner 2003; Larcher et al. 2010). However, exceptions are obvious. The temperature of the *S. alpina* leaf shown in Fig. 7.1 is initially very low when the leaf is exposed to full sunlight. Comparable light/temperature extremes occur also in other alpine plant species when a clear day follows a cold night with snowfall. The snow can remain at the leaf surface while light energy slowly melts the snow (Fig. 7.2). In this situation the leaf experiences temperatures at around 0°C while the PFD is high. Moser et al. (1977) described such combinations of high irradiation and low temperature for several alpine species. The other extreme, temperatures up to 40°C and higher, were measured in summer at high PFD (Körner 2003; Larcher et al. 2010).

Fig. 7.2 *R. glacialis* plants after snowfall in July at the Col du Galibier (French Alps). The leaves are covered by snow while strong sunlight is absorbed (Photo by Peter Streb)

The temperature optimum for photosynthesis in high mountain ecotypes is slightly lower than the temperature optimum in low altitude ecotypes but not very different from lowland species. It was found to be 16–25°C for most species (Körner and Diemer 1987; Rawat and Purohit 1991; Goldstein et al. 1996; Cabrera et al. 1998; Manuel et al. 1999; Körner 2003; Germino and Smith 2001; Zhang and Hu 2008; Kleier and Rundel 2009). At low temperatures the photosynthetic sugar formation in alpine ecotypes of *Trifolium repens* is higher than that in low elevation ecotypes (Mächler and Nösberger 1977).

Most frequently alpine plants, at least in the Alps, experience PFDs lower than approximately 600 μmol $m^{-2}$ $s^{-1}$ together with air temperatures lower than the optimum (Körner 2003). The temperature optimum of photosynthesis in ambient $CO_2$ decreases when the PFD decreases (Moser et al. 1977; Körner 1982; Terashima et al. 1993; Körner 2003). The same effect was, however, also well described for non-alpine plant species (Berry and Björkman 1980; Leegood and Edwards 1996). How can this shift in temperature optimum by light be explained? Under limiting light (usually less than 200 μmol $m^{-2}$ $s^{-1}$) the effect of temperature on net carbon assimilation in C3 species is low. At high temperatures, the dark respiration increases and affects the net carbon assimilation. Furthermore, the solubility of oxygen relative to $CO_2$ increases, and the specificity factor of $CO_2$ for Rubisco decreases (Leegood and Edwards 1996). This effect increases photorespiration relative to carbon assimilation. If light saturates photosynthesis, maximum assimilation increases to an optimum with rising temperatures, because higher temperatures accelerate enzymatic reactions. Furthermore, the PFD saturating photosynthesis increases up to the optimum temperature. Beyond this optimum, photosynthesis decreases because Rubisco is deactivated at high temperatures, and damage to the photosystems might limit electron transport activity (Berry and Björkman 1980). Thus the thermal optimum of $CO_2$ assimilation may change

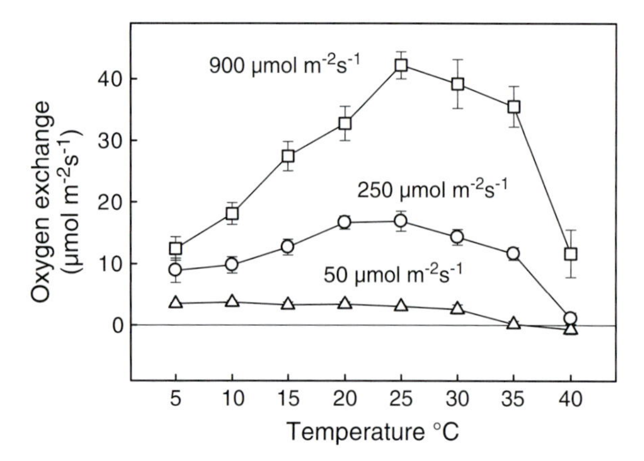


**Fig. 7.3** Gross oxygen evolution (oxygen evolution – dark respiration) of *R. glacialis* leaves measured at saturating $CO_2$ at various PFDs as a function of leaf temperature. Measurements were done with a Hansatech D2 leaf chamber connected to a thermostated water bath. The temperature of the leaf was determined after introducing a thermocouple touching an *R. glacialis* leaf into the chamber. The setting of the thermostat and the respective light and temperature conditions were recorded. Subsequently, the thermocouple was replaced by the oxygen electrode and thermostat settings were adjusted to the desired leaf temperature. Measurements started at 25°C. Thereafter the temperature was decreased to 5°C and subsequently stepwise increased up to 40°C. For every leaf temperature dark respiration was measured before the PFD density was stepwise increased starting with the lowest PFD (Streb and Laureau, unpublished)

continuously during the day with changing PFDs as shown *in situ* for *Carex curvula* and *Rheum nobile* (Körner 1982; Terashima et al. 1993).

Oxygen evolution measurements at saturating $CO_2$, which suppresses photorespiration, show the temperature and light response of leaves of the alpine species *R. glacialis* (Fig. 7.3). Since respiration also increases with temperature (Atkin and Tjoelker 2003) and might impair the thermal optimum at low light, gross oxygen evolution (oxygen evolution – dark respiration) was calculated, assuming that respiration is not decreased by light. Under limiting light (50 µmol $m^{-2}$ $s^{-1}$) gross oxygen evolution is independent from leaf temperature between 5°C and 30°C and decreased at higher temperatures. Little differences in the thermal optimum were observed at 250 or 900 µmol $m^{-2}$ $s^{-1}$ PFD. The slow decrease of gross oxygen evolution at temperatures higher than 25°C might be explained by successive deactivation of Rubisco activase and the rapid decrease at temperatures above 35°C by additional impairment of PSII function under all light conditions (Larcher et al. 1997; Streb et al. 2003c; Haldimann and Feller 2004).

Alpine plants are regularly exposed to conditions of temperature extremes at low (Figs. 7.1 and 7.2) and at high temperatures together with exceptionally high irradiance. Low as well as high temperatures will then affect photosynthetic activity, and alpine plants have to cope with excess excitation energy under these two contrasting conditions.

The temperature extremes, where net photosynthesis was measured in alpine species, ranged from −3 (e.g. *G. reptans*) to −7°C (e.g. *R. glacialis*) at low temperatures and 38°C (*G. reptans, R. glacialis, S. alpina*) up to 47°C (*Acrctostaphylos uva-ursi*) at high temperatures and depend on the light conditions (Moser et al. 1977; Larcher et al. 1997; Körner 2003; Streb et al. 2003c).

### 7.4.2 Photosynthesis in Alpine Plants as Influenced by Low $CO_2$ Partial Pressure

The air pressure decreases with altitude by approximately 10%/1,000 m and so does the partial pressure of $CO_2$ (Körner 2003). Because the concentration of a gas in the liquid phase (i.e. in the chloroplast) is proportional to its partial pressure in the gas phase (Henry's law), it was suggested that alpine plants suffer from $CO_2$ limitation especially at high altitude. Furthermore, the $CO_2$ concentration at a given atmospheric pressure depends on the absolute temperature. The effect of low temperature on the $CO_2$ concentration at high compared to low altitude is, however, small (7% for a difference of 20°C) as compared to the effect of lower air pressure compared to higher air pressure (30% for a difference of 3,000 m altitude). It should be noted that the oxygen partial pressure decreases as well and that the $CO_2/O_2$ ratio at the level of Rubisco is not changed by altitude (Friend and Woodward 1990).

In addition, carbon assimilation (*A*) depends on the $CO_2$ concentration in the chloroplast which is determined by $CO_2$ uptake and the conductance for $CO_2$ diffusion to the chloroplast (Box 7.1, Eq. 7.1). The $CO_2$ diffusion occurs first in the gas phase by successively passing the boundary layer, which depends on the wind speed, the leaf structure and size, the stomata and the substomatal cavities, and the intercellular air space. Second, $CO_2$ enters the cell walls, passing a liquid phase to the carboxylation site in the chloroplasts (Evans and Loreto 2000). Accordingly, the total resistance ($R_{total}$) to $CO_2$ diffusion is considered as the sum of four different resistances (Box 7.1, Eq. 7.2) with $g_{total} = 1/r_{total}$. The $CO_2$ diffusion in the gas phase depends on the diffusion coefficient of $CO_2$ which is a function of temperature and of the partial pressure of $CO_2$ (Box 7.1, Eq. 7.3). From this equation an approximately 1.3 fold higher diffusion coefficient of $CO_2$ can be calculated for 3,000 m.a.s.l., counteracting the lower $CO_2$ concentration at high altitude (Terashima et al. 1995).

The stomatal conductance depends on the stomatal density, the stomatal pore area and the mean length of diffusion through the epidermis (Meidner and Mandsfield 1968). The stomatal density increases with altitude in temperate mountains (Friend and Woodward 1990; Körner 2003) and this could facilitate $CO_2$ diffusion. Accordingly, a higher stomatal conductance was measured in four crop species grown at high altitude in the Himalaya as compared to lower altitude grown counterparts (Kumar et al. 2005). However, a decrease in stomatal conductance with altitude was measured in *Metrosideros polymorpha* (Vitousek et al. 1990; Meinzer et al. 1992), *Buddleja davidii* (Shi et al. 2006) and *P. cuspidatum* (Kogami et al. 2001).

Figure 7.4 shows variations of net $CO_2$ assimilation of leaves of *R. glacialis* as a function of $C_i$ at 2,100 m.a.s.l., 25°C and a PFD of 600 μmol $m^{-2}\,s^{-1}$. From these results a stomatal limitation of 40% can be calculated according to Farquhar and Sharkey (1982) (Box 7.1, Eq. 7.4). This is in the same order of magnitude as can be estimated from results of Terashima et al. (1993) obtained for *Rheum nobile* and Shi et al. (2006) obtained in *Buddleja davidii* and much higher than the 10% stomatal limitation in non-drought stressed herbaceous lowland-plants.

The conductance to $CO_2$ diffusion in the substomatal cavity ($g_{ias}$) depends on the porosity and tortuosity of the leaf, i.e. the leaf thickness and the surface of the mesophyll cells which contacts the intracellular air space. While the intracellular air volume was similar in high altitude and low altitude plant species, the mesophyll area contacting the intracellular air space was higher in more than 20 different alpine species measured by Körner (2003). Kogami et al. (2001) determined, however, a decrease in porosity of the mesophyll with altitude in *P. cuspidatum*.

**Box 7.1**

Equations used in the text

*A*: Assimilation; $g_{CO2}$: total conductance for $CO_2$ diffusion; $C_a$: $CO_2$ concentration outside the leaf; $C_c$: $CO_2$ concentration in the chloroplast; $D_{CO2}$: diffusion coefficient for $CO_2$; *L*: length of the diffusion path

$$A = g_{CO_2} \times (c_a - c_c) = \left(\frac{D_{CO2}}{L}\right) \times (c_a - c_c) \qquad (7.1)$$

$r_a$: resistance of the boundary layer; $r_s$: resistance of the stomata; $r_{ias}$: resistance of the substomatal cavities; $r_i$: resistance of the internal liquid phase; $r_{ias} + r_i = r_m$

$$R_{total} = r_a + r_s + r_{ias} + r_i \qquad (7.2)$$

$D_{CO2}$: diffusion coefficient of $CO_2$ $D_{CO2}{}^{0}$: diffusion coefficient of $CO_2$ under standard conditions ($T^0$ 273 K; $P^0$ 1,013 hPa)

$$D_{CO2} = D_{CO2}{}^{0} \times \left(\frac{T}{T^0}\right)^2 \times \left(\frac{P^0}{P}\right) \qquad (7.3)$$

*A*: assimilation at a given internal partial pressure of $CO_2$ ($C_i$), $A^0$: theoretical assimilation if $C_i$ were equal to the external $CO_2$ partial pressure ($C_a$), i.e. if stomata did not limit assimilation

$$R_s(\%) = \frac{A^0 - A}{A^0} \times 100 \qquad (7.4)$$

The $CO_2$ must further enter the cell wall and pass the liquid phase of the cell. The decrease in temperature at higher altitude is expected to increase the viscosity and therefore to slow down diffusion. The diffusion of $CO_2$ in the cell is probably facilitated by $CO_2$ pores and carbonic anhydrase (Evans et al. 2009). However, as mentioned by Evans et al. (2009), more than 50% of the total diffusion resistance may be attributed to the dissolution of $CO_2$ and the passage of the cell wall. The increase in cell wall thickness by

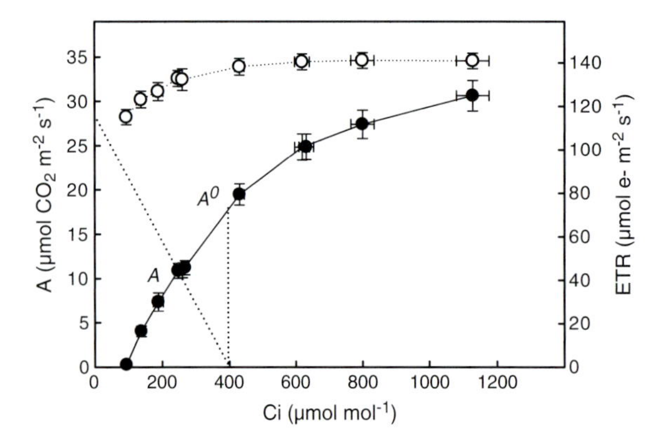


**Fig. 7.4** Carbon assimilation (A) and photosynthetic electron transport (ETR) of an *R. glacialis* leaf measured as the function of internal $CO_2$ partial pressure ($C_i$) at 2,100 m.a.s.l., 25°C and 600 µmol $m^{-2}$ $s^{-1}$. The measurements allow the calculation of stomatal limitation (Box 1, Eq. 7.4) (Cornic and Streb, unpublished)

44% in high-mountain as compared to lowland *P. cuspidatum* described by Kogami et al. (2001) may therefore be crucial in controlling the diffusion of $CO_2$ in this species.

The data in Fig. 7.4 (ETR and A), obtained from *R. glacialis* leaves, were used to estimate a mesophyll conductance of 0.111 mol $m^{-2}$ $s^{-1}$ at 380 ppm $CO_2$, applying the calculation published by Epron et al. (1995). The mesophyll conductance is therefore lower than the estimated stomatal conductance ($0.141 \pm 0.01$, $n = 8$) and corresponds to a chloroplastic $CO_2$ molar ratio of 150 µmol $mol^{-1}$ (Cornic and Streb unpublished). Similar results were reported for other high mountain plants (Kogami et al. 2001; Sakata and Yokoi 2002; Shi et al. 2006). These values are smaller than those for cultivated crop plants but similar to trees (Evans and Loreto 2000). In addition, increased mesophyll resistance was reported for some species growing at high elevation (Kogami et al. 2001; Sakata and Yokoi 2002; Shi et al. 2006). Furthermore, Kogami et al. (2001) described a ten times higher conductance of the internal air space as compared to conductance in the liquid phase, and Shi et al. (2006) described a rapidly varying mesophyll conductance similar to the stomatal conductance in response to environmental conditions. Thus, net $CO_2$ assimilation could be limited more by a decrease in $CO_2$ conductance than by lower external $CO_2$ partial pressures, as already concluded by Friend and Woodward (1990).

In the dry matter of high-mountain plants the $\delta^{13}C$ increases with elevation (Körner et al. 1991), indicating less discrimination against $^{13}C$. This effect can results from higher diffusion resistance of $CO_2$ or higher carboxylation efficiency, which correlates with higher Rubisco contents. Both can induce a low $CO_2$ content at the carboxylation site. Higher carboxylation efficiency correlating with the leaf mass area and the nitrogen content per unit leaf area was reported for alpine plants by Körner and Diemer (1987). Nevertheless, Kogami et al. (2001); Sakata and Yokoi (2002) provided evidences favouring a lower mesophyll conductance to explain the high $\delta^{13}C$ in alpine species. While Vitousek et al. (1990) found a strong correlation between the leaf mass area and the foliar $\delta^{13}C$ in *Metrossideros polymorpha* and attributed this to a higher mesophyll resistance, Cordell et al. (1999), investigating the same species, concluded that the carboxylation efficiency is higher. Moreover, Shi et al. (2006) assigned the less negative $\delta^{13}C$ neither to changes in diffusion conductance nor to changes in carboxylation efficiency and suggested that the low $C_c$ is affected by low temperature effects only. Obviously more work is needed to understand the $^{13}C$ discrimination at high elevation, and these investigations may also show biodiversity in various species.

Reports on carbon assimilation efficiency in alpine plants were contrasting, as shown by several authors and for several plant species (Friend and Woodward 1990; Körner 2003; Sakata and Yokoi 2002; Kumar et al. 2005, 2006; Shi et al. 2006). In *Typha latifolia* ecotypes higher Rubisco contents conferred higher assimilation at high $CO_2$. While in this species the specificity factor of Rubisco ($S_{CO2/O2}$) remained unchanged, the species *Mimulus cardinalis* from high altitude has a lower Km for $CO_2$ than a lowland clone (Friend and Woodward 1990), which is in agreement with results of Sakata and Yokoi (2002) obtained in *Reynoutria japonica*. Castrillo (1995) did not find different Km values for Rubisco in different altitude populations of *Espeletia schutzii* but higher maximum activities at higher elevation.

The higher photosynthetic capacity reported in several studies for alpine plants as calculated based on leaf surface may simply result from their greater leaf thickness (Körner 2003). When the photosynthetic capacity of alpine plants at $CO_2$ saturation was compared to lowland plants on a chlorophyll basis, no significant difference was observed (Streb et al. 1997).

The $CO_2/O_2$ ratio in chloroplasts and the Rubisco specificity factor determine the rate of photorespiration relative to carbon assimilation in alpine plants. As $S_{CO2/O2}$ increases with decreasing temperature (Jordan and Ogren 1984), it can be expected that photorespiration is lower at high altitude. Photorespiration does occur in alpine species, allowing high rates of electron transport even when stomata are closed (Heber et al. 1996). Even higher rates of photorespiration were reported in alpine as compared to lowland ecotypes of *Trifolium repens* when grown and measured under identical conditions (Mächler et al. 1977). Mächler and Nösberger (1978) suggested that higher photorespiration in alpine ecotypes is induced by lower growth temperature. Photorespiration was clearly temperature-dependent and higher at 22°C than at 6°C at natural $C_i$ in *R. glacialis* leaves (Streb et al. 2005). However, at low temperatures electron transport to carboxylation and to photorespiration was nearly identical. High rates of photorespiration at low $C_i$ can also be concluded from Fig. 7.4 when comparing the rate of electron transport and carbon assimilation. The observation that the abundance of chloroplast protrusions was higher in alpine plants might also be interpreted as an indication of enhanced metabolite exchange for photorespiration resulting from the increased chloroplast surface (Lütz and Engel 2007; Buchner et al. 2007). Photorespiration in alpine plants may therefore serve as security valve when consuming excess energy under conditions of high PFD that exceed the capacity of carbon assimilation and to maintain high electron transport rates. This was demonstrated in leaves of *R. glacialis*, where excess electrons from photosynthesis were consumed partially by photorespiration (Streb et al. 2005). When photorespiration was blocked with phosphinothricin, an inhibitor of glutamine synthase, the primary electron acceptor of PSII ($Q_A$) became more reduced, and non-photochemical fluorescence (qN) quenching was increased relative to *R. glacialis* leaves in the absence of the inhibitor (Streb et al. 1998). However, other alpine plants like *S. alpina* responded less to the inhibition of photorespiration by phosphinothricin, suggesting that the importance of photorespiration differs in various alpine species (Streb et al. 1998).

### 7.4.3 C4 and CAM Plants in Alpine Environments

Only a few species performing C4 or CAM type photosynthesis were investigated in alpine environments, and their abundance is low (Körner 2003). Contrary to C3 plants, no shift in the thermal optimum of assimilation at various PFDs is expected in C4 plants, since their photorespiratory activity is suppressed. Measurements of the $\delta^{13}C$ in leaves of plants growing in alpine meadows in Tibet (Yi et al. 2003) or in the organic soil matter of grasslands and forests from Papua New Guinea (Bird et al. 1994) gave values close to −26‰ at 3,000 m.a.s.l., indicating the absence of C4 and CAM photosynthesis in these environments. In a comprehensive literature survey, Sage et al. (1999) reported that the proportion of C4 species relative to C3 species decreases strongly with altitude. According to Sage et al. (1999), transition from C4 to C3 species occurs when mean minimum air temperatures in the growing season are lower than 15°C, and C4 species are almost absent at minimum air temperatures between 8°C and 10°C. In most C4 species photosynthesis is adversely affected at chilling temperature and light (Leegood and Edwards 1996). However, some cold-tolerant C4 plants, including alpine species, were described (P'yankov et al. 1992; Pittermann and Sage 2000; Naidu et al. 2003). In *Miscanthus giganteus* the temperature response of carbon assimilation was similar in cold and warm-grown leaves, while it was strongly suppressed in cold-grown maize leaves (Naidu et al. 2003). The temperature optimum of carbon assimilation was, however, slightly shifted to higher temperatures in warm as compared to cold-grown leaves. P'yankov et al. (1992) could not find a different temperature optimum in C4 and C3 species of the Pamir. In *Miscanthus giganteus* the protein contents of pyruvate orthophosphate dikinase was elevated and those of phosphoenol pyruvate carboxylase and Rubisco unchanged in plants grown at cold as compared to growth at warm temperatures, which was not the case in maize leaves cultivated under the same conditions (Naidu et al. 2003). Furthermore, the kinetic properties of Rubisco ($K_{cat}$, $v_{max}$ and activation energy) were not significantly different in cold and warm-grown *Miscanthus giganteus*, and the catalytic site of Rubisco was identical to that of *Zea mays*

(Wang et al. 2008). By contrast, Pittermann and Sage (2000) concluded from a comparison of high-elevation and low elevation *Bouteloua gracilis* that photosynthesis at low temperatures is limited by Rubisco activity in this species.

Some evidence for CAM metabolism is found in studies on the European Alps and the Andes (Osmond et al. 1975; Wagner and Larcher 1981; Keeley and Keeley 1989). *Several Sempervivum* species from dry areas in the German and Italian Alps present both an increase of acidity during the night and $\delta^{13}C$ values of about $-17$‰ characteristic of CAM metabolism. Sedum species, showing a moderate increase in acidity at the end of the night, have $\delta^{13}C$ values usually observed in C3 plants (Osmond et al. 1975). Wagner and Larcher (1981) showed that CAM metabolism in *Sempervivum montanum* was facultative and depended on the prevailing night and day temperatures. CAM metabolism was also described for high elevation tropical cactus in central Peru (4,000–4,700 m.a.s.l.) (Keeley and Keeley 1989). Nevertheless, it should be noted that maximum and minimum air temperatures at this investigation site were neither very high (maximum 19.1°C) nor very low (−0.7°C). However, more work on the characterization of alpine CAM metabolism is necessary, in particular about facultative CAM metabolism and inducing stress conditions.

## 7.5 Photoinhibition at Temperature Extremes

### 7.5.1 Photoinactivation

Most plants exposed to high PFDs at low or high temperature show symptoms of photodamage (Berry and Björkman 1980). Such photodamage occurs because enzymatic reactions and transport processes are sensitive to temperature, while light absorption is not (Falk et al. 1996; Öquist and Huner 2003). If the temperature decreases by 10°C most enzymatic activities decrease by a factor of approximately 2, but also substrate affinities and activation energies change with temperature (Leegood and Edwards 1996). The different temperature sensitivity of individual enzymes can lead to the accumulation of metabolites, especially sugars and sugar phosphates and to a limitation of phosphate as observed after low temperature stress in non-acclimated plant species but less in cold-acclimated species (Hurry et al. 2000; Streb et al. 2003b). On the other hand, sugar accumulation and phosphate limitation may trigger acclimation processes in the plant (Hurry et al. 2000; Rolland et al. 2002). In the alpine environment the lower $CO_2$ supply (Chapter 4.2) can aggravate the imbalance between light absorption and utilisation by limiting carbon assimilation.

Among the early effects of light-induced damage, photoinhibition of the PSII reaction centre protein D1 is the most widely studied effect (Aro et al. 1993). However, the enzyme catalase is also inactivated by light, and photoinhibition of the PSI reaction centre was reported in selected plant species (Terashima et al. 1994; Feierabend 2005; Scheller and Haldrup 2005). Prolonged exposure to high light intensities under stress conditions leads to more general photooxidative damage, such as chlorophyll loss and damage to other proteins and to membranes (Feierabend et al. 1992; Suzuki and Mittler 2006). The inactivated D1 protein and the inactivated catalase protein are replaced by new protein synthesis (Aro et al. 1993; Feierabend 2005). Low or high temperature may either enhance the rate of inactivation or (and) slow down the rate of replacement, leading to a net loss of proteins during light stress (Melis 1999; Murata et al. 2007; Takahashi and Murata 2008). The PSII repair cycle was described by Aro et al. (1993) and involves several steps such as triggering the D1 protein for degradation, transport of the damaged PSII reaction centre from appressed to non-appressed membranes, degradation of the D1 protein and synthesis of new proteins, as well as reorganisation of the PSII complex, transport to the appressed membranes and its activation (Aro et al. 1993). Common methods to investigate photoinhibition of PSII are the decrease of the Fv/Fm ratio in dark-adapted leaves, the loss of quantum yield of gas exchange, a lowering of the photosynthetic capacity and the net loss of the D1 protein of PSII (Krause and Cornic 1987; Aro et al. 1993; Baker 2008). Since regulatory photosynthetic down-regulation may overlap with photoinhibitory damage, care must be taken to separate the two effects of light and environmental stress (Baker 2008).

The catalase protein is directly inactivated by the absorption of blue light or indirectly through oxidative reactions originating in high-intensity red light in the chloroplast (Shang and Feierabend 1999; Feierabend 2005). A chloroplast-mediated inactivation of catalase was also observed at low light intensities in algae. In this case the catalase inactivation indicates reducing conditions in the chloroplast and may trigger the transmittance of a hydrogen peroxide-induced signal effecting gene expression in the nucleus (Shao et al. 2008). On the other hand, catalase serves as an important antioxidant enzyme scavenging $H_2O_2$ produced during photorespiration (Feierabend 2005).

Chloroplast-mediated in vitro catalase inactivation by red light was weaker for chloroplasts isolated from the alpine plant species *R. glacialis* than for chloroplasts isolated from lowland plants (Streb et al. 1998), suggesting less reducing conditions in the chloroplasts of this alpine species. Besides catalase, the D1 protein is also inactivated by reducing conditions at the PSII acceptor side, i.e. when the plastoquinone pool and the primary stable electron acceptor ($Q_A$) become reduced (Aro et al. 1993; Huner et al. 1993; Melis 1999).

Low and high temperatures enhance the inactivation of PSII and of catalase (Feierabend et al. 1992). At both high and low temperatures the formation of reactive oxygen species in light is accelerated, which may lead to potential general cellular damage and block repair processes (Wise 1995; Apel and Hirt 2004; Murata et al. 2007).

### 7.5.2 Photoinactivation in Alpine Plants

Reports about photoinhibition in alpine plants are controversial. Rawat and Purohit (1991) reported that photosynthesis in alpine herbs in the Himalaya is adversely affected by high PFD and low or high temperatures. Similarly, carbon assimilation was significantly reduced in tropical alpine *Lobelia rhynchopetalum* at high PFD over 2,000 μmol $m^{-2}$ $s^{-1}$ (Fetene et al. 1997). Williams et al. (2003) measured significant photoinhibition in alpine shrubs of the Tasmanian mountains in spring, which was partially dynamic and partially chronic. Alpine herbs are highly sensitive to photoinhibition in the dehydrated state, while mosses are protected (Heber et al. 2000). However, even in winter no significant photoinhibition was measured in several other alpine plant species from the Alps, North America and the Himalaya with the variable fluorescence method and by gas exchange measurements (Heber et al. 1996; Streb et al. 1997; Germino and Smith 2000; Hacker and Neuner 2006; Kumar et al. 2006). Leaves of alpine plants artificially exposed to low temperatures at 1,000 μmol $m^{-2}$ $s^{-1}$ showed only minor inactivation of PSII and of catalase and no sign of chlorophyll damage (Streb et al. 1997). However, at the same PFD at 38°C the Fv/Fm ratio and the oxygen evolution capacity declined in *S. alpina* and *R. glacialis* leaves, suggesting that these alpine plants are more sensitive to high-temperature-induced photoinhibition than to low-temperature-induced photoinhibition (Streb et al. 2003c). In darkness most alpine plants are equally sensitive to high temperature as lowland plants or even more resistant (Gauslaa 1984; Larcher et al. 1997; Körner 2003; Streb et al. 2003c; Larcher et al. 2010). High-temperature-induced photoinhibition could be one reason why some alpine plant species are unable to grow at lower elevation with high temperature events (Streb et al. 2003c; Zhang and Hu 2008).

The high tolerance to PSII photoinhibition in *S. alpina* and *R. glacialis* leaves is not a specific property of their D1 protein, since in isolated chloroplasts or thylakoids electron transport rates were similarly inactivated by light as in preparations from susceptible plants (Streb et al. 1997). The reduced light sensitivity of a catalase isozyme of the alpine species *H. alpina*, however, resulted from a specifically modified structure of the enzyme protein (Streb et al. 1997; Engel et al. 2006). Its primary structure differs in six unusual amino acid substitutions from the majority of all other catalases (Engel et al. 2006), indicating that the modification of its structure confers light tolerance to this catalase.

Recently Takahashi and Murata (2008) suggested that environmental stresses, such as cold and heat, accelerate photoinhibition by blocking the repair of damaged proteins rather than by enhancing their inactivation. However, when the resynthesis of the D1 protein was blocked by lincomycin or chloramphenicol or the resynthesis of catalase was blocked by cycloheximide, alpine species like *S. alpina*, *H. alpina* and *R. glacialis* still show less photoinactivation of PSII and of catalase than non-alpine species (Streb et al. 1997). Furthermore, the turnover of the D1 protein in intact leaves of *R. glacialis* and *S. alpina* was markedly retarded as compared to the D1 turnover in non-alpine species under identical conditions (Shang and Feierabend 1998). These results suggest that the investigated alpine species are tolerant to photoinhibitory conditions in that they slow down the rate of inactivation rather than enhancing repair capacities. Other alpine plant species, such as *Eriophorum angustifolium*, simply avoid photoinhibitory conditions by retarding the greening of the leaves until temperature conditions are favourable (Lütz 1996). On the other hand, leaf closure of *Saxifraga paniculata* leaves prevents overheating and water loss and thereby protects the plant from photoinhibitory damage by reducing the absorption of light (Neuner et al. 1999).

### 7.5.3 Reducing Conditions in Chloroplasts of Alpine Plants

In cold-acclimated plants the ability to keep the relative reduction state of the photosynthetic electron transport chain low correlates well with their higher tolerance to photoinactivation of PSII and catalase (Huner et al. 1998; Streb and Feierabend 1999; Streb et al. 1999; Öquist and Huner 2003). Several parameters reflecting the redox state of the chloroplasts were investigated in alpine plants. The relative reduction state of the primary stable electron acceptor of PSII ($Q_A$) was estimated based on chlorophyll fluorescence. In two alpine snowbank species investigated by Germino and Smith (2000), the relative reduction state of $Q_A$ at varying temperatures and light intensities did not correlate with photoinhibition as measured by the Fv/Fm ratio. A strong reduction of $Q_A$ was measured at low temperatures and high light values in *G. montanum* leaves, but this species is highly tolerant to photoinhibition (Manuel et al. 1999). By contrast, in an Andes ecotype of *C. quitensis* higher relative $Q_A$ reduction correlated with a stronger low-temperature-induced photoinhibition as compared to an Antarctic ecotype (Bravo et al. 2007). Divergent results were also obtained for *S. alpina* and *R. glacialis* leaves. While the relative reduction state of $Q_A$ in leaves of *R. glacialis* was kept in a strongly oxidised state at various light intensities and temperatures, those of *S. alpina* became much more reduced. Similarly, the relative electron transport rate, as calculated from chlorophyll fluorescence assays, was much higher in *R. glacialis* leaves than in those of *S. alpina* (Streb et al. 1998; Streb et al. 2003c). The initial activity of NADPH-malate deshydrogenase, an indicator of reducing conditions in the chloroplast (Backhausen et al. 2000), was low in leaves of *R. glacialis* in full sunlight at midday at the natural growing site, suggesting the absence of strong reducing conditions (Streb et al. 2005). Furthermore, ascorbate was oxidised in *R. glacialis* leaves in an alpine environment during the course of the day, while the reduced ascorbate content increased more than the dehydroascorbate content in leaves of *S. alpina* (Streb et al. 1997). In conclusion, the different results about reducing conditions in alpine plants suggest that the mere occurrence of reducing conditions alone does not allow to predict photoinhibitory damage in these plants and that some alpine species such as *S. alpina* tolerate reducing conditions, while other species such as *R. glacialis* avoid reducing conditions.

## 7.6 Protection Strategies Against High PFD and Temperature Extremes in Alpine Plants

### 7.6.1 Carbon Assimilation and Photorespiration

How can alpine plants protect their photosynthetic machinery against photoinactivation? The simplest mechanism would be that alpine plants avoid an imbalance between absorbed light energy and its utilisation by sufficiently high carbon assimilation and photorespiration even under stress conditions. In plants exposed to low temperatures and high PFD, the availability of phosphate for photosynthetic ATP formation can become a limiting factor (Falk et al. 1996; Strand et al. 1999; Hurry et al. 2000). This phosphate limitation may result from reduced sucrose synthesis. As a consequence hexose-phosphates and PGA accumulate, and the ATP/ADP ratio decreases (Hurry et al. 1994, 1995; Savitch et al. 1997). This was shown in comparative experiments with pea leaves and alpine plants. While characteristics of Pi limitation were observed in pea leaves, no such limitation was seen in alpine plants, suggesting that photosynthesis at low temperature is not impaired by phosphate limitation in *R. glacialis* and *S. alpina* (Streb et al. 2003b). The alpine species with the highest electron transport rate at PSII (*R. glacialis*) was therefore tested with gas exchange and chlorophyll fluorescence assays in order to determine whether transported electrons are entirely consumed by carbon assimilation, respiration and photorespiration. While photorespiration consumes most of the electrons not used for carbon assimilation, approximately 15–20 μmol $m^{-2} s^{-1}$ electrons are distributed to other oxygen-consuming reactions at a high internal partial pressure of $CO_2$ and at 6°C and 22°C, respectively (Streb et al. 2005, see also Fig. 7.4). The primary products of carbon assimilation are sucrose and fructose in *R. glacialis* leaves, which are not respired immediately but obviously transported to storage organs (Nogués et al. 2006). Despite the fact that malate and ranunculin represent more than 50% of the total soluble leaf carbon, both compounds were not labelled during short-time labelling experiments in light (Streb et al. 2003a; Nogués et al. 2006). The low initial activity of NADPH-malate deshydrogenase and the respiratory coefficient of approximately 1 further suggest that electrons are not transported to mitochondria via the malate valve nor oxidised by respiration (Streb et al. 2005; Nogués et al. 2006). In conclusion, no other exceptional metabolites were synthesised in *R. glacialis* leaves, which could explain higher rates of electron consumption beyond sucrose synthesis and Calvin cycle activity, but photorespiration remains an important electron and ATP sink even at low temperatures. Similarly, Kumar et al. (2006) measured higher oxygenase activity of Rubisco at high altitude but similar accumulation of phosphoglycerate in barley and wheat leaves investigated at different altitudes.

### 7.6.2 Non-radiative Dissipation of Excess Excitation Energy

In order to balance light absorption, electron transport and carbon assimilation under stress conditions, plants can dissipate absorbed light energy as heat at the level of PSII (Niyogi 1999; Ort and Baker 2002). This energy dissipation is frequently measured by non-photochemical fluorescence quenching (NPQ or qN) (van Kooten and Snel 1990). More than 75% of absorbed light energy can be dissipated by this pathway (Niyogi 1999). Part of the NPQ depends on the synthesis of zeaxanthin and the presence of the PsbS protein (Demmig-Adams and Adams 2006). However, also zeaxanthin-independent NPQ was described (Niyogi 1999; Finazzi et al. 2004).

High amounts of xanthophyll cycle pigments with a high de-epoxidation state and concomitant high levels of NPQ were observed in a great variety of plant species from various origins, suggesting that NPQ and xanthophyll cycle pigments are functionally related in order to prevent overexitation of the photosynthetic apparatus and to prevent plants from photooxidative damage (Demmig-Adams and Adams 1992a; 1992b; 2006; Demmig-Adams 1998). Furthermore, overwintering conifers and cold-acclimated plants can retain their de-epoxidation state even in darkness and display reduced photochemical efficiency after exposure to light, which supports the view that the xanthophyll cycle participates in the protection against photooxidative damage (Öquist and Huner 2003; Demmig-Adams and Adams 2006; Zarter et al. 2006a, b). Moreover, zeaxanthin synthesis in cold-acclimated rye leaves is more rapid than in non-cold-acclimated rye leaves when leaves are exposed to high PFD at low temperatures, although these leaves show the same amount of xanthophyll cycle pigments and zeaxanthin content after prolonged exposure (Streb et al. 1999).

#### 7.6.2.1 The Xanthophyll Cycle and Non-photochemical Fluorescence Quenching in Alpine Plants

The xanthophyll cycle and non-photochemical fluorescence quenching were also measured in various alpine and antarctic plants. In six Tasmanian alpine species studied by Williams et al. (2003), the de-epoxidation state was higher in spring than in autumn leaves. In some species the de-epoxidation state was, however, twice as high as in others. Similar strong variations in xanthophyll cycle pigments were measured by Wildi and Lütz (1996) at high altitude. Although the de-epoxidation state of the xanthophyll cycle pigments was higher in cold-acclimated than in non-cold-acclimated *C. quitensis* from the Andes, the total pool of xanthophyll cycle pigments was higher in the non-cold-acclimated plants. However, both, the de-epoxidation state and the total pool of xanthophylls were higher after growth at high as compared to low PFD (Bravo et al. 2007). Leaves of *S. alpina*, *S. pusilla* or *H. alpina* had significantly higher levels and higher de-epoxidation states of xanthophyll cycle pigments than leaves of *R. glacialis* both in darkness and after light exposure in their natural environment as well as under experimental conditions (Wildi and Lütz 1996; Streb et al. 1997, 1998). The higher xanthophyll cycle activity and level of *S. alpina* leaves as compared to those of *R. glacialis* leaves correlated with higher non-photochemical fluorescence quenching at PFDs up to 1.000 μmol $m^{-2}$ $s^{-1}$ and moderate temperature. At higher PFDs or at low or high temperatures, however, the values of qN were similar in *S. alpina* and *R. glacialis* leaves despite the strong differences in the xanthophyll cycle pigments and their de-epoxidation state (Streb et al. 1998, 2003c), suggesting the occurrence of zeaxanthin-independent non-photochemical quenching in *R. glacialis* leaves.

In the presence of DTT, the xanthophyll cycle activity and NPQ decreased (Demmig-Adams et al. 1990). Incubation of leaves in the presence of DTT diminished the de-epoxidation of violaxanthin and non-photochemical fluorescence quenching in *R. glacialis* and *S. alpina* leaves at moderate temperatures (Streb et al. 1998). However, DTT was less efficient in blocking the xanthophyll de-epoxidation at high temperatures (Streb et al. 2003c). At low temperatures the lowered zeaxanthin level had only minor effects on qN as compared to ambient temperature (Streb et al. 1998). Furthermore, at ambient temperature in the presence of DTT qN did not respond to increasing PFDs in *S. alpina* leaves, while qN increased well with PFD in *R. glacialis* leaves (Streb et al. 1998). In general, relative reduction states of $Q_A$ were higher and PSII electron transport rates lower in leaves of *S. alpina* and *R. glacialis* in the presence of DTT as compared to control leaves, suggesting that the xanthophyll cycle is involved in non-photochemical fluorescence quenching in both plants. However, zeaxanthin-depleted *S. alpina* leaves showed higher relative $Q_A$ reduction, lower rates of PSII electron transport and higher extents of photoinhibition in comparison to zeaxanthin-depleted *R. glacialis* leaves (Streb et al. 1998). The NPQ is thought to protect the tropical alpine plant *Lobelia rhynchopetalum* from photoinhibition, when closed stomata and low temperature in the morning affect photosynthetic assimilation (Fetene et al. 1997).

Altogether, the xanthophyll cycle-mediated non-photochemical fluorescence quenching is of importance for alpine plants and keeps the reduction state of $Q_A$ low, reduces electron transport rates and protects plants from photoinhibition. However, non-photochemical fluorescence quenching is also important in the absence of zeaxanthin synthesis, suggesting that the relationship between zeaxanthin contents and protection by non-photochemical fluorescence quenching is complex in alpine plants. Furthermore, the occurrence of xanthophyll cycle dependent non-photochemical fluorescence quenching is ubiquitous among higher plants and functions also in non-alpine and even tropical plants with similar efficiency. Therefore non-photochemical quenching alone cannot be sufficient to protect alpine plants from cold-induced photoinhibition. Non-photochemical fluorescence quenching is more rapidly induced in cold-acclimated plants, and because higher zeaxanthin contents are retained, a sudden increase of PFD can diminish the initial photooxidative damage of alpine or cold-acclimated plants. Nevertheless, the fact that electron transport rates are comparatively high under stress conditions in *R. glacialis* leaves suggests that protection against overreduction and photoinhibition is mainly achieved by processes downstream of the PSII reaction centre in this plant.

### 7.6.3 PTOX and Cyclic Electron Transport Around PSI

In the alpine plant *G. montanum* measurements of the PSII and PSI redox state at high PFD and low temperature suggested the occurrence of cyclic electron transport around PSI (Manuel et al. 1999). This was interpreted to increase the pH gradient at the thylakoid membrane and to regulate energy dissipation at the level of PSII in order to protect this plant from destructive photoinhibition (Manuel et al. 1999). The PSI cyclic electron transport can increase the ATP/NADPH ratio as compared to linear electron flow and modulate the NPQ at the level of PSII (Joliot and Joliot 2002; Makino et al. 2002; Rumeau et al. 2007; Laisk et al. 2007).

Recently a plastid terminal oxidase (PTOX) with similarities to the mitochondrial alternative oxidase was described (Carol and Kuntz 2001). PTOX is involved in carotenoid synthesis during chloroplast development and in chlororespiration (Carol and Kuntz 2001; Peltier and Cournac 2002; Nixon and Rich 2006). Since PTOX can accept electrons from PSII via the plastoquinone pool and from cyclic PSI electron transport via several plastoquinone reductases, PTOX activity is assumed to act as a security valve to keep the plastoquinone pool oxidised as well as to prevent overreduction in the chloroplast and photoinhibition (Carol and Kuntz 2001; Peltier and Cournac 2002; Nixon and Rich 2006; Quiles 2006; Rumeau et al. 2007; Shahbazi et al. 2007; Tallon and Quiles 2007; Diaz et al. 2007).

In most plant species tested, the PTOX protein is present in low concentrations only, and electron consumption can potentially account for much less than 1% of total electron transport activity (Ort and Baker 2002). Under high PFD and heat or low temperature treatment, PTOX levels are, however, up-regulated in various plant species (Quiles 2006; Tallon and Quiles 2007; Rumeau et al. 2007; Busch et al. 2008). All alpine plant species tested for their PTOX content also show elevated levels as compared to lowland plants or even to PTOX overexpressing *Lycopersicon esculentum* (Streb et al. 2005). In leaves of *G. montanum* the leaf PTOX content increased with the altitude at which the plants were collected (Streb et al. 2005). Together with cyclic electron flow around PSI as reported by Manuel et al. (1999), a high PTOX content and the supposed high PTOX activity could function as security valve under high PFD and additional environmental stress. One could speculate that excess light energy absorbed by PSII is channelled to oxygen via PTOX, while excess PFD absorbed by PSI is cycled around PSI (Fig. 7.5).

The highest PTOX content was measured in *R. glacialis* leaves, and in this species a high proportion of the electron transport rates could not be attributed to carbon assimilation and photorespiration (Streb et al. 2005). Furthermore, the PTOX content of *R. glacialis* leaves declined when plants were grown at low elevation for several weeks, while their sensitivity to photoinhibition increased (Streb et al. 2003, 2005). In addition, the content of the NDH H protein, which is part of the NADPH-plastoquinone-reductase catalysing cyclic electron transport around PSI via NADPH, differed in acclimated and deacclimated *R. glacialis* leaves in a similar way as the PTOX protein content (Streb et al. 2005). Considering the ability of *R. glacialis* leaves to avoid reducing conditions in the chloroplast (see Chapter 7.5.3) and further considering the high sensitivity of *R. glacialis* leaves to oxidative stress conditions (see Chapter 7.6.4.2), a similar protective function of PTOX and cyclic electron transport can be assumed as proposed for *G. montanum*.

However, at present the potential protective role of PTOX (Fig. 7.5) is still under discussion. Pine needles increase their PTOX content during acclimation to winter conditions (Busch et al. 2008). Pine needles with elevated PTOX content and transgenic Arabidopsis leaves with increased PTOX showed similar relative reduction states of $Q_A$ as leaves with lower PTOX content (Rosso et al. 2006; Busch et al. 2008), suggesting that the PTOX protein does not consume excess electrons from the plastoquinone pool.

Evidence of a potential protective function of PTOX (Fig. 7.5) comes from experiments with *Chlamydomonas rheinhardtii* (Cournac et al. 2002),

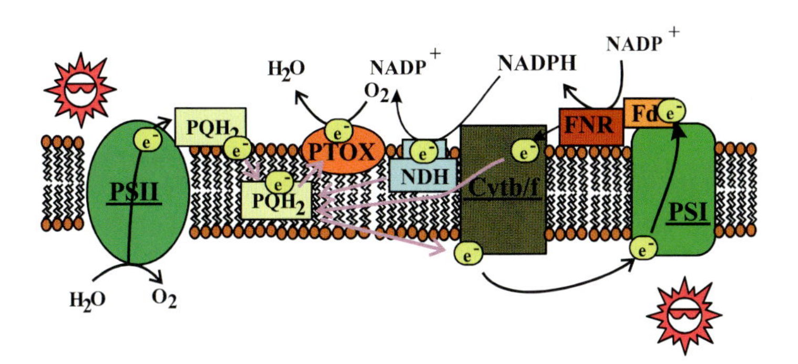


**Fig. 7.5** Proposed protective function of PTOX and cyclic electron transport to keep the oxidation state of the electron transport chain high. PTOX accepts electrons from plastoquinone and keeps the oxidation state of the plastoquinone pool high. Excess absorbed light energy entering the electron transport chain from PSII is transferred to oxygen via PTOX to form water. PSI can donate electrons via ferredoxin (Fd) and the cyt b6/f complex to the plastoquinone pool. These electrons may return to PSI, thus increasing the pH gradient and inducing NPQ at the level of PSII. Excess NADPH is oxidised by the NADPH-plastoquinone oxidoreductase (NDH) transferring the electrons to the plastoquinone pool and further to PTOX to reduce oxygen to water

transgenic plants lacking catalase and ascorbate peroxidase activity (Rizhsky et al. 2002), *Avena sativa* and two *Brassica* species (Quiles 2006; Diaz et al. 2007; Tallon and Quiles 2007), green and white leaf sections of the *Arabidopsis immutans* mutant (Aluru et al. 2007) and tomato plants (Shahbazi et al. 2007). In PTOX overexpressing tobacco plants, however, photoinhibition was lower at low PFD but higher at high PFD than in the WT. The authors suggest that PTOX mediates superoxide production and that its protecting activity might depend on SOD activities (Heyno et al. 2009).

The varying observations in different plant species might be explained by the dual function of PTOX (Shahbazi et al. 2007), by a potential lack of cofactors and (or) cyclic electron transport or by incorrect assembly in transgenic plants. Pine needles down-regulate their photosynthetic activity during winter and mainly emit absorbed light energy as heat (Öquist and Huner 2003). In this case PTOX activity may mainly serve to increase their carotenoid content. In alpine plants however, the PTOX protein is needed to compensate stress periods during active photosynthesis. Nevertheless, the protective action of PTOX in combination with cyclic electron transport around PSI in alpine plants remains to be further substantiated.

Interestingly, different species from a high altitude location showed higher rates of mitochondrial alternative oxidase activity as compared to the same species from a low altitude (Kumar et al. 2007). Possibly both PTOX and mitochondrial alternative oxidase are up-regulated at high altitude.

## 7.6.4 Antioxidants and Antioxidative Enzymes

Excess absorbed light energy not dissipated or consumed by other protective reactions may lead to the formation of ROS (Asada 1996; Apel and Hirt 2004). Up to a threshold concentration, ROS and their detoxification products can act as signals transmitted from the chloroplast to the nucleus to induce acclimation (Apel and Hirt 2004; Foyer and Noctor 2005). However, excessive ROS formation generates oxidative damage reflected in photoinhibition, pigment bleaching, lipid peroxidation, more general protein damage and inhibition of repair capacities, and finally leads to cell death (Asada 1996; Moller et al. 2007; Takahashi and Murata 2008). Since oxidative stress is induced at high PFD and low temperature (Wise 1995; Asada 1996; Ort and Baker 2002), control of ROS formation and ROS detoxification might be of key importance for the survival of alpine plants. Accordingly, enhanced antioxidative protection by elevated activities of antioxidative enzymes and elevated contents of lipid-soluble and water-soluble antioxidants were often interpreted to contribute to temperature and light stress tolerance (Noctor and Foyer 1998; Ort and Baker 2002).

### 7.6.4.1 Singlet Oxygen

The formation of singlet oxygen at the level of PSII is thought to participate in irreversible damage to the D1 protein during photoinhibition and in initiating lipid peroxidation (Aro et al. 1993; Niyogi 1999; Krieger-Liszkay et al. 2008; Triantaphylides et al. 2008). Besides carotenoids, α-tocopherol is also involved in singlet oxygen scavenging and defence against lipid peroxidation (Niyogi 1999; Munné-Bosch 2005; Kruk et al. 2005), while zeaxanthin appears to have a higher antioxidant capacity as compared to other xanthophylls (Havaux et al. 2007). The high resistance of most alpine plants to low-temperature induced photoinhibition (Chapter 5.2) suggests that singlet oxygen is either efficiently scavenged or its formation is avoided. However, the highly variable carotenoid and zeaxanthin contents in alpine plants (Chapter 3.2) and similarly high variations of the α-tocopherol content (Wildi and Lütz 1996; Streb et al. 1998, 2003) suggest divergent scavenging mechanisms. At least alpine species with low carotenoid and α-tocopherol content can therefore be supposed to have the ability to avoid singlet oxygen formation

### 7.6.4.2 Other Reactive Oxygen Species

Mainly at the level of PSI, electrons can be transferred to oxygen by forming superoxide in the Mehler reaction. Superoxide itself can give rise to the formation of other reactive oxygen species such as the perhydroxyl radical, $H_2O_2$ and the extremely toxic hydroxyl radical (Asada 1996; Apel and Hirt 2004). In the water-water-cycle photosynthetically formed superoxide is finally oxidised to water by a cascade of different enzyme reactions involving the two most important soluble antioxidants ascorbate and glutathione as well as SOD and APx (Asada 1999). However, also outside the chloroplast antioxidant defence reactions, like catalase activity, are important. It is therefore conceivable that an increase in antioxidative scavenging capacity, in particular of components of the ascorbate-glutathione cycle, could alleviate photooxidative damage.

As already shown for other protective mechanisms, antioxidant contents and antioxidant enzyme activities also vary markedly in different alpine species. Very high antioxidant contents were measured in high-altitude ecotypes of *S. alpina* and *S. pusilla*, medium levels in *Poa laxa*, *Carex curvula*, and *G. reptans*, and only low concentrations of major antioxidants were found in leaves of *H. alpina*, *R. glacialis*, *Taraxacum alpinum*, *Dryas octopetala* and *Tanacetum alpinum* (Wildi and Lütz 1996; Streb et al. 1997, 1998, 2003a). Similar differences were measured in Anatolian mountain species (Öncel et al. 2004). With few exceptions, the contents of antioxidants increase with altitude (Wildi and Lütz 1996; Wang et al. 2009). In leaves of *S. alpina* and *Polygonum viviparum* ascorbate contents are among the highest values published in the literature (Streb et al. 1997, 1998, 2003a; Wang et al. 2009, see also Bligny and Aubert this volume). In *S. alpina* leaves, ascorbate is even the second major soluble carbon metabolite next to sucrose (Streb et al. 2003a). However, ascorbate contents within chloroplasts are similar in species with a high concentration of ascorbate in the leaves (*S. alpina*) and those with a relatively low total ascorbate content in the leaves (*R. glacialis* and *H. alpina*). In *R. glacialis* and *H. alpina* leaves, almost the entire cellular ascorbate is present in the chloroplasts, while in *S. alpina* leaves most of the ascorbate content is found outside the chloroplasts (Streb et al. 1997, Bligny and Aubert, this volume).

Similar to the antioxidant contents, the activities of major antioxidant enzymes such as SOD, glutathione reductase and catalase are also high in *S. alpina* but low in *R. glacialis* leaves (Streb et al. 1997). In *Polygonum viviparum* from the Tianshsan mountain, the activities of several antioxidative enzymes increased with altitude (Wang et al. 2009). In roots of four perennial grasses the activity of antioxidative enzymes increased during acclimation to winter conditions and again during rewarming in spring (Zhou and Zhao 2004). In cell cultures of the alpine species *Chorispora bungeana* antioxidant enzyme activities increased during freezing events, suggesting that antioxidantive enzyme activity is involved in freezing resistance (Guo et al. 2006).

Species with high and low antioxidant content and antioxidant enzyme activity were tested for their oxidative stress tolerance after the application of paraquat in light (Streb et al. 1998). Leaves of *S. alpina* with the highest antioxidant scavenging capacity were much more tolerant to paraquat-induced oxidative stress than other alpine or non-alpine plant species, while leaves of *R. glacialis* were among the most sensitive under the tested species (Streb et al. 1998). The results show that antioxidative protection differs widely in alpine plants. Individual species are highly protected against oxidative stress, while others are more sensitive.

## Conclusions

Perennial alpine plant species from temperate mountains have to assimilate sufficient carbon during a short vegetation period to guarantee the survival of a long winter period. The vegetation period is frequently challenged by extremes of light and temperature and characterised by a lower $CO_2$ supply as compared to lower altitude. Furthermore, microclimatic constraints such as high wind speed and low water supply can affect the vitality of the plants. Alpine plants maintain high rates of photosynthesis under varying environmental stress conditions, and carbon use efficiency is higher than in lowland plants. Furthermore, alpine plants survive extreme temperature and light events and show less photoinhibition and less cellular damage than lowland plants. Morphological characteristics of alpine plants, such as small size, cushions, rosettes and tussocks, ameliorate the microclimatic condition at the leaf level. Several alpine plant species show characteristics of acclimation to high light intensities and low temperature and avoid the limitation of phosphate supply under stress conditions. However, excess light absorption occurs under certain environmental conditions, and alpine plants use different defence strategies to prevent overexcitation of the photosynthetic apparatus, ROS formation, and possess a high capacity for ROS detoxification (Fig. 7.6). Although not every possible defence strategy was tested in all alpine species investigated, special features of individual alpine species suggest great biodiversity of tolerance mechanisms. Figure 7.6

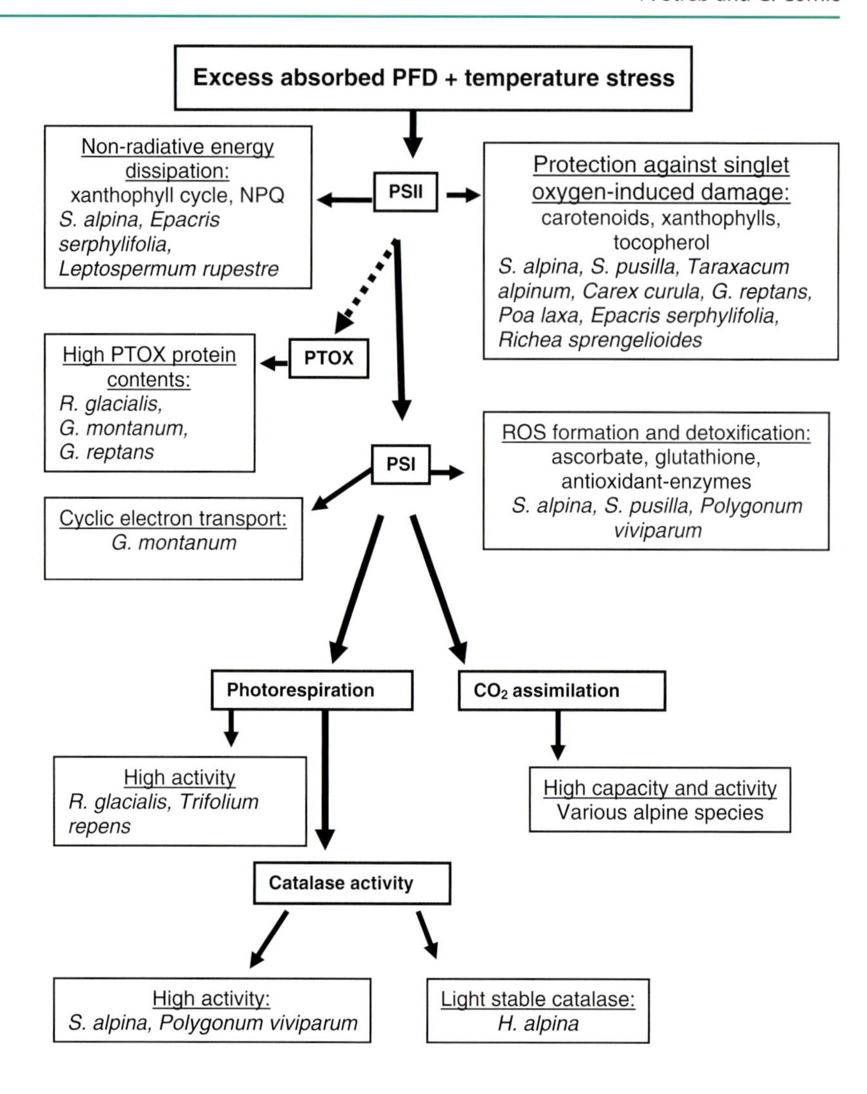


**Fig. 7.6** Potential protection strategies of alpine plants against combinations of light stress and temperature extremes. For each species mentioned in the figure, experimental evidence demonstrating the relevance of the respective protection strategy was obtained, as indicated in the boxes

highlights protection strategies and the respective species for which evidence is presented in this review. From Fig. 7.6 it is further evident that, where tested, alpine species do not simply rely on one protective strategy but use multiple mechanisms simultaneously. This could enable alpine species to replace the failure of one protective level by another, and this may distinguish alpine species from non-adapted lowland species, which often apply similar mechanisms of protection, sometimes with similar efficiency. *S. alpina* leaves are an example of very high non-photochemical quenching combined with high xanthophyll cycle activity and content possibly protecting the plant from overreduction under most conditions. Under reducing conditions an exceptionally high antioxidative protection capacity will keep damage low. *R. glacialis* consumes excess absorbed light energy by carbon assimilation and photorespiration. Reducing conditions may be avoided by the action of PTOX as safety valve. High PTOX contents have so far been found in all alpine plant species tested and may be characteristic of alpine species, but their functional significance has yet to be demonstrated. The discovery of a light-stable catalase in *H. alpina* leaves shows that modifications of protein properties can also be employed to better tolerate alpine constraints.

**Acknowledgements** We are grateful to Prof Dr. J. Feierabend for critical reading of the manuscript. We acknowledge the Station Alpine du Lautaret for their work facilities.

## References


Aluru MR, Stessman DJ, Spalding MH, Rodermel SR (2007) Alterations in photosynthesis in *Arabidopsis* lacking immutans, a chloroplast terminal oxidase. Photosynth Res 91:11–23

Anderson JM, Osmond CB (1987) Shade-sun responses: compromises between acclimation and photoinhibition. In: Kyle DJ, Osmond CB, Arntzen CJ (eds) Photoinhibition. Elsevier, New York, pp 1–37

Apel K, Hirt H (2004) Reactive oxygen species: metabolism, oxidative stress, and signal transduction. Ann Rev Plant Biol 55:373–399

Aro EM, Virgin I, Andersson B (1993) Photoinhibition of photosystem II. Inactivation, protein damage and turnover. Biochim Biophys Acta 1143:113–134

Asada K (1996) Radical production and scavenging in the chloroplasts. In: Baker NR (ed) Advances in photosynthesis, Vol. 5: photosynthesis and the environment. Kluwer Academic Publishers, Dordrecht, pp 123–150

Asada K (1999) The water-water cycle in chloroplasts: scavenging of active oxygens and dissipation of excess photons. Ann Rev Plant Physiol Plant Mol Biol 50:601–639

Atkin OK, Tjoelker MG (2003) Thermal acclimation and the dynamic response of plant respiration to temperature. Trends Plant Sci 8:343–351

Backhausen JE, Kitzmann C, Horton P, Scheibe R (2000) Electron acceptors in isolated intact spinach chloroplasts act hierarchically to prevent over-reduction and competition for electrons. Photosynth Res 64:1–13

Bailey S, Walters RG, Jansson S, Horton P (2001) Acclimation of *Arabidopsis thaliana* to the light environment: the existence of separate low light and high light responses. Planta 213:794–801

Baker NR (2008) Chlorophyll fluorescence: a probe of photosynthesis in vivo. Ann Rev Plant Biol 59:89–113

Berry J, Björkman O (1980) Photosynthetic response and adaptation to temperature in higher plants. Ann Rev Plant Physiol 31:491–543

Bird M, Haberle S, Chivas A (1994) Effect of the altitude on the carbon-isotope composition of forest and grassland soils from Papua New Guinea. Global Biogeochem Cycles 8:13–22

Bravo LA, Saavedra-Mella FA, Vera F, Guerra A, Cavieres LA, Ivanov A, Huner NPA, Corcuera LJ (2007) Effect of cold acclimation on the photosynthetic performance of two ecotypes of *Colobanthus quitensis* (Kunth) Bartl. J Exp Bot 58:3581–3590

Buchner O, Holzinger A, Lütz C (2007) Effects of temperature and light on the formation of chloroplast protrusions in leaf mesophyll cells of high alpine plants. Plant Cell Environ 30:1347–1356

Busch F, Hüner NPA, Ensminger I (2008) Increased air temperature during simulated autumn conditions impairs photosynthetic electron transport between photosystem II and photosystem I. Plant Physiol 147:402–414

Cabrera HM, Rada F, Cavieres L (1998) Effects of temperature on photosynthesis of two morphologically contrasting plant species along an altitudinal gradient in the tropical high Andes. Oecologia 114:145–152

Carol P, Kuntz M (2001) A plastid terminal oxidase comes to light: implications for carotenoid biosynthesis and chlororespiration. Trends Plant Sci 6:31–36

Cartellieri E (1940) Über die Transpiration und Kohlensäureassimilation an einem hochalpinen Standort. Sitzungsberichte der mathematisch-naturwissenschaftlichen Klasse Abteilung I 149:95–143

Castrillo M (1995) Ribulose-1.5-bis-phosphate carboxylase activity in altitudinal populations of *Espeletia schultzii* Wedd. Oecologia 101:193–196

Cordell S, Goldstein G, Meinzer FC, Handley LL (1999) Allocation of nitrogen and carbon in leaves of *Metrosideros polymorpha* regulates carboxylation capacity and $\delta^{13}C$ along an altitudinal gradient. Funct Ecol 13:811–818

Cournac L, Latouche G, Cerovic Z, Redding K, Ravenel J, Peltier G (2002) In vivo interactions between photosynthesis, mitorespiration, and chlororespiration in *Chlamydomonas reinhardtii*. Plant Physiol 129:1921–1928

Demmig-Adams B (1998) Survey of thermal energy dissipation and pigment composition in sun and shade leaves. Plant Cell Physiol 39:474–482

Demmig-Adams B, Adams WW III (1992a) Photoprotection and other responses of plants to high light stress. Ann Rev Plant Physiol Plant Mol Biol 99:599–626

Demmig-Adams B, Adams WW III (1992b) Carotenoid composition in sun and shade leaves of plants with different life forms. Plant Cell Environ 15:411–419

Demmig-Adams B, Adams WW III (2006) Photoprotection in an ecological context: the remarkable complexity of thermal energy dissipation. New Phytol 172:11–21

Demmig-Adams B, Adams WW III, Heber U, Neimanis S, Winter K, Krüger A, Czygan FC, Bilger W, Björkman O (1990) Inhibition of zeaxanthin formation and of rapid changes in radiationless energy dissipation by dithiothreitol in spinach leaves and chloroplasts. Plant Physiol 92:293–301

Diaz M, De Haro V, Munoz R, Quiles MJ (2007) Chlororespiration is involved in the adaptation of *Brassica* plants to heat and high light intensity. Plant Cell Environ 30:1578–1585

Diemer M, Körner Ch (1996) Lifetime leaf carbon balances of herbaceous perennial plants from low and high altitudes in the central Alps. Funct Ecol 10:33–43

Engel N, Schmidt M, Lütz C, Feierabend J (2006) Molecular identification, heterologous expression and properties of light-insensitive plant catalases. Plant Cell Environ 29:593–607

Epron D, Godard D, Cornic G, Genty B (1995) Limitation of net $CO_2$ assimilation rate by internal resistances to $CO_2$ transfer in the leaves of two tree species (*Fagus sylvatica* L. and *Castanea sativa* Mill.). Plant Cell Environ 18:43–51

Evans JR, Loreto F (2000) Acquisition and diffusion of $CO_2$ in higher plant leaves. In: Leegood RC, Sharkey TD, von Caemmerer S (eds) Photosynthesis: physiology and metabolism, vol 9. Kluwer Academic Publishers, Dordrecht/Boston/London, pp 332–351

Evans JR, Kaldenhoff R, Genty B, Terashima I (2009) Resistances along the $CO_2$ diffusion pathway inside leaves. J Exp Bot 60:2235–2248

Falk S, Maxwell DP, Laudenbach DE, Huner NPA (1996) Photosynthetic adjustment to temperature. In: Baker NR (ed) Photosynthesis and the environment, vol 5. Kluwer Academic Publishers, Dordrecht, pp 367–385

Farquhar GD, Sharkey TD (1982) Stomatal conductance and photosynthesis. Ann Rev Plant Physiol 33:317–345

Feierabend J (2005) Catalases in plants: molecular and functional properties and role in stress defence. In: Smirnoff N (ed) Antioxidants and reactive oxygen species in plants. Blackwell Publishers, Oxford, pp 101–140

Feierabend J, Schaan C, Hertwig B (1992) Photoinactivation of catalase occurs under both high- and low-temperature stress conditions and accompanies photoinhibition of PSII. Plant Physiol 100:1554–1561

Fetene M, Nauke P, Lüttge U, Beck E (1997) Photosynthesis and photoinhibition in a tropical alpine giant rosette plant, *Lobelia rhynchopetalum*. New Phytol 137:453–461

Finazzi G, Johnson GN, Dallosto L, Joliot P, Wollman F-A, Bassi R (2004) A zeaxanthin-independent nonphotochemical quenching mechanism localized in the photosystem II core complex. Proc Natl Acad Sci USA 101:12375–12380

Foyer CH, Noctor G (2005) Oxidant and antioxidant signalling in plants: a re-evaluation of the concept of oxidative stress in a physiological context. Plant Cell Environ 28:1056–1071

Friend AD, Woodward FI (1990) Evolutionary and ecophysiological responses of mountain plants to the growing season environment. Adv Ecol Res 20:59–124

Gauslaa Y (1984) Heat resistance and energy budget in different scandinavian plants. Hol Ecol 7:1–78

Germino MJ, Smith WK (2000) High resistance to low-temperature photoinhibition in two alpine, snowbank species. Physiol Plant 110:89–95

Germino MJ, Smith WK (2001) Relative importance of microhabitat, plant form and photosynthetic physiology to carbon gain in two alpine herbs. Funct Ecol 15:243–251

Goldstein G, Drake DR, Melcher P, Giambelluca TW, Heraux J (1996) Photosynthetic gas exchange and temperature-induced damage in seedlings of the tropical alpine species *Argyroxiphium sandwicense*. Oecologia 106:298–307

Guo F-X, Zhang M-X, Chen Y, Zhang W-H, Xu S-J, Wang J-H, An L-Z (2006) Relation of several antioxidant enzymes to rapid freezing resistance in suspension cultured cells from alpine *Chorispora bungeana*. Cryobiology 52:241–250

Hacker J, Neuner G (2006) Photosynthetic capacity and PSII efficiency of the evergreen alpine cushion plant *Saxifraga paniculata* during winter at different altitudes. Arct Antarct Alp Res 38:198–205

Haldimann P, Feller U (2004) Inhibition of photosynthesis by high temperature in oak (*Quercus pubescence L.*) leaves grown under natural conditions closely correlates with a reversible heat-dependent reduction of the activation state of ribulose-1,5-bisphosphate carboxylase/oxygenase. Plant Cell Environ 27:1169–1183

Havaux M, Dall'Osto L, Bassi R (2007) Zeaxanthin has enhanced antioxidant capacity with respect to all other xanthophylls in *Arabidopsis* leaves and functions independent of binding of PSII antennae. Plant Physiol 145:1506–1520

Heber U, Bligny R, Streb P, Douce R (1996) Photorespiration is essential for the protection of the photosynthetic apparatus of C3 plants against photoinactivation under sunlight. Bot Acta 109:307–315

Heber U, Bilger W, Bligny R, Lange OL (2000) Phototolerance of lichens, mosses and higher plants in an alpine environment: analysis of photoreactions. Planta 211:770–780

Henrici M (1918) Chlorophyllgehalt und Kohlensäure-Assimilation bei Alpen- und Ebenen-Pflanzen. Verh naturforsch Ges Basel 30:43–134

Heyno E, Gross CM, Laureau C, Culcasi M, Pietri S, Krieger-Liszkay A (2009) Plastid alternative oxidase (PTOX) promotes oxidative stress when overexpressed in tobacco. J Biol Chem 284:31174–31180

Huner NPA, Elfman B, Krol M, McIntosh A (1984) Growth and development at cold-hardening temperatures. Chloroplast ultrastructure, pigment content, and composition. Can J Bot 62:53–60

Huner NPA, Öquist G, Hurry VM, Krol M, Falk S, Griffith M (1993) Photosynthesis, photoinhibition and low temperature acclimation in cold tolerant plants. Photosynth Res 37:19–39

Huner NPA, Öquist G, Sarhan F (1998) Energy balance and acclimation to light and cold. Trends Plant Sci 3:224–230

Hurry VM, Malmberg G, Gardeström P, Öquist G (1994) Effects of a short-term shift to low temperature and of long-term cold hardening on photosynthesis and ribulose-1.5-bisphosphate carboxylase/oxygenase and sucrose phosphate synthase activity in leaves of winter rye (*Secale cereale* L.). Plant Physiol 106:983–990

Hurry VM, Keerberg O, Pärnik T, Gardeström P, Öquist G (1995) Cold-hardening results in increased activity of enzymes involved in carbon metabolism in leaves of winter rye (*Secale cereale* L.). Planta 195:554–562

Hurry V, Strand Å, Furbank R, Stitt M (2000) The role of inorganic phosphate in the development of freezing tolerance and the acclimatization of photosynthesis to low temperature is revealed by the pho mutants of *Arabidopsis thaliana*. Plant J 24:383–396

Joliot P, Joliot A (2002) Cyclic electron transfer in plant leaf. Proc Natl Acad Sci USA 99:10209–10214

Jordan DB, Ogren WL (1984) The $CO_2/O_2$ specificity of ribulose 1.5-bisphosphate carboxylase/oxygenase. Planta 161:308–313

Keeley JE, Keeley SC (1989) Crassulacean acid metabolism (CAM) in high elevation tropical cactus. Plant Cell Environ 12:331–336

Kleier C, Rundel P (2009) Energy balance and temperature relations of *Azorella compacta*, a high-elevation cushion plant of the central Andes. Plant Biol 11:351–358

Kogami H, Hanba YT, Kibe T, Terashima I, Masuzawa T (2001) $CO_2$ transfer conductance, leaf structure and carbon isotope composition of *P. cuspidatum* leaves from low and high altitudes. Plant Cell Environ 24:529–538

Körner Ch (1982) $CO_2$ exchange in the alpine sedge *Carex curvula* as influenced by canopy structure, light and temperature. Oecologia 53:98–104

Körner C (2003) Alpine plant life. Functional plant ecology of high mountain ecosystems. Springer Verlag, Berlin/Heidelberg

Körner Ch, Diemer M (1987) In situ photosynthetic responses to light, temperature and carbon dioxide in herbaceous plants from low and high altitude. Funct Ecol 1:179–194

Körner Ch, Farquhar GD, Wong SC (1991) Carbon isotope discrimination by plants follows latitudinal and altitudinal trends. Oecologia 88:30–40

Krause GH, Cornic G (1987) $CO_2$ and $O_2$ interactions in photoinhibition. In: Kyle DJ, Osmond CB, Arntzen CJ (eds) Photoinhibition. Elsevier, New York, pp 169–196

Krieger-Liszkay A, Fufezan C, Trebst A (2008) Singlet oxygen production in photosystem II and related protection mechanism. Photosynth Res 98:551–564

Kruk J, Holländer-Czytko H, Oettmeier W, Trebst A (2005) Tocopherol as singlet oxygen scavenger in photosystem II. J Plant Physiol 162:749–757

Kumar N, Kumar S, Ahuja PS (2005) Photosynthetic characteristics of *Hordeum, Triticum, Rumex* and *Trifolium* species at contrasting altitude. Photosynthetica 43:195–201

Kumar N, Kumar S, Vats SK, Ahuja PS (2006) Effect of altitude on the primary products of photosynthesis and the associated enzymes in barley and wheat. Photosynth Res 88:63–71

Kumar N, Vyas D, Kumar S (2007) Plants at high altitude exhibit higher component of alternative respiration. J Plant Physiol 164:31–38

Laisk A, Eichelmann H, Oja V, Talts E, Scheibe R (2007) Rates and roles of cyclic and alternative electron flow in potato leaves. Plant Cell Physiol 48:1575–1588

Larcher W, Wagner J, Lütz C (1997) The effect of heat on photosynthesis, dark respiration and cellular ultrastructure of the arctic-alpine psychrophyte *Ranunculus glacialis*. Photosynthetica 34:219–232

Larcher W, Kainmüller C, Wagner J (2010) Survival types of high mountain plants under extreme temperatures. Flora 205:3–18

Leegood RC, Edwards GE (1996) Carbon metabolism and photorespiration: temperature dependence in relation to other environmental factors. In: Baker NR (ed) Photosynthesis and the environment, vol 5. Kluwer Academic Publishers, Dordrecht, pp 191–221

Logan BA, Barker DH, Demmig-Adams B, Adams WW III (1996) Acclimation of leaf carotenoid composition and ascorbate levels to gradients in the light environment within an Australian rainforest. Plant Cell Environ 19:1083–1090

Lütz C (1996) Avoidance of photoinhibition and examples of photodestruction in high alpine *Eriophorum*. J Plant Physiol 148:120–128

Lütz C, Engel L (2007) Changes in chloroplast ultrastructure in some high-alpine plants: adaptation to metabolic demands and climate. Protoplasma 231:183–192

Mächler F, Nösberger J (1977) Effect of light intensity and temperature on apparent photosynthesis of altitudinal ecotypes of *Trifolium repens* L. Oecologia 31:73–78

Mächler F, Nösberger J (1978) The adaptation to temperature of photorespiration and of the photosynthetic carbon metabolism of altitudinal ecotypes of *Trifolium repens* L. Oecologia 35:267–276

Mächler F, Nösberger J, Erismann KH (1977) Photosynthetic $^{14}CO_2$ fixation products in altitudinal ecotypes of *Trifolium repens* L. with different temperature requirements. Oecologia 31:79–84

Makino A, Miyake C, Yokota A (2002) Physiological functions of the water-water cycle (Mehler reaction) and the cyclic electron flow around PSI in rice leaves. Plant Cell Physiol 43:1017–1026

Manuel N, Cornic G, Aubert S, Choler P, Bligny R, Heber U (1999) Protection against photoinhibition in the alpine plant *Geum montanum*. Oecologia 119:149–158

Maxwell K, Johnson GN (2000) Chlorophyll fluorescence – a practical guide. J Exp Bot 51:659–668

Medek DE, Ball MC, Schortemeyer M (2007) Relative contributions of leaf area ratio and net assimilation rate to change in growth rate depend on growth temperature: comparative analysis of subantarctic and alpine grasses. New Phytol 175:290–300

Meidner H, Mandsfield TA (1968) Physiology of stomata. McGraw-Hill, Maidenhair/England

Meinzer FC, Rundel PW, Goldstein G, Sharifi MR (1992) Carbon isotope composition in relation to leaf gas exchange and environmental conditions in Hawaiian *Metrosideros polymorpha* populations. Oecologia 91:305–311

Melis A (1999) Photosystem-II damage and repair cycle in chloroplasts: what modulates the rate of photodamage in vivo? Trends Plant Sci 4:130–135

Moller IM, Jensen PE, Hansson A (2007) Oxidative modifications to cellular components in plants. Ann Rev Plant Biol 58:459–481

Mooney HA, Billings WD (1961) Comparative physiological ecology of arctic and alpine populations of *Oxyria digyna*. Ecol Mon 31:1–29

Moser W, Brzoska W, Zachhuber K, Larcher W (1977) Ergebnisse des IBP-Projekts "Hoher Nebelkogel 3184 m" Sitzungsberichte der Österreichischen Akademie der Wissenschaften (Wien). Mathematisch-Naturwissenschaftliche Klasse. Abteilung I 186:387–419

Munné-Bosch S (2005) The role of alpa-tocopherol in plant stress tolerance. J Plant Physiol 162:743–748

Murata N, Takahashi S, Nishiyama Y, Allakhverdiev SI (2007) Photoinhibition of photosystem II under environmental stress. Biochim Biophys Acta 1767:414–421

Murchie EH, Horton P (1998) Contrasting patterns of photosynthetic acclimation to the light environment are dependent on the differential expression of the responses to altered irradiance and spectral quality. Plant Cell Environ 21:139–148

Naidu SL, Moose SP, Al-Shoaibi AK, Raines CA, Long SP (2003) Cold tolerance of C4 photosynthesis in *Miscanthus x giganteus*: adaptation in amounts and sequence of C4 photosynthetic enzymes. Plant Physiol 132:1688–1697

Neuner G, Braun V, Buchner O, Taschler D (1999) Leaf rosette closure in the alpine rock species *Saxifraga paniculata* mill.: significance for survival of drought and heat under high irradiation. Plant Cell Environ 22:1539–1548

Nixon PJ, Rich PR (2006) Chlororespiratory pathways and their physiological significance. In: Wise RR, Hoober JK (eds) The structure and function of plastids. Springer, Netherlands, pp 237–251, Chapter 12

Niyogi KK (1999) Photoprotection revisited: genetic and molecular approaches. Ann Rev Plant Physiol Plant Mol Biol 50:333–359

Noctor G, Foyer CH (1998) Ascorbate and glutathione: keeping active oxygen under control. Ann Rev Plant Physiol Plant Mol Biol 49:249–279

Nogués S, Tcherkez G, Streb P, Pardo A, Baptist F, Bligny R, Ghashghaie J, Cornic G (2006) Respiratory carbon metabolites in the high mountain plant species *Ranunculus glacialis*. J Exp Bot 57:3837–3845

Öncel I, Yurdakulol E, Keles Y, Kurt L, Yildiz A (2004) Role of antioxidant defense system and biochemical adaptation on stress tolerance of high mountain and steppe plants. Acta Oecol 26:211–218

Öquist G, Huner NPA (2003) Photosynthesis of overwintering evergreen plants. Ann Rev Plant Biol 54:329–355

Ort DR, Baker NR (2002) A photoprotective role of $O_2$ as an alternative electron sink in photosynthesis? Cur Op Plant Biol 5:193–198

Osmond B, Ziegler H, Stichler W, Trimborn P (1975) Carbon isotope discrimination in alpine succulent plants supposed to be capable of crassulacean acid metabolism (CAM). Oecologia 18:209–217

P'yankov VI, Voznesenskaya EV, Kuz'min AN, Demidov ED, Vasil'ev AA, Dzyubenko OA (1992) C4 photosynthesis in alpine species of the pamirs. Sov Plant Physiol 39:421–430

Peltier G, Cournac L (2002) Chlororespiration. Ann Rev Plant Biol 53:523–550

Pfannschmidt T (2003) Chloroplast redox signals: how photosynthesis controls its own genes. Trends Plant Sci 8:33–41

Pittermann J, Sage RF (2000) Photosynthetic perfomance at low temperature of *Bouteloua gracilis* Lag., a high-altitude C4 grass from the Rocky mountains. USA Plant Cell Environ 23:811–823

Quiles MJ (2006) Stimulation of chlororespiration by heat and high light intensity in oat plants. Plant Cell Environ 29:1463–1470

Rawat AS, Purohit AN (1991) $CO_2$ and water vapour exchange in four alpine herbs at two altitudes and under varying light and temperature conditions. Photosynth Res 28:99–108

Rizhsky L, Hallak-Herr E, van Breusegem F, Rachmilevitch S, Barr JE, Rodermel S, Inze D, Mittler R (2002) Double antisense plants lacking ascorbate peroxidase and catalase are less sensitive to oxidative stress than single antisense plants lacking ascorbate peroxidase or catalase. Plant J 32:329–342

Rolland F, Moore B, Sheen J (2002) Sugar sensing and signaling in plants. Plant Cell 14(Suppl):185–205

Rosso D, Ivanov AG, Fu A, Geisler-Lee J, Hendrickson L, Geisler M, Stewart G, Krol M, Hurry V, Rodermel SR, Maxwell DP, Hüner NPA (2006) IMMUTANS does not act as a stress-induced safety valve in the protection of the photosynthetic apparatus of *Arabidopsis* during steady-state photosynthesis. Plant Physiol 142:574–585

Rumeau D, Peltier G, Cournac L (2007) Chlororespiration and cyclic electron flow around PSI during photosynthesis and plant stress response. Plant Cell Environ 30:1041–1051

Sage RF, Wedin DA, Li M (1999) The biogeography of C4 photosynthesis: patterns and controlling factors. In: Sage RF, Monson RK (eds) Plant biology. Academic, San Diego, pp 313–373

Sakata T, Yokoi Y (2002) Analysis of the $O_2$ dependency in leaf-level photosynthesis of two *Reynoutria japonica* populations growing at different altitudes. Plant Cell Environ 25:65–74

Savitch LV, Gray GR, Huner NPA (1997) Feedback-limited photosynthesis and regulation of sucrose-starch accumulation during cold acclimation and low-temperature stress in a spring and winter wheat. Planta 201:18–26

Scheller HV, Haldrup A (2005) Photoinhibition of photosystem I. Planta 221:5–8

Shahbazi M, Gilbert M, Labouré A-M, Kuntz M (2007) Dual role of the plastid terminal oxidase in tomato. Plant Physiol 145:691–702

Shang W, Feierabend J (1998) Slow turnover of the D1 reaction center protein of photosystem II in leaves of high mountain plants. FEBS Lett 425:97–100

Shang W, Feierabend J (1999) Dependence of catalase photoinactivation in rye leaves on light intensity and quality and characterization of a chloroplast-mediated inactivation in red light. Photosynth Res 59:201–213

Shao N, Beck CF, Lemaire SD, Krieger-Liszkay A (2008) Photosynthetic electron flow affects $H_2O_2$ signaling by inactivation of catalase in *Chlamydomonas reinhardtii*. Planta 228:1055–1066

Shi Z, Liu S, Liu X, Centritto M (2006) Altitudinal variation in photosynthetic capacity, diffusional conductance and delta $^{13}C$ of butterfly bush (*Buddleja davidii*) plants growing at high elevations. Physiol Plant 128:722–731

Strand A, Hurry V, Henkes S, Huner N, Gustafsson P, Gardeström P, Stitt M (1999) Acclimation of *Arabidopsis* leaves developing at low temperatures. Increasing cytoplasmic volume accompanies increased activities of enzymes in the Calvin cycle and in the sucrose-biosynthesis pathway. Plant Physiol 119:1387–1397

Streb P (1994) Lichtschäden und Streßwirkungen in Blättern und antioxidative Schutzmechanismen. Dissertation am Fachbereich Biologie der J.W. Goethe Universität

Streb P, Feierabend J (1999) Significance of antioxidants and electron sinks for the cold-hardening-induced resistance of winter rye leaves to photo-oxidative stress. Plant Cell Environ 22:1225–1237

Streb P, Feierabend J, Bligny R (1997) Resistance to photoinhibition of photosystem II and catalase and antioxidative protection in high mountain plants. Plant Cell Environ 20:1030–1040

Streb P, Shang W, Feierabend J, Bligny R (1998) Divergent strategies of photoprotection in high-mountain plants. Planta 207:313–324

Streb P, Shang W, Feierabend J (1999) Resistance of cold-hardened winter rye leaves (*Secale cereale* L.) to photo-oxidative stress. Plant Cell Environ 22:1211–1223

Streb P, Aubert S, Gout E, Bligny R (2003a) Reversibility of cold- and light-stress tolerance and accompanying changes of metabolite and antioxidant levels in the two high mountain plant species *Soldanella alpina* and *Ranunculus glacialis*. J Exp Bot 54:405–418

Streb P, Aubert S, Gout E, Bligny R (2003b) Cold- and light-induced changes of metabolite and antioxidant levels in two high mountain plant species *Soldanella alpina* and *Ranunculus glacialis* and a lowland species *Pisum sativum*. Physiol Plant 118:96–104

Streb P, Aubert S, Bligny R (2003c) High temperature effects on light sensitivity in the two high mountain plant species *Soldanella alpina* (L) and *Ranunculus glacialis* (L). Plant Biol 5:432–440

Streb P, Josse E-M, Gallouët E, Baptist F, Kuntz M, Cornic G (2005) Evidence for alternative electron sinks to photosynthetic carbon assimilation in the high mountain plant species *Ranunculus glacialis*. Plant Cell Environ 28:1123–1135

Suzuki N, Mittler R (2006) Reactive oxygen species and temperature stresses: a delicate balance between signaling and destruction. Physiol Plant 126:45–51

Takahashi S, Murata N (2008) How do environmental stresses accelerate photoinhibition? Trends Plant Sci 13:178–182

Tallon C, Quiles MJ (2007) Acclimation to heat and high light intensity during the development of oat leaves increases the NADH DH complex and PTOX levels in chloroplasts. Plant Sci 173:438–445

Terashima I, Masuzawa T, Ohba H (1993) Photosynthetic characteristics of a giant alpine plant, *Rheum nobile* Hook. f. et Thoms. and of some other alpine species measured at 4300 m, in the eastern Himalaya, Nepal. Oecologia 95:194–201

Terashima I, Funayama S, Sonoike K (1994) The site of photoinhibition in leaves of *Cucumis sativus* L. at low temperature is photosystem I, not photosystem II. Planta 193:300–306

Terashima I, Masuzawa T, Ohba H, Yokoi Y (1995) Is photosynthesis suppressed at higher elevations due to low $CO_2$ pressure? Ecology 76:2663–2668

Triantaphylides Ch, Krischke M, Hoeberichts FA, Ksas B, Gresser G, Havaux M, Van Breusegem F, Mueller MJ (2008) Singlet oxygen is the major reactive oxygen species involved in photooxidative damage to plants. Plant Physiol 148:960–968

van Kooten O, Snel JFH (1990) The use of chlorophyll fluorescence nomenclature in plant stress physiology. Photosynth Res 25:147–150

Vitousek PM, Field CB, Matson PA (1990) Variation of foliar $\delta^{13}C$ in Hawaiian *Metrosideros polymorpha*: a case of internal resistance? Oecologia 84:362–370

Wagner J, Larcher W (1981) Dependence of $CO_2$ gas exchange and acid metabolism of the alpine CAM plant *Sempervivum montanum* on temperature and light. Oecologia 50:88–93

Walters RG (2005) Towards an understanding of photosynthetic acclimation. J Exp Bot 56:435–447

Wang D, Naidu SL, Portis AR Jr, Moose SP, Long SP (2008) Can the cold-tolerance of C4 photosynthesis in Miscanthus x giganteus relative to Zea mays be explained by differences in activities and thermal properties of Rubisco? J Exp Botany 59:1779–1787

Wang Y, He W, Huang H, An L, Wang D, Zhang F (2009) Antioxidative responses to different altitudes in leaves of alpine plant *Polygonum viviparum* in summer. Acta Physiol Plant 31:839–848

Wildi B, Lütz C (1996) Antioxidant composition of selected high alpine plant species from different altitudes. Plant Cell Environ 19:138–146

Williams EL, Hovenden MJ, Close DC (2003) Strategies of light energy utilisation, dissipation and attenuation in six co-occurring alpine heath species in Tasmania. Funct Plant Biol 30:1205–1218

Wise RR (1995) Chilling-enhanced photooxidation: the production, action and study of reactive oxygen species produced during chilling in the light. Photosynth Res 45:79–97

Yi XF, Yang YQ, Zhang XA, Li LX, Zhao L (2003) No C4 plants found at the Haibei alpine meadow ecosystem research station in Qinghai, China: evidence from stable carbon isotope studies. Acta Bot Sinica 45:1291–1296

Zarter CR, Adams WW III, Ebbert V, Adamska I, Jansson S, Demmig-Adams B (2006a) Winter acclimation of PsbS and related proteins in the evergreen *Arctostaphylos uva-ursi* as influenced by altitude and light environment. Plant Cell Environ 29:869–878

Zarter CR, Demmig-Adams B, Ebbert V, Adamska I, Adams WW III (2006b) Photosynthetic capacity and light harvesting efficiency during the winter-to-spring transition in subalpine conifers. New Phytol 172:283–292

Zhang S-B, Hu H (2008) Photosynthetic adaptation *Meconopsis integrifolia* Franch. and *M. horridula* var. racemosa Prain. Bot Stud 49:226–233

Zhou R, Zhao H (2004) Seasonal pattern of antioxidant enzyme system in the roots of perennial forage grasses grown in alpine habitat, related to freezing tolerance. Physiol Plant 121:399–408

# Specificities of Metabolite Profiles in Alpine Plants

# 8

Richard Bligny and Serge Aubert

## Abbreviations

| | |
|---|---|
| Glcn-6-P | gluconate 6-phosphate |
| L-AA | L-ascorbic acid |
| m asl | metres above sea level |
| MeG | methyl-$\beta$-D-glucopyranoside |
| PCA | perchloric acid |
| PP-pathway | pentose phosphate pathway |
| ROS | reactive oxygen species |
| SOD | superoxide dismutase |

Frequently Mentioned Plant Species

*Soldanella alpina*
*Geum montanum*
*Xanthoria elegans*
*Acer pseudoplatanus*
*Arabidopsis thaliana*

## 8.1 Introduction

Given that plants cannot escape their environment, they have evolved many strategies to survive, grow, and reproduce, including the capability to synthesise over 200,000 specialized and highly variable metabolites (Yonekura-Sakakibara and Saito 2009). In the severe alpine environment plants experience particularly low and high temperature extremes, intense solar radiation under clear conditions, strong wind effects, and variable mean dates for snow melting depending on slope and exposure (Körner 2003). Demanding environmental conditions have long been shown to exert a profound influence on the soluble metabolite composition of plants, although plants from high elevation habitats have been poorly analysed (Harborne 1982; Alonso-Amelot 2008). In the case of an alpine plant like *Arnica montana*, which grows at altitudes ranging from 600 to 2,200 m in the Tyrolean Alps, it was shown, for example, that secondary metabolite profiles differ according to the altitude, including strong enhancements of flavonoids and phenolic acids at high altitude (Spitaler et al. 2006; Zidorn 2009). More generally, in a great number of species the alpine climate characteristics, particularly the decrease of mean temperatures, positively influence the plant concentrations of soluble carbohydrates (reviewed by Körner 2003). Accordingly, in the culture cells of the laboratory model plants Arabidopsis (*A. thaliana* L.) and sycamore (*Acer pseudoplatanus* L.) we observed that the decrease of the nutrient media temperature from 20°C to 10°C causes cells to accumulate on average twice as many soluble sugars and sugar-P,

R. Bligny
Laboratoire de Physiologie Cellulaire & Végétale, Unité Mixte de Recherche, Institut de Recherche en Technologies et Sciences pour le Vivant, Grenoble cedex 9, France

S. Aubert (✉)
Station Alpine Joseph Fourier, Unité Mixte de Service, Université Joseph Fourier, Grenoble cedex 9, France
e-mail: serge.aubert@ujf-grenoble.fr

C. Lütz (ed.), *Plants in Alpine Regions*,
DOI 10.1007/978-3-7091-0136-0_8, © Springer-Verlag/Wien 2012

and also nucleotides, mainly due to the strong decrease of cell growth and respiration rates (unpublished).

This chapter aims to focus on possible adaptive responses in alpine plants suggested by the unexpected accumulation of specific metabolites. We have chosen to focus on two groups of plants hyper-accumulating either ascorbate (belonging to *Primulaceae*) or methyl-β-D-glucopyranoside (belonging to *Rosaceae*). We will also show how the accumulation of gluconate 6-P may contribute to the quasi instantaneous reviviscence of dry alpine lichens after rewetting.

## 8.2 Metabolomic Analysis

The analysis of metabolites has long been the cornerstone of biological research with the aim being to gain insights into the interaction of living organisms with their environment. The identification and quantitative measurement of metabolites present in plants is generally done according to methods that differ in their sensitivity, the number of compounds analysed, and the level of structural information provided (Sumner et al. 2003). In this domain metabolite profiling and fingerprinting are two general approaches used to analyse small molecules based on complementary techniques. According to Last et al. (2007), metabolite profiling is the analysis of known metabolites belonging to predetermined classes (Roessner et al. 2001), whereas fingerprinting detects hundreds of only partially identified molecules. In the first case, mass spectrometry (MS) coupled with gas or liquid chromatography (GC and LC) or with capillary electrophoresis offers sensitivity and accuracy. In the second case, techniques like high-performance liquid chromatography (HPLC) and NMR spectroscopy, though less sensitive, can be used for rapid profiling of a large number of samples. These techniques are non-destructive, and they can further provide structural information on metabolites (Fraser et al. 2000; Ratcliffe and Shachar-Hill 2001). NMR, for example, provides metabolic fingerprinting with interesting chemical specificity for compounds containing nuclei with non-zero magnetic moments like $^{1}H$, $^{31}P$, $^{13}C$, or $^{15}N$, allowing the identification and quantification of a great number of metabolites in complex mixtures such as cell extracts (Fan 1996; Roberts 2000; Colquhoun 2007). The examples developed in this chapter should highlight the advantages of NMR when seeking a general approach to data mining in non-targeted metabolomics.

Another invaluable advantage provided by NMR spectroscopy is that until now it has been the only method permitting the non-invasive analysis of cell metabolite localisation and the study of their dynamics in the different subcellular compartments (Ratcliffe 1994; Aubert et al. 1996, 1999). This method provides information on the absolute concentration of the more abundant metabolites and eliminates the need for extractions and sample preparation procedures (Bligny and Douce 2001). As described below, we utilised *in vitro* and *in vivo* $^{13}C$-NMR methods to quantify ascorbate in a series of alpine and lowland plants, and to characterise the subcellular compartmentation of this metabolite.

## 8.3 Where Do Some *Primulaceae* Accumulate So Much L-Ascorbic Acid?

It is well-known that the increasing light energy received and absorbed by plants at high altitude may generate the formation of an excess of reactive oxygen species (ROS) in chloroplasts (Asada 1996; Apel and Hirt 2004), leading to oxidative damages such as DNA, carbohydrate, and protein degradation, lipid peroxidation and photoinhibition (Moller et al. 2007; Takahashi and Murata 2008). To alleviate photooxidative damage and to protect their photosynthetic machinery, alpine plants have adopted a number of biochemical strategies (Streb and Cornic; Lütz et al.; Lütz and Seidlitz, this book). These strategies include the oxidation of superoxides via cascades of enzyme reactions as shown for the ascorbate-glutathione cycle first proposed by Foyer and Halliwell (1976) and reviewed by Noctor and Foyer (1998). Most high altitude ecotypes actually contain more antioxidants than those growing at lower elevation due to the combined effect of lower temperature and higher light intensity (Wildi and Lütz 1996; Streb and Feierabend 1999; 2003). Among them, L-ascorbic acid (L-AA, vitamin C) is the major antioxidant found in plants. Very high concentrations of L-AA were found in some plants, for example in the young leaves of *Soldanella alpina* growing at 2,500 m in the French Hautes-Alpes (Streb et al. 2003), and also in young leaves of *Polygonum viviparum* growing at 3,900 m in the Tianshan Mountains in China (Wang et al. 2009). The question arises as to whether L-AA accumulation is a characteristic of alpine plants shared by many genus or even families, or whether it is restricted to a few species.

### 8.3.1 Accumulation of L-AA in the Foliage of Lowland and Alpine Plants

L-AA was measured in plant tissue perchloric acid extracts (PCA) using $^{13}$C-NMR for *in vitro* analysis as shown in Fig. 8.1. Although L-AA is a universal constituent of green plants, very variable amounts of this vitamin were found in lowland and alpine plant leaves (Table 8.1). In many plants, including the arctic-alpine *Polygonum viviparum* and the subantarctic *Pringlea antiscorbutica*, L-AA concentrations were below the threshold of $^{13}$C-NMR detection or close to a few μmol/g tissue wet wt, whereas it was more abundant in the leaves of various plants belonging to *Rosaceae* and *Primulaceae* families, reaching 46–50 μmol/g wet wt in young *S. alpina* leaves, for example. For the sake of comparison, the highest L-AA concentration of any fresh fruit known, 180 μmol/g wet wt, was found in the Australian bush plum *Terminalia ferdinandiana* (Cunningham et al. 2009). Apparently there is no clear relation between the huge variations in L-AA and the taxonomical position of plants. There is also huge variation in response to high light intensities (Foyer 1993), to the age of leaves and to the organ in question in *S. alpina* (Table 8.1), as well as between fruit and vegetables (Davey et al. 2000). Therefore, considering tissues as a whole, what does a threshold for L-AA concentration mean?

A major question arises from a biochemical point of view, namely in which cell compartment the metabolite accumulates and at what concentration in each compartment. Indeed, when cells contain large

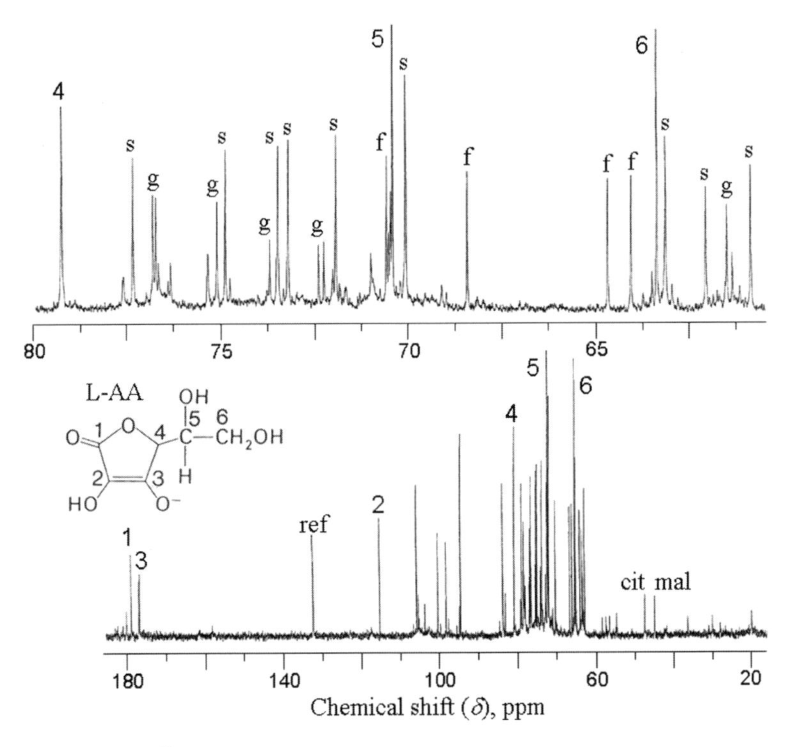


Fig. 8.1 *In vitro* proton-decoupled $^{13}$C-NMR spectrum of a young *Soldanella alpina* leaf PCA extract adjusted at pH 7.5. The extract is prepared as described by Aubert et al. (1996) from 10 g of 15-d old *S. alpina* leaves harvested during the first week of July at 2,450 m asl near the Lautaret pass, at 10 am. This spectrum recorded at 20°C on a Bruker AMX 400 WB spectrometer tuned at 100.6 MHz is the result of 1,000 transients with a 6-s repetition time (1 h). Peak assignments are as follows: L-AA, L-ascorbic acid, the different carbons of L-AA are numbered on the molecule and on the corresponding peaks of resonance; g, glucose; f, fructose; cit, citrate; mal, malate; ref, reference (250 μmol of maleate)

**Table 8.1** L-ascorbic acid content of selected plant leaves and sometimes fruits, flowers, and roots when indicated

| | | | μmol/g wet wt | % soluble sugars (mol/mol) |
|---|---|---|---|---|
| Dicots | *Actinidiaceae* | *Actinidia sinensis* Planch. (fr, lp) | 20–22 | 25–28 |
| | *Campanulacées* | *Campanula scheuchzeri* | nd | – |
| | *Caryophyllaceae* | *Cerastium latifolium* | nd | – |
| | | *Dianthus superbus* | nd | – |
| | | *Silene acaulis* | nd | – |
| | *Chenopodiaceae* | *Spinacia oleracea* L. | nd | – |
| | | *Chenopodium bonus henricus* | 6.0–7.1 | 12–14 |
| | *Compositae* | *Achillea millefolium* | 4.2–4.9 | 15–17 |
| | | *Berardia subacaulis* | 1.0–2.0 | 3.0–4.0 |
| | | *Centaurea uniflora* | 4.3–5.2 | 7.5–8.6 |
| | | *Doronicum grandiflorum* | 3.2–4.1 | 5.0–6.5 |
| | | *Homogyne alpina* | 7.0–9.0 | – |
| | | *Taraxacum officinalis* | nd | – |
| | | *Tussilago farfara* L. | nd | – |
| | | *Stemmacantha rhapontica* | 8.5–9.5 | 19–21 |
| | *Crassulaceae* | *Sedum anacampseros* | 6.3–7.3 | 13–15 |
| | | *Sempervivum montanum* | 4.1–5.2 | 10–12 |
| | *Crucifereae* | *Arabidopsis thaliana* | 3.0–4.0 | 20–25 |
| | | *Brassica broccoli* | 5.5–6.5 | 13–16 |
| | | *Brassica repanda* | 4.9–6.0 | 8.1–10 |
| | | *Coincya richerii* | 1.2–2.5 | 6.0–8.0 |
| | | *Pringlea antiscorbutica* R. Br. | nd | – |
| | *Ericaceae* | *Arctostaphylos uva-ursi* | nd | – |
| | | *Rhododendron ferrugineum* | 8.0–10 | 8.5–11 |
| | *Fabaceae* | *Oxytropis campestris* (L.) DC | nd | – |
| | | *Phaseolus vulgaris* | 7.2–8.8 | 16–17 |
| | | *Pisum sativum* L. | 6.0–7.5 | – |
| | | *Trifolium pratense* | 5.5–6.5 | 10–12 |
| | *Fagaceae* | *Fagus silvatica* | 5.8–7.0 | 9.0–11 |
| | *Gentianaceae* | *Gentiana acaulis* | nd | – |
| | | *Gentiana lutea* | 3.4–4.7 | 4.0–5.2 |
| | | *Gentiana punctata* | 3.8–5.0 | 4.2–5.5 |
| | *Geraniaceae* | *Geranium ibericum* | nd | – |
| | *Gesneriaceae* | *Ramondia myconi* | nd | – |
| | *Oenotheraceae* | *Epilobium montanum* | nd | – |
| | *Umbelifereae* | *Eryngium alpinum* | nd | – |
| | | *Heracleum sphondylium* | nd | |
| | | *Laserpitium siler* L. | 4.4–5.6 | 6.4–7.5 |
| | | *Meum atamanticum* | nd | – |
| | | *Petroselinum sativum* Mill. | 8.0–10 | 14–18 |
| | *Papaveraceae* | *Papaver alpina* | nd | – |
| | *Plantaginaceae* | *Plantago alpina* | 1.5–2.5 | 2.5–3.5 |
| | *Plumbaginaceae* | *Statice montana* | nd | – |
| | *Polemoniaceae* | *Polemonium caeruleum* | nd | – |
| | *Polygonaceae* | *Polygonum viviparum* | nd | – |
| | | *Rumex acetosella* | nd | – |
| | | *Rumex scutatus* | nd | – |
| | *Primulaceae* | *Primula beesiana* | 22–26 | 30–35 |
| | | *Primula officinalis* L. (lp) | 36–40 | 45–49 |
| | | *Primula vulgaris* L. (lp) | 28–32 | 32–36 |

(continued)

**Table 8.1** (continued)

| | | | μmol/g wet wt | % soluble sugars (mol/mol) |
|---|---|---|---|---|
| | | *Soldanella alpina* L. | 46–50 | 41–45 |
| | | *S. alpina* (old leaves) | 7.5–9.0 | 4.5–6.0 |
| | | *S. alpina* (flowers) | 5.0–6.0 | 6.5–7.5 |
| | | *S. alpina* (roots) | nd | – |
| | *Ranunculaceae* | *Aconitum napellus* | 5.0–6.0 | 6.2–7.3 |
| | | *Pulsatilla alpina* | 9.0–10 | 10–11 |
| | | *Ranunculus acris* L. | nd | – |
| | | *Ranunculus glacialis* L. | 3.5–4.5 | 13–15 |
| | | *Ranunculus pyrenaicus* | nd | – |
| | *Rosaceae* | *Acaena microphylla* Hook. fil | 6.0–7.0 | 10–11 |
| | | *Alchemilla alpina* | 10–12 | 11–13 |
| | | *Alchemilla pentaphyllea* L. | 2.5–3.2 | 2.8–3.5 |
| | | *Dryas octopetala* L. | 9.7–10.5 | 7.8–8.5 |
| | | *Fragaria vesca* | 6.3–7.2 | 10–11 |
| | | *Geum coccineum* Sibth. & Sm | 17–20 | 9.0–11 |
| | | *Geum montanum* L. | 10–11 | 7.5–8.5 |
| | | *Geum pyrenaicum* Miller | 15–17 | 8.2–9.5 |
| | | *Geum reptans* L. | 18–20 | 11–13 |
| | | *Geum rivale* L. | 2.0–2.5 | 8.5–10 |
| | | *Geum urbanum* | nd | – |
| | | *Potentilla grandiflora* L. | 7.5–8.0 | 5.3–6.7 |
| | | *Potentilla pentaphyllea* L. | 4.5–5.6 | 3.3–5.4 |
| | | *Rosa canina* L. | 3.5–4.2 | 7.8–8.8 |
| | | *Rubus fruticosus* | 19–21 | 17–19 |
| | | *Sanguisorba minor* L. | nd | – |
| | | *Sibbaldia procumbens* L. | 4.1–5.2 | 11–14 |
| | | *Woronowia elegans* (Albov) Juz. | 2.4–3.2 | 2.0–2.8 |
| | | *Cercocapnos* sp. | 14–18 | 28–33 |
| | | *Cotoneaster* sp. | nd | – |
| | | *Kerria japonica* | 8.6–9.5 | 35–40 |
| | | *Malus* sp. | 12–16 | 30–35 |
| | | *Prunus* sp. | nd | – |
| | *Salicaceae* | *Salix serpyllifolia* | nd | – |
| | *Saxifragaceae* | *Saxifraga aizoïdes* | 6.5–8.0 | 8.4–10 |
| | | *Saxifraga rotondifolia* | 7.0–8.0 | 10–12 |
| | *Scrophulariaceae* | *Linaria alpina* | nd | – |
| | | *Pedicularis verticillata* | nd | – |
| | | *Scutellaria alpina* | 4.5–5.7 | 9.5–11 |
| | *Solanaceae* | *Lycopersicum esculentum* | 5.0–6.0 | 2.0–3.0 |
| | | *Nicotiana sylvestris* | 3.3–4.2 | 2.3–3.3 |
| | | *Solanum tuberosum* L. | 6.8–8.1 | 10–12 |
| Monocots | *Amaryllidaceae* | *Narcissus poeticus* | nd | – |
| | | *Narcissus pseudonarcissus* L. | 8.3–9.2 | 3.1–4.2 |
| | *Araceae* | *Arum maculatum* L. | 6.0–7.0 | 5.6–6.6 |
| | *Cypereceae* | *Carex foetida* | nd | – |
| | | *Kobresia myosuroides* | 8.0–9.0 | 5.0–6.0 |
| | *Iridaceae* | *Crocus vernus* | 5.8–6.8 | 6.5–7.5 |
| | | *Iris* sp | 30–33 | 35–38 |
| | *Juncaceae* | *Luzula lutea* | nd | – |
| | | *Luzula nivalis* | 3.1–4.0 | 5.7–6.5 |

(continued)

**Table 8.1** (continued)

| | | | μmol/g wet wt | % soluble sugars (mol/mol) |
|---|---|---|---|---|
| | *Liliaceae* | *Tulipa* sp | nd | – |
| | *Orchidaceae* | *Himantoglosum hircinum* L. | 2.2–3.0 | 7.3–8.2 |
| | | *Oncidium* sp | 5.0–6.0 | 4.0–5.0 |
| | *Poaceae* | *Alopecurus gerardii* | nd | – |
| | | *Festuca spadicea* | 2.4–3.1 | 3.5–4.3 |
| | | *Secale cereale*[a] L. | 12–16 | – |
| Gymnosperms | *Pinaceae* | *Abies alba* | nd | – |
| | | *Pinus sylvestris* L. | nd | – |
| Pteridophytes | | *Athyrium* sp | nd | – |
| Bryophytes | | *Bryum* sp | nd | – |

The high altitude plant leaves selected in this table were harvested in the area of the Lautaret (2,100 m) and Galibier (2,600 m) passes, Hautes-Alpes, France, and the lowland plant (lp, as indicated) leaves near Grenoble (200–500 m). Young mature leaves (10 g) were sampled during the first days of July between 11 am and 12 pm, under 1,500–2,500 μmol/m$^2$/s PPFD and 15–30°C environmental conditions; old leaves (when mentioned) were sampled at the end of September. In some of the analysed species, nd means that the amount of L-AA present in PCA extracts is below the threshold of $^{13}$C-NMR detection under our NMR measurement conditions (Aubert et al. 1996), i.e. about 2 μmol for 1-h accumulation time, which enables L-AA concentrations of ca 0.2 μmol/g tissue wet wt to be detected

[a]acc. to Streb and Feierabend (1999)

amounts of metabolites, such as soluble sugars or organic acids, most of them are located in the large central vacuole for osmotic reasons (Aubert et al. 1996). We thus hypothesised that this could also be the case for L-AA when it reaches 50 μmol/g wet wt in *S. alpina* leaves. To substantiate this hypothesis we utilised the non-invasive *in vivo* NMR technique as shown below.

### 8.3.2 Cellular and Subcellular Localization of L-AA in *S. alpina* Leaves

While L-ascorbic acid has very important metabolic and antioxidant functions in plants and animals, its synthesis by the Smirnoff-Wheeler pathway was only identified in plants in the late 1990s (Wheeler et al. 1998). However, though multiple roles of L-AA in the physiology and metabolism of plant cells are now well known (reviewed by Davey et al. 2000 and Conklin 2001), the studies on its subcellular localization remain inconclusive. For example, whereas in species containing relatively low amounts of L-AA, such as the lowland plant *Spinacia oleracea* or the alpine *Ranunculus glacialis,* most of this metabolite is localized in the chloroplast, in *S. alpina* accumulating high amounts of L-AA it was principally outside the chloroplast (Streb et al. 1997). However, it was not determined in which other cytoplasmic compartment (cytosol, mitochondria, or the vacuole) most of the L-AA was present. Taking into account that the volume of the cytoplasm occupies nearly 10% of the cell volume in chlorophyllus parenchyma cells, the calculated concentration of L-AA in the cytoplasm of young *S. alpina* leaves would average 400–500 mM. Since such a high concentration for a soluble metabolite in the cytoplasm seems unlikely, this means that the vacuole must contain large stores of L-AA in these leaves.

To substantiate this hypothesis we took advantage of the fact that: (1) the pKa of the $C_3$ hydroxyl of L-AA is acidic (pKa = 4.2), meaning that the $^{13}$C-chemical shift of this carbon is pH-dependent between roughly 3.0 and 5.5; (2) the vacuolar pH (vac-pH) measured from the chemical shift of the vacuolar phosphate (vac-Pi) by *in vivo* $^{31}$P-NMR is 4.9–5.1, whereas that of the cytoplasm is close to 7.5. To unambiguously discriminate between vacuolar and cytoplasmic L-AA pools, freshly harvested young *S. alpina* leaves were utilised because they contain high amounts of L-AA (Table 8.1). Leaves were cut into small pieces and placed in an NMR tube as described by Aubert et al. (1996). The portion of spectrum centred on the position of the $C_3$ shows a predominant peak at 175.5 ppm flanked by a shoulder at 176.0 ppm (Fig. 8.2). The main peak was attributed to vacuolar L-AA, and the shoulder was tentatively attributed to the cytosolic L-AA. In order to confirm these attributions, cells were perfused

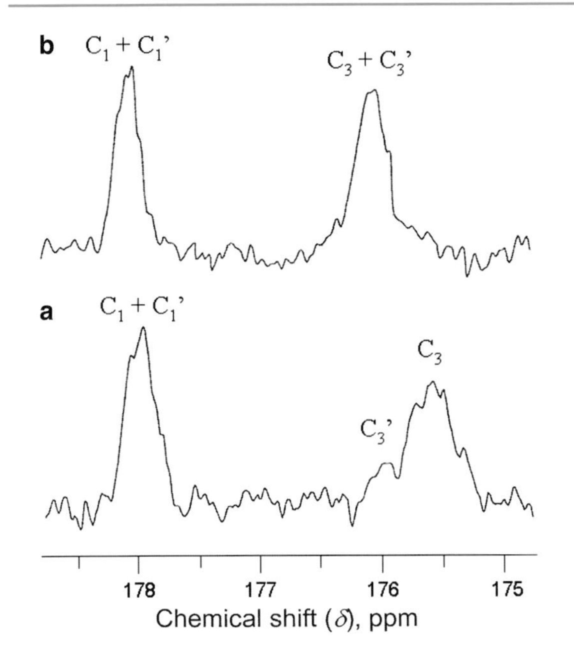


Fig. 8.2 Compartmentation of L-AA between cytoplasm and vacuole, determined in *Soldanella alpina* leaves by *in vivo* $^{13}$C-NMR. Young *S. alpina* leaves (8 g) were cut into 4 mm$^2$ pieces and perfused in the 25 mm NMR tube as described by Aubert et al. (1996). The spectra are recorded at 20°C on a Bruker AMX 400 WB spectrometer equipped with a 25 mm probe tuned at 100.6 MHz, and they are the result of 3,600 transients with a 1-s repetition time (1 h). Only the portion of spectra centred on the $C_1$ and $C_3$ carbons of L-AA is shown. Samples were successively perfused with a standard nutrient medium adjusted at pH 6.2 (**a**) and at pH 9.0 in the presence of 1 mM $NH_4^+$ (**b**). Peak assignments are as follows: $C_1$, and $C'_1$ correspond to the first carbon of L-AA molecules respectively present in the vacuole and in the cytoplasm, respectively; $C_3$, and $C'_3$ correspond to the third carbon of L-AA molecules respectively present in the vacuole and in the cytoplasm, respectively. Note that in the range of pHs considered (4.9–7.5) the first carbon of L-AA does not shift, whereas the third does

with a medium containing 1 mM $NH_4^+$ at pH 9.0, which increases the vacuolar pH from 5.0 to about 7.0 within 15 min (Aubert et al. 1999). As shown in Fig. 8.2, the alkalisation of the vacuole shifted the $C_3$ signals of L-AA from 175.5 to 176.0 ppm, overlapping the peak attributed to the cytoplasmic L-AA. The $C_1$ (at 178.0 ppm) did not significantly shift as well as the signals corresponding to the other carbons of L-AA (not shown on the graph). Referring the chemical shifts of the $C_3$ carbon (at 175.5 ppm) to a calibration curve indicates that most of the intracellular L-AA was localised in a cell compartment at pH ~ 5.0, i.e. the vacuole. In the leaves utilised in this experiment the L-AA concentration was measured as 20 μmol/wet wt, which means that its vacuolar concentration was close to this value. In some cases it may reach a concentration as high as 50 mM. The concentration of the cytoplasmic L-AA giving a signal at 176.0 ppm when leaves are incubated at pH 6.2 with a standard medium was much lower, corresponding to about 1 μmol/g FW. Assuming from microscopic observation that the cytoplasm to cell ratio is close to 0.10 in the chlorophylous parenchyma cells of *S. alpina* leaves, one can estimate that the cytoplasmic L-AA concentration is ca. 10 mM. Furthermore, if most of the cytoplamic L-AA was present in the chloroplasts (Davey et al. 2000), which account for ca. 40% of the cytoplasmic volume in this material, the concentration of L-AA in chloroplasts should be close to 25 mM. This value is of the same order of magnitude as those measured in the chloroplasts of other plant materials (Foyer 1993; Davey et al. 2000). Experiments performed throughout a summer season led to the conclusion that the concentration of L-AA is relatively constant in the cytoplasm, whereas it varies considerably in the vacuole according to the age of the leaves. This suggests that the vacuolar L-AA may compensate for cytoplasmic L-AA's need to fulfil essential functions in photosynthetic metabolism, resulting in its variable concentration based on tissue age and the alpine plant in question.

These results also raise the question of the transport of L-AA across the tonoplast, since this monovalent anion is theoretically unable to permeate membranes at physiological pH and is synthesised in the cytoplasm, with a terminal step catalysed by a mitochondrial enzyme (Mapson et al. 1954), and can accumulate massively in the vacuole. In addition the strong decrease of L-AA in old leaves indicates that it effluxes out of the vacuole to be reused by the cell metabolism in the cytoplasm, suggesting the existence of a carrier-mediated transport in the tonoplast. Indeed, high ascorbate contents decrease in *S. alpina* leaves during winter, suggesting relocation of ascorbate from the vacuole to the metabolism of the cell in order to guarantee the survival of the leaves or to protect flowers in early spring (Streb, unpublished). Finally, an improved understanding of the mechanisms of L-AA storage in the vacuole could help to selectively increase the concentration of this vitamin in crop plants and thus increase their nutritional value as well as their tolerance to oxidative stresses.

## 8.4 Methyl-β-D-Glucopyranoside (MeG) Synthesis and Accumulation in *Geum montanum* Leaves: A Possible Role in Methanol Detoxification?

Methanol (MeOH) is the second most abundant volatile organic compound massively released into the atmosphere (Guenther et al. 1995; Sharkey 1996). MeOH is produced by young expanding leaves as well as during the degradation of pectin and lignin (Lewis and Yamamoto 1990; Fall and Benson 1996). Pectin methylesterases catalyse the demethylation of pectin during insect herbivory, contributing to the elicitation of defence responses and plant-herbivore interactions (von Dahl et al. 2006; Körner et al. 2009). Different pectic polysaccharides, such as rhamnogalacturonans, contain methylated glucosides, for example 2-*O*-methyl fucose and 2-*O*-methyl xylose (reviewed by Bacic et al. 1988). *O*-methyl sugars are also known to occur in secondary metabolites, such as cardenolides containing 3-*O*-methyl-glucose, 2,3-di-*O*-methyl glucose and various methylethers of deoxy-glucose and deoxy-galactose (Connolly and Hill 1991). In these cases, the sugar polymers are formed by a glucosidic linkage involving the carbon C1 of carbohydrates, excluding the methylation of this carbon.

In the open air the methanol emitted via stomata is rapidly diluted. On the other hand, its complete miscibility with water may concentrate MeOH, for example when it is produced by the degradation of the litter of alpine plants during spring. At that moment overwintering leaves are flattened on the ground by melting snow and they are permanently in contact with water. In plants the assimilation of methanol occurs through the folate-mediated single-carbon metabolism via the formation of methyl-tetrahydropteroylpolyglutamate ($CH_3H_4PteGlu_n$) and *S*-adenosylmethionine (Gout et al. 2000), with a risk of interferences with the C-1 pathway regulation. At this level, the *de novo* synthesis of methyl-*β*-D-glucopyranoside (MeG), which does not involve the formation of $CH_3H_4PteGlu_n$ (Gout et al. 2000), could help cells to avoid this risk. The MeG is synthesized via an unspecific transglycosylation process catalysed by a hydrolase in the presence of a glycosyl acceptor such as methanol, ethanol, or glycerol: (Glc-O-X + $CH_3OH \leftrightarrow$ Glc-O-$CH_3$ + X). Surprisingly, MeG was identified for the first time in the early 1980s in white clover foliage (Smith and Phillips 1981), but its intracellular compartmentation and cellular function were investigated much later (Gout et al. 2000). Interestingly, it should be mentioned that at the beginning of the last century, an abundant glycoside was detected by Bonnier and Douin (1911–1935) in ethanolic extracts of different plant species belonging to the genus *Geum*. This glycoside was probably MeG, as suggested by the evidence described below.

### 8.4.1 Accumulation of MeG in the Foliage of Lowland and Alpine Plants

Like ascorbate, MeG has been measured in different plants species and plant organs from the $^{13}$C-NMR spectra of PCA extracts. As shown for *Geum montanum* leaves collected at the Col du Lautaret (Fig. 8.3), among the carbon resonances of the major solutes accumulated in this plant, such as sucrose, glucose, fructose, and malate, seven resonance peaks centred at 58.00, 61.58, 70.35, 73.99, 76.69, 76.71, and 104.00 ppm correspond to the seven carbons of MeG (Gout et al. 2000; Aubert et al. 2004). The peaks centred at 58.00 and 104.00 ppm (Fig. 8.3, insets) correspond to the added methyl group and to the $C_1$ of glucose, respectively.

The accumulation of MeG in different alpine and lowland plant leaves and the synthesis of $^{13}$C-MeG from [$^{13}$C] methanol are shown in Table 8.2. It appears that various members of the subfamily *Rosoideae* of *Rosaceae* accumulate more than 1 μmol MeG/g wet wt. A substantial accumulation is observed in nine out of the ten genus tested, but not in five species of the genus *Alchemilla*. Among plants belonging to a given genus, MeG accumulates much more in some species than in others. For example, the leaves of only seven species of *Geum*, among ten tested species, accumulate MeG. The species of other subfamilies (*Spiraeoideae*, *Maloideae*, and *Prunoideae*) do not accumulate MeG above 1 μmol/g wet wt. Similarly, MeG is not detected in the leaf extracts of 250 other plants belonging to the main families of French flora. Only two exceptions were observed:

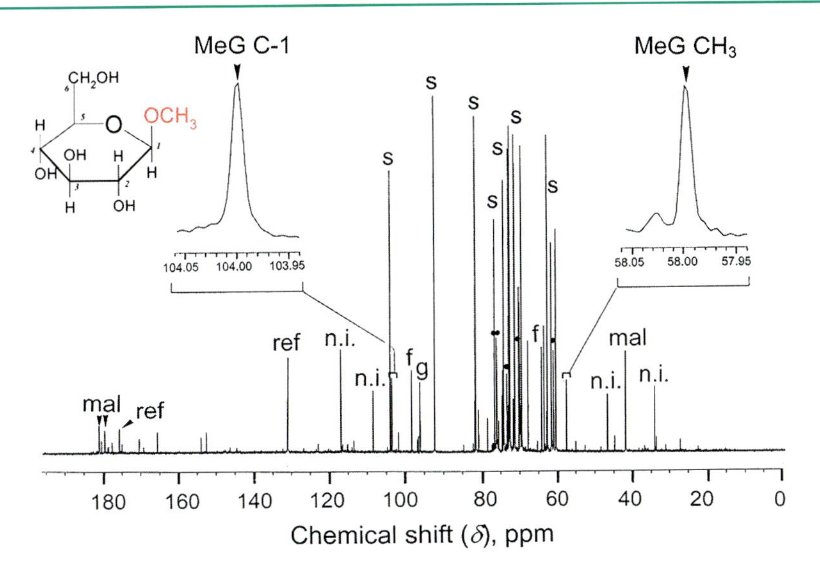


**Fig. 8.3** *In vitro* proton-decoupled $^{13}$C-NMR spectrum of a *Geum montanum* leaf PCA extract adjusted at pH 7.5. The PCA extract was prepared as described by Aubert et al. (1996) from 10 g of adult (1 month old) *G. montanum* leaves harvested during the first week of July at the Lautaret pass (2,100 m asl) at 10 am. Insets: expanded scales showing the resonance peaks of the methyl group (58.00 ppm) and $C_1$ (104.00 ppm) of MeG. The positions of the five other MeG resonance peaks are indicated by solid circles. This spectrum was recorded as mentioned in the legend of Fig. 8.1. Peak assignments are as follows: MeG, methyl-*β*-D-glucopyranoside; s, sucrose; g, glucose; f, fructose; mal, malate; n.i., not identified; ref, reference (100 µmol of maleate)

*Oxytropis campestris* and *Pisum sativum* (*Fabaceae*). Interestingly MeG is abundant in white clover, also belonging to the *Fabaceae* family, where it was discovered (Smith and Phillips 1981). Table 8.2 also indicates that all the dicots tested, as well as non-photosynthetic suspension-cultured cells, show an activity of methanol incorporation to glucose detected in the presence of [$^{13}$C] methanol. As expected, the *Rosaceae* in which MeG is detected in the absence of added methanol synthesise MeG at a much higher rate (up to 6.5–7.5 µmol/d/g wet wt) compared with the non-accumulating species (usually less than 1 µmol/d/g wet wt in other tested *Rosaceae* and dicots), suggesting that the availability of methanol is the main limiting factor for its synthesis in plants naturally accumulating this glycoside. However, the rate of MeG accumulation in many dicots and monocots is not measurable by $^{13}$C-NMR, or is very low, and the gymnosperms, pteridophyts, and mosses tested do not accumulate MeG or incorporate $^{13}$C-methanol in MeG at all, indicating that the enzyme catalysing the synthesis of this glycoside may vary significantly according to the plant under analysis.

### 8.4.2 Cellular and Subcellular Localization of MeG in *G. montanum*

Although MeG is detected in *G. montanum* leaves, rhizome and roots, it was abundant only in leaves (Fig. 8.2). The concentration of MeG increases in *G. montanum* leaves in the course of ageing. It may account for up to 20% of soluble carbohydrates before senescence in overwintering leaves. However, in contrast to non-methylated soluble carbohydrates MeG is not reallocated to roots and rhizomes, since these storage organs contain only low levels of MeG throughout the vegetation season (<3 µmol/g wet wt in rhizomes and barely detectable in roots). By contrast, sucrose, glucose, and fructose decrease in old leaves, but increase in rhizomes and roots, indicating that they are reallocated before winter in subterranean organs (Fig. 8.4). Very high concentrations of MeG are also measured in the overwinter leaves of *Dryas octopetala*, whereas the concentration of this glycoside remains very low in the subterranean organs of the plant. These results suggest that MeG is not further metabolised after its synthesis in the cytoplasm. In agreement, and in contrast to the sugars consumed to sustain respiration, it does not decline in leaves kept in

**Table 8.2** Methyl-β-D-glucopyranoside content of selected alpine and lowland plant leaves

| | | | | μmol/g wet wt | % soluble sugars (mol/mol) | $^{13}$C-methanol incorporation μmol/d/g wet wt |
|---|---|---|---|---|---|---|
| Dicots | | | | | | |
| | *Rosaceae* | | | | | |
| | | *Rosoideae* | *Acaena microphylla* | 3.4–4.0 | 3.5–4.0 | 2.4–2.8 |
| | | | *Alchemilla pentaphyllea L.* | nd | – | 0.10–0.12 |
| | | | *Dryas octopetala L.* | 35–40 | 28–32 | |
| | | | *Fragaria vesca* | 1.3–2.2 | 2.7–3.3 | |
| | | | *Geum coccineum* | 14–17 | 5.6–6.0 | |
| | | | *Geum heterocarpum Boiss.* | 3.5–4.0 | 3.4–3.8 | |
| | | | *Geum montanum L.* | 16–20 | 12–14 | 5.5–6.5 |
| | | | *Geum pyrenaicum* | 30–35 | 15–18 | |
| | | | *Geum reptans L.* | 25–29 | 12–15 | |
| | | | *Geum rivale L.* | 4.0–4.6 | 16–20 | |
| | | | *Geum trifolia* | 11–14 | 10–12 | |
| | | | *Potentilla grandiflora L.* | 15–20 | 11–13 | 6.5–7.5 |
| | | | *Rosa canina* | 3.5–4.0 | 7.8–9.0 | |
| | | | *Rubus fruticosus* | 6.9–8.0 | 6.5–7.4 | |
| | | | *Sanguisorba minor* | 15–18 | 10–12 | |
| | | | *Sibbaldia procumbens L.* | 18–22 | 11–14 | |
| | | | *Waldensteinia fragarioides* | 16–21 | 36–40 | |
| | | | *Woronowia elegans* | 16–20 | 17–20 | |
| | | *Prunoideae* | *Prunus sp.* | nd | – | 0.04–0.08 |
| | *Aceraceae* | | *Acer pseudoplatanus*[a] | nd | – | 1.2–1.4 |
| | *Apiaceae* | | *Daucus carota*[a] | nd | – | 0.05 |
| | *Caryophyllaceae* | | *Silene alba*[a] | nd | – | 0.04 |
| | *Chenopodiaceae* | | *Spinacia oleracea* | nd | – | 0.26–0.32 |
| | *Fabaceae* | | *Oxytropis campestris* | 9.5–12 | 34–40 | |
| | | | *Pisum sativum* | 2.2–3.0 | – | 0.56–0.68 |
| | *Ranunculaceae* | | *Ranunculus acris* | nd | – | 0.16–0.19 |
| | *Solanaceae* | | *Solanum tuberosum* | nd | – | 0.76–0.84 |
| Monocots | | | | | | |
| | *Amaryllidaceae* | | *Narcissus pseudonarcissus* | nd | – | 0.40–0.50 |

(continued)

**Table 8.2** (continued)

| | | | μmol/g wet wt | % soluble sugars (mol/mol) | $^{13}$C-methanol incorporation μmol/d/g wet wt |
|---|---|---|---|---|---|
| | *Araceae* | *Arum maculatum* | nd | – | nd |
| | *Iridaceae* | *Iris sp* | nd | – | 0.005–0.01 |
| | *Liliaceae* | *Tulipa sp* | nd | – | 0.01–0.02 |
| | *Orchidaceae* | *Himantoglosum hircinum* | nd | – | 0.003–0.005 |
| | *Poaceae* | *Zea mays* | nd | – | nd |
| | | *Zea mays*[a] | nd | – | nd |
| Gymnosperms | | *Cedrus sp* | nd | | nd |
| Pteridophytes | | *Athyrium* | nd | – | nd |
| Bryophytes | | *Bryum sp* | nd | – | nd |

The high altitude and lowland plants selected in this table were harvested as indicated in Table 8.1 and were principally chosen among those naturally containing $^{13}$C-NMR-detectable MeG, and/or those in which MeG was detected when leaves were incubated in the presence of [$^{13}$C]methanol. The names of 250 other plants distributed in the main families of the French flora, in which MeG was not detected, are not mentioned. In particular, MeG was not detected in ten *Rosaceae* from the subfamilies *Spiraeoideae* and *Maloideae*. The absence of detection (nd) has the same meaning as indicated in the legend of Table 8.1
[a]Suspension-cultured cells

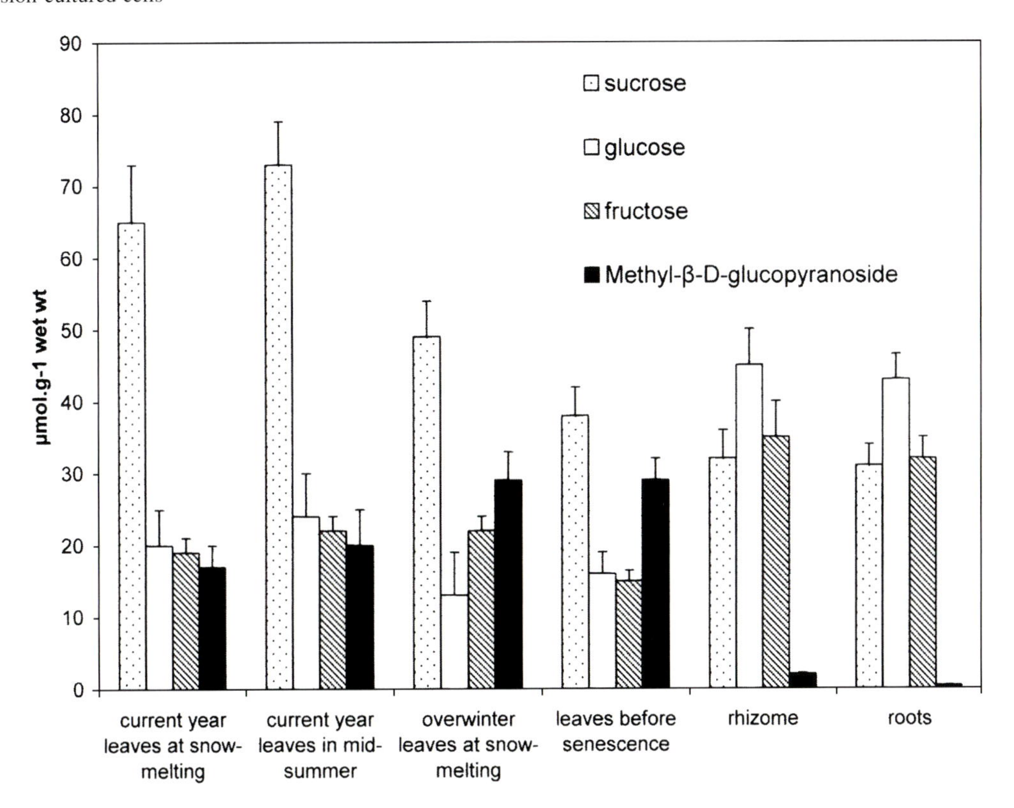


**Fig. 8.4** Seasonal variation and organ-specific distribution of MeG, sucrose, fructose, and glucose in *Geum montanum* leaves. Quantification was done from metabolite profiles obtained by $^{13}$C-NMR analysis. Values are given as means ± SE (n = 3)

the dark during several days or in culture cells transferred to a sugar-free nutrient medium after a period of MeG synthesis in the presence of glucose and MeOH (Aubert et al. 2004).

In contrast to L-AA, which can be localised in plant cells using *in vivo* $^{13}$C-NMR, this technique cannot be utilised here because the chemical shift of MeG does not depend on the pH as it does for L-AA. Consequently, to determine the subcellular location of MeG between the cytoplasm and vacuole, it is necessary to analyse purified vacuoles. However, since it is difficult to purify intact organelles from *G. montanum*, the preparation of vacuoles was done from culture cells having accumulated MeG (Aubert et al. 2004). Comparisons with organic acids, such as malate or citrate known to be present almost exclusively in the vacuole (Gout et al. 1993), led to the conclusion that MeG is stored in this cell compartment. This pointed to the question of localisation of MeG synthesis in plant cells: in the vacuole where it is stored, in the apoplasm, or in the cytoplasm. Experiments done with [1-$^{13}$C] glucose and [$^{13}$C]MeOH indicate that MeG is synthesised in the cytoplasm (Aubert et al. 2004) and not in the vacuole where it accumulates, nor in the cell wall, where methanol can be produced according to pectin de-methylation processes (Nemecek-Marshall et al. 1995; Körner et al. 2009). Furthermore, the absence of MeG synthesis in purified cell mitochondria or plastids incubated with labelled methanol and glucose (Aubert et al. 2004) suggests that MeG is synthesised in the cytosol, prior to transfer to the vacuole.

### 8.4.3 Physiological Significance of MeG Accumulation in Plants

The reason why some plant species belonging to *Rosaceae* and sometimes to *Fabaceae* accumulate MeG remains unclear. Of course, like many other methylated molecules (quaternary ammonium, tertiary sulfonium, methyl-inositols, etc.), MeG could be involved in osmotic stress tolerance. In fact, MeG promotes the opening of cut rose flowers via the lowering of the apoplastic and the increase of the symplastic osmotic water potential in the petals (Ichimura et al. 1999). In addition, *G. montanum* is highly resistant to water deficit (Manuel et al. 1999). In this plant, however, MeG already accumulates in young leaves when the snow melts in the absence of water or salt-induced osmotic stress. MeG could also be involved in hydroxyl radical scavenging as reported by Popp and Smirnoff (1995) for methyl-inositols. In this respect, MeG accumulates principally in those alpine *Rosaceae*, which are directly exposed to high light irradiation, resulting in the production of more ROS (Asada 1994; Streb et al. 1998). However, with MeG being stored in the vacuole, it is likely that it would not permit the protection of the cytoplasmic cell components and structures. Moreover, although abundant in some species, MeG cannot be considered as a storage compound for carbon metabolism because it is not reallocated in growing tissues during the vegetative season in contrast to sucrose or glucose, and it is not further metabolised to prevent cells from autophagy upon carbon starvation. Finally, MeG could also be involved in freezing tolerance, as are other soluble sugars (for a review of cell physiology of plants growing in cold environments, see Lütz 2010). Interestingly, *Geum montanum* leaves survive freezing events down to −20°C, in contrast to other alpine species (Streb, unpublished).

An alternative hypothesis is that the synthesis of MeG in the cytosol may help limit the accumulation of a potentially toxic concentration of methanol in the cytoplasm of cells exposed to this alcohol. Indeed, methanol is a solvent of membrane lipids, thus representing a danger to the integrity of cell structures. However, no significant effects of 5 mM MeOH on the membrane integrity of culture cells were observed (Gout et al. 2000). By contrast, its oxidation produces formaldehyde, a very reactive component considered to be highly toxic to cell metabolism. The metabolisation of methanol via the folate-mediated single-carbon metabolism (Gout et al. 2000) could also interfere with the regulation of this pathway as outlined above. Further work is needed to investigate the existence of a correlation between the accumulation of MeG in the leaves of the alpine plants accumulating this glycoside and the amount of methanol produced in the leaves or incorporated into leaf tissues as a product of pectin or lignin demethylation.

## 8.5 The Role of Polyols and of the Gluconate 6-Phosphate Accumulated During Desiccation and Reviviscence of the Lichen *Xanthoria elegans* (Link) Thalli

Lichens are a symbiotic association between a photosynthetic autotrophic green alga and/or cyanobacterium and a heterotrophic fungus. They are among the most resistant of living organisms, growing in extreme environments including the deserts and frigid areas of all five continents. In the alpine environment, lichens are particularly exposed to harsh fluctuations of water supply, light intensity, and temperature (Kappen 1988; Körner 2003).

The metabolic activity of lichen thalli indicated by gas exchanges may remain significant at temperatures as low as −20°C, but it becomes undetectable when their water content decreases below 15% (Lange 1980; Schroeter et al. 1991), which may happen at high altitude on sun-exposed rocks. However, in many cases lichens take up sufficient moisture from atmospheric water vapour to stay above this low hydration threshold, enabling significant metabolic activity (Lange 1980). Indeed, hydration is facilitated by high concentrations of polyols in both photo- and mycobiont (Rundel 1988) that lower water potential (Lange et al. 1990). Immersion in water or rewetting in moist air triggers the reviviscence of dry lichens (Smith and Molesworth 1973). Respiration and photosynthesis promptly resume normal activities upon hydration (Bewley 1979), indicating that cell damage induced by drying is rapidly repaired (Farrar and Smith 1976). However, desiccated membranes may be leaky, exposing the cell to losses of organic and inorganic solutes during rewetting (Dudley and Lechowicz 1987; Longton 1988). Consequently, cell survival requires a rapid resealing of membrane disruptions (McNeil and Steinhardt 1997; McNeil et al. 2003), and a prompt restarting of metabolic activity is expected.

When photosynthetic water oxidation is stopped by dehydration, intense solar radiation is another environmental parameter that threatens lichens due to the increased production of potentially damaging ROS. In addition, following re-hydration, a burst of intracellular production of ROS was also reported (Weissman et al. 2005). In the plant kingdom, the ability to withstand desiccation and light stress involves various mechanisms reviewed by Hoekstra et al. (2001), Rascio and La Rocca (2005), and Heber et al. (2005). One specific feature of dry poikilohydric organisms is that they do not emit light-induced fluorescence, indicating a loss of the PSII function (Lange et al. 1989). These organisms minimise damage by activating exciton transfer to quenchers and converting excess energy to heat (Bilger et al. 1989; Heber et al. 2000). In *Ramalina lacera* the superoxide dismutase, catalase, glutathione reductase, and glucose 6-P dehydrogenase activities are enhanced both in photo- and mycobiont (Weissman et al. 2005).

In desiccation-tolerant higher plants, the repair of cell damage upon re-hydration, and the subsequent restoration of cell function, involves various complex inductive mechanisms including gene transcriptions and an increased protein turnover (Oliver et al. 2004), therefore involving a delay ranging from a few hours to several days (Gaff 1997). By contrast, we observed that the respiration and the photosynthetic activity of different lichens harvested in high mountain areas, including *Cladonia rangiferina* and *portentosa*, *Cetraria islandica* and *nivalis*, *Peltigera aphthosa*, *Evernia prunastri*, *Collema tenax*, and *Xanthoria elegans*, restarts almost immediately after rewetting. The swiftness of this recovery suggested that it relies on the preservation/accumulation of key pools of metabolites during dehydration. It was hypothesised that metabolic intermediates, such as respiratory substrates, nucleosides, and pyridine nucleotides do not significantly leak from the photo- and mycobiont during dehydration/hydration cycles in lichens. To confirm this hypothesis the respiration and photosynthetic activities and the metabolite profiles of *X. elegans* thalli were assessed in parallel during such cycles.

### 8.5.1 Metabolite Profiling of *X. elegans* Thalli During Dehydration/Hydration Cycles

*X. elegans* is a saxicolous lichen of the *Teloschistaceae* family. It contains an ascomycetous fungus (*Teloschistes* sp) and a unicellular green alga (*Trebouxia* sp) (Helms 2003). The thallus of this saxicolous lichen grows on rocky surfaces where it undergoes frequent hydration/dehydration cycles depending on atmospheric conditions. When exposed to the summer sun, *X. elegans* thalli can lose most of its constitutive water in a dry atmosphere within approximately 30 min. They become wet again as soon as they receive water, like many other lichens (Smith and Molesworth 1973; Farrar and Smith 1976; Larson 1981). The thalli utilised for the assays described below were collected on limestone rocks, at 2,800 m, above the Galibier pass (Hautes-Alpes), an area characterised by a relatively continental and dry climate.

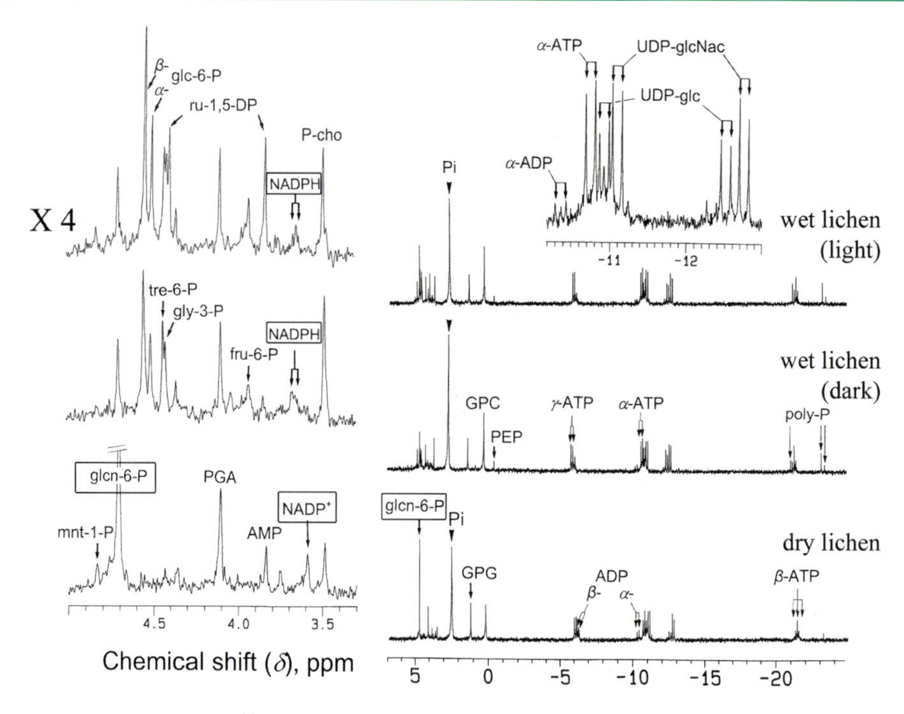


**Fig. 8.5** *In vitro* proton-decoupled $^{31}$P-NMR spectra of *Xanthoria elegans* thalli PCA extracts adjusted at pH 7.5. Extracts are prepared from 4 g of thalli (on a dry weight basis) as described by Aubert et al. (1996). Thalli were collected dry in the light (**a**), wet in the dark (**b**), or wet in the light (**c**). The spectra recorded at 20°C on a Bruker AMX 400 WB spectrometer tuned at 161.93 MHz are the result of 4,096 scans with 3.6-s repetition intervals (4 h). Peak assignments are as follows: mnt-1-P, mannitol 1-P; glcn-6-P, gluconate 6-P; glc-6-P, glucose 6-P; tre-6-P trehalose 6-P; gly-3-P, glycerol 3-P; PGA, phosphoglycerate; fru-6-P, fructose 6-P; ru-1,5-DP, ribulose 1,5-diphosphate; AMP, adenosine monophosphate; P-cho, P-choline; GPG, glycerophosphoglycerol; GPC, glycerophosphocholine; PEP, phosphoenolpyruvate; UDP-glc, uridine 5′-diphosphate-α-D-glucose; UDP-glcNAc, uridine 5′-diphospho-N-acetylglucosamine; poly-P, polyphosphates

The metabolic profile of dry *X. elegans* thalli (Fig. 8.5 and Table 8.3) indicates that the two most abundant P-compounds measured by $^{31}$P-NMR are inorganic phosphate and gluconate 6-P, an intermediate of the pentose phosphate pathway (PP-pathway). Other identified P-compounds are mannitol 1-P, glycerate 3-P, AMP, NADP$^+$, P-choline, the two phosphodiesters glycerylphosphoryl-glycerol and -choline (GPG and GPC), nucleosides (mainly ATP and ADP), and nucleoside diphosphate sugars, UDP-glc and UDP-glcNAc. Polyphosphates are also detected in PCA extracts. However, since they largely precipitate during PCA extraction, their concentration is calculated from *in vivo* NMR spectra (Aubert et al. 2007).

By contrast, wet thalli contain significant pools of various sugar phosphates including glucose 6-P, trehalose 6-P, fructose 6-P, glyceraldehyde 3-P and phosphoenolpyruvate. Conversely, the concentration of gluconate 6-P is low. In addition, wet thalli contain NADPH (NADP$^+$ was not detected), P-Cho, GPC, and ATP (but less ADP and AMP), and their UDP-glc pool is threefold higher. PGA, GPG, and UDP-glcNAc pools are similar in both dry and wet thalli. The metabolite profiles of wet thalli harvested in the light resemble those of wet thalli harvested in the dark, except for the presence of an important double peak corresponding to the BBC cycle intermediate ribulose 1,5-diphosphate located in the chloroplasts of the algal partner. $^{13}$C-NMR spectra analyses show fewer differences in relation to the water status of thalli. Typically, dry and wet thalli spectra exhibit major resonance peaks corresponding to polyols, namely arabitol, mannitol, and ribitol (Aubert et al. 2007). The polyol concentrations of wet lichens do not differ significantly with light or dark conditions during the harvest.

**Table 8.3** Metabolic profiles of dry and wet *X. elegans* thalli

| Metabolite | Dry lichens (light) | Wet lichens (dark) | Wet lichen (light) |
|---|---|---|---|
| sucrose | 27 ± 3 | 29 ± 3 | 27 ± 3 |
| trehalose | 9 ± 2 | 5 ± 2 | 8 ± 2 |
| ribitol | 110 ± 10 | 90 ± 9 | 105 ± 10 |
| arabitol | 360 ± 30 | 240 ± 20 | 340 ± 30 |
| mannitol | 240 ± 20 | 290 ± 30 | 250 ± 20 |
| glutamate | 45 ± 5 | 40 ± 4 | 43 ± 5 |
| glutamine | 120 ± 12 | 71 ± 7 | 125 ± 12 |
| choline | 85 ± 9 | 80 ± 8 | 88 ± 9 |
| betaine | 170 ± 15 | 150 ± 14 | 160 ± 15 |
| polyphosphates | 6.5 ± 1 | 6.4 ± 1 | 6.2 ± 1 |
| Pi | 2.5 ± 0.4 | 2.7 ± 0.4 | 3.6 ± 0.5 |
| mannitol 1-P | 0.06 ± 0.01 | 0.02 ± 0.01 | 0.50 ± 0.01 |
| gluconate 6-P | 0.91 ± 0.09 | 0. 18 ± 0.02 | 0.20 ± 0.02 |
| glucose 6-P | nd | 0.52 ± 0.05 | 0.81 ± 0.08 |
| trehalose 6-P | nd | 0.22 ± 0.02 | 0.22 ± 0.02 |
| glycerol 3-P | nd | 0.15 ± 0.02 | 0.17 ± 0.02 |
| ru-1,5-DP | nd | nd | 0.27 ± 0.03 |
| PGA | 0.27 ± 0.03 | 0.25 ± 0.03 | 0.27 ± 0.03 |
| fructose 6-P | nd | 0.80 ± 0.01 | 0.13 ± 0.02 |
| AMP | 0.12 ± 0.02 | 0.04 ± 0.01 | nd |
| NADPH | 0.02 ± 0.01 | 0.08 ± 0.01 | 0.08 ± 0.01 |
| $NADP^+$ | 0.05 ± 0.01 | nd | nd |
| P-choline | 0.15 ± 0.02 | 0.40 ± 0.04 | 0.33 ± 0.03 |
| GPG | 0.30 ± 0.03 | 0.25 ± 0.02 | 0.25 ± 0.02 |
| GPC | 0.28 ± 0.03 | 0.47 ± 0.05 | 0.47 ± 0.05 |
| PEP | nd | 0.07 ± 0.01 | 0.06 ± 0.01 |
| ATP | 0.38 ± 0.04 | 0.48 ± 0.05 | 0.53 ± 0.05 |
| ADP | 0.14 ± 0.02 | 0.08 ± 0.01 | 0.05 ± 0.01 |
| UDP-glc | 0.13 ± 0.02 | 0.33 ± 0.04 | 0.33 ± 0.04 |
| UDP-glcNAc | 0.39 ± 0.04 | 0.43 ± 0.04 | 0.46 ± 0.05 |

Dry thalli are collected in the light and wet thalli in the dark (at dawn) or in the light. Metabolites were identified and quantified from PCA extracts, using maleate and methylphosphonate as internal standards for $^{13}C$- and $^{31}P$-NMR analyses, respectively. Abbreviations are as indicated in the legend of Fig. 8.5; nd, not detected. Values given as μmol/g lichen dry wt were obtained from a series of independent experiments and are means ± SD ($n = 5$)

### 8.5.2 Time Course Changes of the P-Compound Profile of *X. elegans* Thalli During Desiccation and Following Hydration

When hydrated lichens were allowed to dry under natural conditions, their metabolic profile reverted to that of initially dry thalli within ca. 30 min. In particular, gluconate 6-P starts to accumulate and glc-6-P and ru-1,5-DP to decrease as soon as the water content of thalli drops below 30–35%. At the same time, NADPH decreases and $NADP^+$ accumulates symmetrically, suggesting either that the cell's requirement for redox power during dehydration was not compensated or/and that production of NADPH was blocked. A decrease of ATP (by nearly 30%), with the corresponding increase of ADP and AMP, is also observed. The accumulation of gluconate 6-P in *X. elegans* during dehydration, when the relative water content of thalli decreases below 30%, could originate from an increase of glucose 6-phosphate dehydrogenase activity, in relation to the production of ROS, as observed in other lichens (Weissman et al. 2005; Kranner and Grill 1994). Indeed, reducing power (NADPH) is required to limit the potentially damaging effect of the ROS burst related to the impaired electron transport chains in

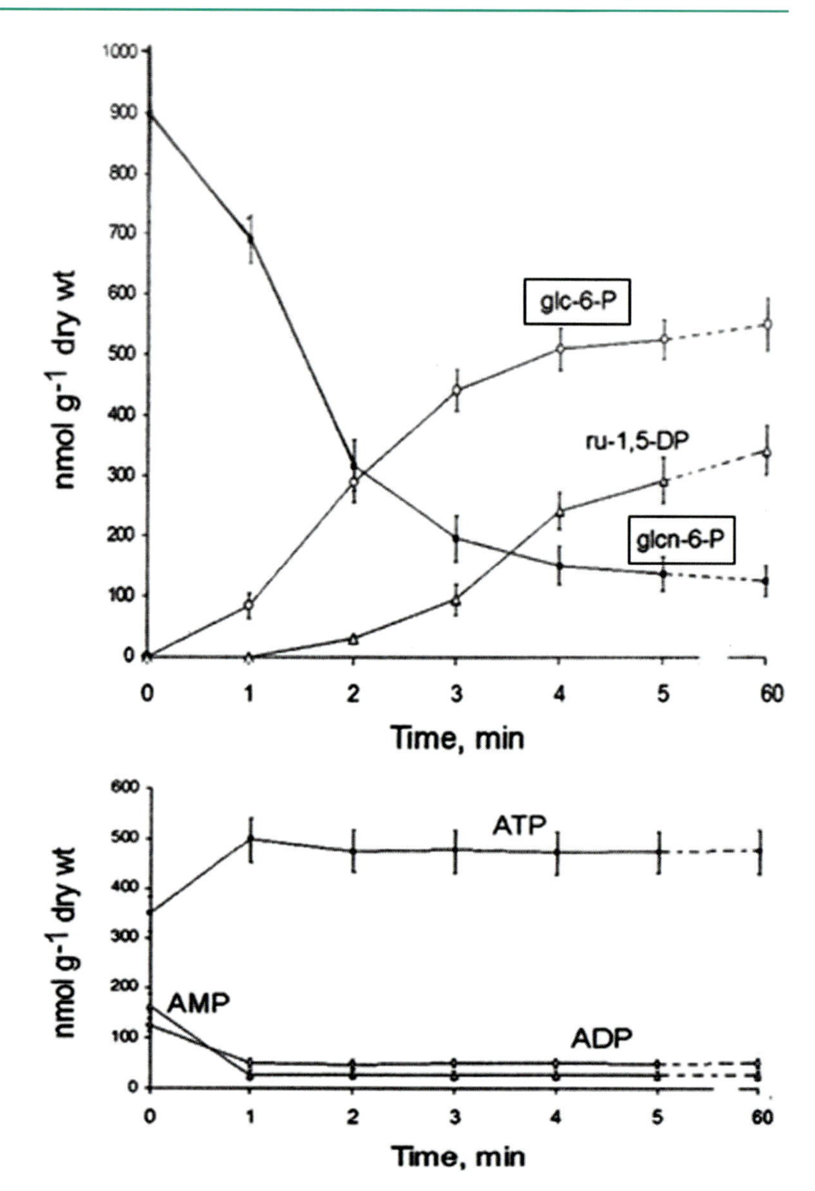


**Fig. 8.6** Time course evolution of gluconate 6-P, glucose 6-P, ribulose 1,5-diphosphate, ATP, ADP, and AMP in *Xanthoria elegans* thalli following rehydration in the light. At time zero, thalli fragments were incubated in a well aerated liquid medium containing 0.2 mM Mops buffer (pH 6.2), at 20°C. Metabolites were quantified from PCA extracts analysed by $^{31}$P-NMR. Values are means ± SD ($n = 5$)

water-stressed cells (Rascio and La Rocca 2005). For example, the maintenance of reduced glutathione for its participation in the ascorbate-glutathione cycle in chloroplasts, via the ascorbate-glutathione cycle (Asada 1994; Foyer et al. 1994), requires NADPH. Similarly, gluconate 6-P accumulates in the alpine plant *S. alpina* and in *Pisum sativum* during cold-induced photoinhibition (Streb et al. 2003).

Following re-hydration, the ATP pool is the first pool of nucleotides to fully recover, reaching a plateau within ca. 1 min (Fig. 8.6), whereas AMP and ADP decrease symmetrically. The pool of glucose 6-P (and also tre-6-P, gly-3-P, and fru-6-P), which was not detected in dry lichen, recovers the value measured in wet lichens during the next 2 min. Interestingly, ru-1,5-DP starts to accumulate after a 2–3 min delay, corresponding to the time taken for photosynthesis to recover, and reaches the concentration measured in wet thalli (Fig. 8.5 and Table 8.3) ca. 5 min later. Conversely, gluconate 6-P decreases to ca. one-fifth of its initial value during the first 3 min before stabilising. $NADP^+$ is reduced to NADPH from the first minutes following rewetting.

Finally, the pools of intermediates involved in membrane lipid synthesis including P-choline and GPC for lecithin and GPG for the phosphatidylglycerol mainly present in plastidial membranes significantly increase during the first minutes following rewetting. This indicates that after a dry period, some membrane repair is required for the physiological activity of lichen to restart.

### 8.5.3 The Role of Polyols and Gluconate 6-P in the Reviviscence of *X. elegans* Thalli

Polyols are potent quenchers of ROS (Jennings et al. 1998) and, as such, they may help protect lichen thalli. In particular, we have observed that mannitol protects ATP from oxidation during desiccation under intense light. More generally, polyols behave as osmoprotectants. In fact, in a similar way to sugars, glutamate, and glycine-betaine, they stabilize proteins and protect intimate cellular structures against the potentially deleterious effect of dehydration (Hoekstra et al. 2001). As well as many other lichens (Vicente and Legaz 1988; Honegger 1991), *X. elegans* contains high amounts of polyols in both photo – (ribitol) and mycobiont (mannitol and arabitol). Although their proportion may change during a dehydration/rehydration cycle, in accordance with previous results obtained by Farrar (1988), their total amount does not fluctuate so much in this lichen. Finally, polyols may constitute alternative metabolic reserves. For example, under longer dark periods, we observed that arabitol is completely metabolised in wet lichens, suggesting that it contributes to sustaining fungal respiration, whereas mannitol remains nearly constant. This last observation tends to prove that mannitol is necessary for the protection of cell structures and metabolites since this polyol is normally metabolised by fungi (Dulermo et al. 2010).

The presence of high concentrations of polyols is nevertheless not sufficient to completely protect the cell structures from ROS-induced peroxidation and de-esterification of glycerolipids that permeate membranes, and from mechanical constraints leading to membrane disruptions during dehydration/rehydration cycles. Hence, the capacity to rapidly reseal disrupted membranes plays a central role in cell survival (McNeil and Steinhardt 1997; McNeil et al. 2003). In this context we observed a threefold increase in P-choline and a doubling of GPC during rehydration (Table 8.3). The increase of these two precursors of phosphatidylcholine synthesis (van der Rest et al. 2002) may be characteristic of the synthesis of phosphatidylcholine-rich membrane systems, like plasma membrane or tonoplast, and thus fulfil a role in the maintenance of cell structural integrity. On the contrary, the stability of GPG suggests that thylakoid membranes, which contain most of chlorophyllous cell's phosphatidylglycerol (Joyard et al. 1993) remains intact in the chloroplasts of photobiont during dehydration/rehydration cycles, thus permitting the rapid restarting of photosynthetic activity.

The cell repair needed following lichen rewetting requires rapid energy production. For this, mitochondria must remain intact in dry lichens, which seems to be the case, since no intermediates of cardiolipin metabolism are detected in PCA extracts (cardiolipin is specific to mitochondria, Bligny and Douce 1980), and they should be supplied with respiratory substrates immediately after rehydration. In the absence of glycolytic intermediates, except PGA likely to be located in the chloroplasts of algae where it permits photosynthesis to restart, the gluconate 6-P accumulated during desiccation was a good candidate. With the aim of testing this hypothesis we inhibited the accumulation of gluconate 6-P in thalli during drying, and we measured the delay preceding a full recovery of respiration and photosynthesis during rehydration. For this, thalli are first incubated for 1 h in the dark in the presence of 5 mM glucosamine (N-glc). N-glc is taken up by cells and phosphorylated to *N*-glucosamine 6-P which is a potent inhibitor of glc-6-P dehydrogenase, blocking the conversion of glucose 6-P into gluconate 6-P (Glaser and Brown 1955). As previously shown in tobacco cells (Pugin et al. 1997), gluconate 6-P and NADPH decreased substantially in the presence of N-glc, while $NADP^+$ increased and glucose 6-P remained constant (Table 8.4). When these thalli are subsequently dehydrated, NADPH is no longer detected, like in the control dry lichen (Table 8.3). However, in contrast to control lichen, gluconate 6-P does not increase (Table 8.4). Under these conditions, lichens treated with glucosamine show a marked delay (1–2 min) after rehydration before respiration recovers to full activity, while no delay is observed in the control lichens. Photosynthesis is similarly delayed. This result shows that the accumulation of gluconate 6-P boosted by the cell need for redox power during

**Table 8.4** Modification of gluconate 6-P, NADPH, and $NADP^+$ induced in *X. elegans* thalli by glucosamine (N-glc) treatment

| Metabolite | Wet thalli | Dry thalli |
|---|---|---|
| *N*-glucosamine 6-P | 0.38 ± 0.04 | 0.37 ± 0.04 |
| glucose 6-P | 0.62 ± 0.06 | nd |
| gluconate 6-P | 0.04 ± 0.01 | 0.03 ± 0.01 |
| NADPH | 0.02 ± 0.01 | nd |
| $NADP^+$ | 0.07 ± 0.01 | 0.08 ± 0.01 |

Lichens were incubated for 1 h in the dark in the presence of 5 mM N-glc (wet thalli), and a fraction of them was subsequently left to dry under natural conditions (dry thalli), prior to PCA extraction. Metabolites were identified and quantified as indicated in Table 8.1. Abbreviations are as indicated in the legends of Fig. 8.3; nd, not detected (<0.05 μmol). Values given as μmol/g lichen dry wt were obtained from a series of independent experiments and are means ± SD ($n = 5$)

dehydration prompted the recovery of cell metabolism after rewetting.

In conclusion, the very rapid recovery of *X. elegans* respiration and photosynthetic activities following rehydration was facilitated by the accumulation of stores of gluconate 6-P during dehydration and by coordinated events associated with preventing oxidative damages and protecting cell components and structures. Gluconate 6-P appeared to accumulate in response to different factors including a need for reducing power delivered by the PP-pathway to limit desiccation-generated ROS and for cell repairs, and the blockade of gluconate 6-P metabolisation by dehydration. The sizeable pools of polyols present in both phyco- and mycobionts serve to protect cells constituents such as nucleotides and to preserve the integrity of intracellular structures. In lichen thalli, like in other poikilohydric organisms such as seeds, progressive dehydration modifies and finally stops metabolic activities. However, in contrast to seeds where dormancy is advantageous to avoid undesirable germination during transiently-favourable conditions (Bewley 1997), the ability of lichen thalli to restart respiration and photosynthesis without delay allows them to take advantage of all reviviscence opportunities offered by the presence of both water and light. This is the case, for example, when winter sun melts the ice to temporarily reveal rocky slopes. In such situations, high net photosynthetic activities observed at low temperatures will enable synthesis of carbohydrates within the minutes following rehydration. Finally, these observations highlight gluconate 6-P accumulation during dehydration as a metabolic adaptation of lichens to the anhydrobiotic cycles imposed by the high-mountain climate.

## Conclusions

Plant cell components can be viewed as the end product of gene expression and environmental constraints. Indeed, plants are able to adjust their metabolism to external perturbations, resulting in altered growth partly because the concentration of intracellular metabolites has changed. In alpine plants, the measurements of most intracellular metabolites can be expected to provide a broad view of the biochemical status of organisms exposed to cold temperatures, high solar radiation, wind and mechanical stresses, unstable soils and nutrient shortage, etc. Under these conditions, the possibility that specific metabolic profiles can be associated with alpine environment was addressed. In other words, does metabolic profiling reveal specific adaptations of alpine plants to their environment? In fact, there are two issues involved. The first is of a semantic nature: does the accumulation of given metabolites modify plant physiology in the sense that it may constitute an adapted response to a given stress? The concomitant accumulation of L-AA in the vacuole of *S. alpina* at high altitude and the low cytoplasmic L-AA concentration shed doubt on the possibility of a close relation between the L-AA concentration in chlorophyllus tissues and the need for protection against harmful oxygen species generated by photosynthesis. The second is that the subcellular flux analysis of concerned pathways should also be considered (Masakapalli et al. 2010). Of course, a set of complementary technologies is available to identify metabolomic profiles and therefore enable a comprehensive view of cell metabolism in different plant tissues to be obtained. Among them, chromatography techniques, mass spectrometry (GC- or LC-MS), and NMR are the most frequently utilised.

However, a major drawback is that, beyond extraction and metabolite analysis techniques, as well as intracellular localisation methods, the sampling of plant material may induce major biases. As highlighted by the data presented in this chapter, it seems that the main difficulty encountered when characterising the metabolite profile of plant tissues is related to sampling. Harvesting plant samples with a view to making comparisons requires the plant location to be chosen carefully based on local climatic conditions before and during sampling, the age of organs, the date and the time. This requires significant and time-consuming work that is rarely done. It might be the reason why the comparison of the metabolic profiles of alpine and lowland plants occasionally shows less difference, for example, in terms of sugars and energy intermediates such as sugar phosphates and nucleotides, or amino-acids, than can be found in plants harvested in different places at the same altitude.

Nevertheless, in some cases the accumulation of a given metabolite can be directly related to the response to a stress. For example, we have shown that the accumulation of gluconate 6-P in lichens during desiccation stages prepares their quasi-immediate reviviscence after rewetting. Apart from this clear-cut example, it has been shown that the higher the altitude and the colder the average temperature, the more abundant sugars, nucleotides and amino-acids are in plants. The accumulation of carbohydrates in plants submitted to cold climates could be a metabolic sink helping to utilise the excess solar energy, thus limiting photoinhibition. The accumulated sugars may also contribute to cryoprotection, as suggested by their synthesis during cold acclimation in woody species (Guy 1990). A comparable accumulation of metabolites involved in energy metabolism is observed in culture cells incubated at a low temperature and in this case, the cell growth becomes negligible long before cell respiration (unpublished data), suggesting that in a cold climate developmental processes may overrule energy metabolism constraints. The effect of temperature on the different processes sustaining cell division and maturation in alpine plants should thus be analysed, in particular those related to the dynamics of cytoskeleton. In Antarctic fishes, for example, the microtubules assemble at very low temperatures, unlike what is observed in homeotherms, due to variations of tubulin primary sequence (Detrich et al. 2000).

Finally, alpine plants frequently accumulate significant quantities of original and specific compounds, the physiological role of which deserves to be further clarified. We gave the example of methyl-$\beta$-D-glucopyranoside in *Geum montanum* and many other *Rosaceae*. One may also cite the ranunculin accumulated in *Ranunculus glacialis* (Streb et al. 2003) and a vast number of compounds involved in secondary metabolism.

**Acknowledgements** We are grateful to Dr. Elisabeth Gout and to Dr. Peter Streb for critical reading of the manuscript, and to Anne-Marie Boisson for the preparation of cell extracts. We acknowledge Pr. Claude Roby and Jean-Luc Le Bail for NMR facilities.

## References

Alonso-Amelot ME (2008) High altitude plants, chemistry of acclimation and adaptation. In: Atta-ur-Rahman (ed) Studies in natural products chemistry, vol 34. Elsevier, Amsterdam, pp 883–981

Apel K, Hirt H (2004) Reactive oxygen species: metabolism, oxidative stress, and signal transduction. Ann Rev Plant Biol 55:373–399

Asada K (1994) Mechanisms for scavenging reactive molecules generated in chloroplasts under light stress. In: Baker NR, Bowyer JR (eds) Photoinhibition of photosynthesis. Bios Scientific Publishers, Oxford, pp 129–142

Asada K (1996) Radical production and scavenging in the chloroplasts. In: Baker NR (ed) Advances in Photosynthesis: photosynthesis and the environment, vol 5. Kluwer Academic Publishers, Dordrecht, pp 123–150

Aubert S, Bligny R, Douce R (1996) NMR studies of metabolism in cell suspensions and tissue cultures. In: Shachar-Hill Y, Pfeffer P (eds) Nuclear magnetic resonance in plant biology. Am Soc Plant Physiol, Rockville, pp 109–144

Aubert S, Hennion F, Bouchereau A, Gout E, Bligny R, Dorne AJ (1999) Subcellular compartmentation of proline in the leaves of the subantarctic Kerguelen cabbage *Pringlea antiscorbutica* R-Br. *in vivo* $^{13}$C-NMR study. Plant Cell Environ 22:255–259

Aubert S, Choler P, Pratt J, Douzet R, Gout E, Bligny R (2004) Methyl-β-D-glucopyranoside in higher plants: accumulation and intracellular localization in *Geum montanum* L. leaves and in model systems studied by $^{13}$C nuclear magnetic resonance. J Exp Bot 406:2179–2189

Aubert S, Juge C, Boisson A-M, Gout E, Bligny R (2007) Metabolic processes sustaining the reviviscence of lichen *Xanthoria elegans* (Link) in high mountain environments. Planta 226:1287–1297

Bacic A, Harris PJ, Stone BA (1988) Structure and function of plant cell walls. In: Preiss J (ed) Encyclopedia of plant physiology, vol 14. Springer, Berlin, pp 297–371

Bewley JD (1979) Physiological aspects of desiccation tolerance. Ann Rev Plant Physiol 30:195–238

Bewley JD (1997) Seed germination and dormancy. Plant Cell 9:1055–1066

Bilger W, Rimke S, Schreiber U, Lange OL (1989) Inhibition of energy transfer to photosystem II in lichens by dehydration: different properties of reversibility with green and blue-green photobionts. J Plant Physiol 134:261–268
Bligny R, Douce R (1980) A precise localization of cardiolipin in plant cells. Biochim Biophys Acta 617:254–263
Bligny R, Douce R (2001) NMR and plant metabolism. Curr Opin Plant Biol 4:191–196
Bonnier G, Douin R (1911–1935) Flore complète illustrée en couleurs de la France, Suisse et Belgique. Paris, Neuchâtel, Bruxelles
Colquhoun I (2007) Use of NMR for metabolic profiling in plant systems. J Pestic Sci 32:200–212
Conklin PL (2001) Recent advances in the role and biogenesis of ascorbic acid in plants. Plant Cell Environ 24:383–476
Connolly JD, Hill RA (1991) Cardenolides. In: Charlwood BV, Harborne DV (eds) Methods in plant biochemistry. Academic, New York, pp 361–368
Cunningham AB, Garnett S, Gorman J, Courtenay K, Boehme D (2009) Eco-Enterprises and *Terminalia ferdinandiana*: "Best Laid Plans" and Australian policy lessons. Econ Bot 63:16–28
Davey MW, Van Montagu M, Inzé D, San Martin M, Kanellis A, Smirnoff N, Benzie IJJ, Strain JJ, Favell D, Fletcher J (2000) Plant L-ascorbic acid: chemistry, function, metabolism, bioavailability, and effects of processing. J Sci Food Agric 80:825–860
Detrich HW, Parker SK, Williams RC, Nogales E, Downing KH (2000) Cold adaptation of microtubule assembly and dynamics. Structural interpretation of primary sequence changes in the alpha- and beta-tubulin of Antarctic fishes. J Biol Chem 275:37038–37047
Dudley S, Lechowicz MJ (1987) Losses of polyol through leaching in subarctic lichens. Plant Physiol 83:813–815
Dulermo T, Rascle C, Billon-Grand G, Gout E, Bligny R, Cotton P (2010) Novel insights into mannitol metabolism in the fungal plant pathogen *Botrytis cinerea*. Biochem J 427:323–332
Fall R, Benson AA (1996) Leaf methanol – the simplest natural product from plants. Trends Plant Sci 1:296–301
Fan TW-M (1996) Metabolic profiling by one- and two-dimentional NMR analysis of complex mixtures. Progr NMR Spectrosc 28:161–169
Farrar JF (1988) Physiological buffering. In: Galun M (ed) Handbook of lichenology II. CRC Press, Boca Raton, pp 101–105
Farrar JF, Smith DC (1976) Ecological physiology of the lichen *Hypogymnia physodes*. III. The importance of the rewetting phase. New Phytol 77:115–125
Foyer CH (1993) Ascorbic acid. In: Alscher RG, Hess JL (eds) Antioxidants in higher plants. CRC Press, Boca Raton, pp 32–57
Foyer CH, Halliwell B (1976) Presence of glutathione and glutatione reductase in chloroplasts: a proposed role in ascorbic acid metabolism. Planta 133:21–25
Foyer CH, Lelandais M, Kunert KJ (1994) Photooxidative stress in plants. Physiol Plant 92:696–717
Fraser PD, Pinto ME, Holloway DE, Bramley PM (2000) Application of high-performance liqid chromatography with photoperiod array detection to the metabolic profiling of plant isoprenoids. Plant J 24:551–558
Gaff DF (1997) Mechanisms of desiccation-tolerance in resurrection vascular plants. In: Basra AS, Basra RK (eds) Mechanisms of environmental stress resistance in plants. Harwood Academic Publishers, The Netherlands, pp 43–58
Glaser BL, Brown DH (1955) Purification and properties of D-glucose 6-phosphate dehydrogenase. J Biol Chem 216:67–79
Gout E, Bligny R, Pascal N, Douce R (1993) $^{13}C$ nuclear magnetic resonance studies of malate and citrate synthesis and compartmentation in higher plant cells. J Biol Chem 268:3986–3992
Gout E, Aubert S, Bligny R, Rébeillé F, Nonomura AR, Benson AA, Douce R (2000) Metabolism of methanol in plant cells. Carbon-13 nuclear magnetic resonance studies. Plant Physiol 123:287–296
Guenther A, Hewitt CN, Erickson D et al (1995) A global model of natural volatile organic compound emissions. J Geophys Res 100:8873–8892
Guy CL (1990) Cold acclimation and freezing stress tolerance: role of protein metabolism. Ann Rev Plant Physiol Plant Mol Biol 41:187–223
Harborne JB (1982) Introduction to ecological biochemistry, 2nd edn. Academic, London, 278 p
Heber U, Bilger W, Bligny R, Lange OL (2000) Phototolerance of lichens, mosses and higher plants in an alpine environment: analysis of photoreactions. Planta 211:770–780
Heber U, Lange OL, Shuvalov VA (2005) Conservation and dissipation of light energy as complementary processes: homoiohydric and poikilohydric autotrophs. J Exp Bot 57:1211–1223
Helms GWF (2003) Taxonomy and symbiosis in associations of *Physciaceae* and *Trebouxia*. Dissertation zur Erlangung des Doktorgrades der Biologischen Fakultät der Georg-August Universität Göttingen 156 p 141 p
Hoekstra FA, Golovina EA, Buitink J (2001) Mechanisms of plant desiccation tolerance. Trends Plant Sci 6:431–438
Honegger R (1991) Functional aspects of the lichen symbiosis. Annu Rev Plant Physiol Plant Mol Biol 42:553–578
Ichimura K, Mukasa Y, Fujiwara T, Kohata K, Goto R, Suto K (1999) Possible roles of methyl glucoside and *myo*-inositol in the opening of cut rose flowers. Ann Bot 83:551–557
Jennings DB, Ehrenshaft M, Pharr DM, Williamson JD (1998) Roles of Mannitol and Mannitol dehydrogenase in active oxygen-mediated plant defense. Proc Natl Acad Sci USA 95:15129–15133
Joyard J, Block MA, Malherbe A, Maréchal E, Douce R (1993) Origin of the synthesis of galactolipids and sulfolipid head groups. In: Moore TS Jr (ed) Lipid metabolism in plants. CRC Press, Boca Raton, pp 231–258
Kappen L (1988) Ecophysiological relationships in different climatic regions. In: Galun M (ed) Handbook of lichenology II. CRC Press, Boca Raton, pp 37–100
Körner C (2003) Alpine plant life. Functional plant ecology of high mountain ecosystems. Springer, Berlin, Heidelberg
Körner E, von Dahl CC, Bonaventure G, Baldwin IT (2009) Pectin methylesterase Na*PME1* contributes to the emission of methanol during insect herbivory and to the helicitation of defence responses in *Nicotiana attenuata*. J Exp Bot 60:2631–2640
Kranner I, Grill D (1994) Rapid change of the glutathione status and the enzymes involved in the reduction of glutathione-

disulfide during the initial stage of wetting of lichens. Crypt Bot 4:203–206

Lange OL (1980) Moisture content and $CO_2$ exchange of lichens. Oecologia 45:82–87

Lange OL, Bilger W, Rimke S, Schreiber U (1989) Chlorophyll fluorescence of lichens containing green and blue green algae during hydration by water vapour uptake and by addition of liquid water. Bot Acta 102:306–313

Lange OL, Pfanz H, Kilian E, Meyer A (1990) Effect of low water potential on photosynthesis in intact lichens and there liberated algal components. Planta 182:467–472

Larson DW (1981) Differential wetting in some lichens and mosses: the role of morphology. Bryologist 84:1–15

Last RL, Jones AD, Shachar-Hill Y (2007) Towards the plant metabolome and beyond. Mol Cell Biol 8:167–174

Lewis NG, Yamamoto E (1990) Lignin: occurrence, biogenesis and biodegradation. Annu Rev Plant Physiol Plant Mol Biol 41:455–496

Longton RE (1988) The biology of polar bryophytes and lichens. Studies in polar research. Cambridge University Press, Cambridge, 391 p

Lütz C (2010) Cell physiology of plants growing in cold environments. Protoplasma. doi:10.1007/s00709-010-0161-5

Manuel N, Cornic G, Aubert S, Choler P, Bligny R, Heber U (1999) Adaptation to high light and water stress in the alpine plant *Geum montanum* L. Oecologia 119:149–158

Mapson LW, Ischerwood FA, Chen YT (1954) Biological synthesis of L-ascorbic acid: the conversion of L-galactono-gamma-lactone into L-ascorbic acid by plant mitochondria. Biochem J 56:21–28

Masakapalli SK, Le Lay P, Huddleston JE, Pollock NL, Kruger NJ, Ratcliffe RG (2010) Subcellular flux analysis of central metabolism in a heterotrophic Arabidopsis cell suspension using steady-state stable isotope labelling. Plant Physiol 152:602–619

McNeil PL, Steinhardt RA (1997) Loss, restoration, and maintenance of plasma membrane integrity. J Cell Biol 137:1–4

McNeil PL, Katsuya M, Vogel SS (2003) The endomembrane requirement for cell surface repair. Proc Natl Acad Sci USA 100:4592–4597

Moller IM, Jensen PE, Hansson A (2007) Oxidative modifications to cellular components in plants. Annu Rev Plant Biol 58:459–481

Nemecek-Marshall M, MacDonald RC, Franzen JJ, Wojciechowski CL, Fall R (1995) Methanol emission from leaves. Enzymatic detection of gas-phase methanol and relation of methanol fluxes to stomatal conductance and leaf development. Plant Physiol 108:1359–1368

Noctor G, Foyer CH (1998) Ascorbate and glutathione: keeping active oxygen under control. Ann Rev Plant Physiol Plant Mol Biol 49:249–279

Oliver MJ, Dowd SE, Zaragoza J, Mauget SA, Payton PR (2004) The rehydration transcriptome of the desiccation-tolerant bryophyte *Tortula ruralis*: Transcript classification and analysis. BMC Genom 5(89):1–19

Popp M, Smirnoff N (1995) Polyol accumulation and metabolism during water deficit. In: Smirnoff N (ed) Environment and plant metabolism: flexibility and acclimation. Bios Scientific Publishers, Oxford, pp 199–215

Pugin A, Frachisse J-M, Tavernier E, Bligny R, Gout E, Douce R, Guern J (1997) Early events induced by the elicitor cryptogein in tobacco cells: involvement of a plasma membrane NADPH oxidase and activation of glycolysis and the pentose phosphate pathway. Plant Cell 9:2077–2091

Rascio N, La Rocca N (2005) Resurrection plants: the puzzle of surviving extreme vegetative desiccation. Crit Rev Plant Sci 24:209–225

Ratcliffe RG (1994) In vivo nuclear magnetic resonance studies of higher plants and algae. Adv Bot Res 20:43–123

Ratcliffe RG, Shachar-Hill Y (2001) Probing plant metabolism with NMR. Annu Rev Plant Physiol Plant Mol Biol 52:499–526

Roberts JKM (2000) NMR adventures in the metabolic labyrinth within plants. Trends Plant Sci 5:30–34

Roessner U, Luedemann A, Brust D, Fiehn O, Linke T, Willmitzer L, Fernie AR (2001) Metabolic profiling allows comprehensive phenotyping of genetically or environmentally modified plant systems. Plant Cell 13:11–29

Rundel PW (1988) Water relations. In: Galun M (ed) Handbook of lichenology II. CRC Press, Boca Raton, pp 17–36

Schroeter B, Jacobsen P, Kappen L (1991) Thallus moisture and microclimatic control of $CO_2$ exchange of *Peltigera aphthosa* (L) Willd on Disco Island (West Greenland). Symbiosis 11:131–146

Sharkey TD (1996) Emission of low molecular mass hydrocarbons from plants. Trends Plant Sci 1:78–82

Smith DC, Molesworth S (1973) Lichen Physiology. XIII. Effects of rewetting dry lichens. New Phytol 72:525–533

Smith AE, Phillips DV (1981) Identification of methyl β-D-glucopyranoside in white clover foliage. J Agric Food Chem 29:850–852

Spitaler R, Schlorhaufer PD, Ellmerer EP, Merfort I, Bortenschlager S, Stuppner H, Zidorn C (2006) Altitudinal variation of secondary metabolite profiles in flowering heads of *Arnica Montana* cv. ARBO. Phytochemistry 67:409–417

Streb P, Feierabend J (1999) Significance of antioxidants and electron sinks for the cold-hardening-induced resistance of winter rye leaves to photo-oxidative stress. Plant Cell Environ 22:1225–1237

Streb P, Feierabend J, Bligny R (1997) Resistance to photoinhibition of photosystem II and catalase and antioxidative protection in high mountain plants. Plant Cell Environ 20:1030–1040

Streb P, Shang W, Feierabend J, Bligny R (1998) Divergent strategies of photoprotection in high-mountain plants. Planta 207:313–324

Streb P, Aubert S, Gout E, Bligny R (2003) Reversibility of cold- and light-stress tolerance and accompanying changes of metabolite and antioxidant levels in the two high mountain plant species *Soldanella alpina* and *Ranunculus glacialis*. J Exp Bot 54:405–418

Sumner LW, Mendes P, Dixon R (2003) Plant metabolomics: large-scale phytochemistry in the functional genomics era. Phytochemistry 62:817–836

Takahashi S, Murata N (2008) How do environmental stresses accelerate photoinhibition? Trends Plant Sci 13:178–182

van der Rest B, Boisson A-M, Gout E, Bligny R, Douce R (2002) Glycerophosphocholine metabolism in higher plant cells. Evidence of a new glyceryl-phosphodiester phosphodiesterase. Plant Physiol 130:244–255

Vicente C, Legaz ME (1988) Lichen enzymology. In: Galun M (ed) Handbook of lichenology I. CRC Press, Boca Raton, pp 239–284

von Dahl CC, Hävecker M, Schlögl R, Baldwin IT (2006) Caterpillar-elicited methanol emission: a new signal in plant-herbivore interactions? Plant J 46:948–960

Wang Y, He W, Huang H, An L, Wang D, Zhang F (2009) Antioxidative responses to different altitudes in leaves of alpine plant *Polygonum viviparum* in summer. Acta Physiol Plant 31:839–848

Weissman L, Garty J, Hochman A (2005) Characterization of enzymatic antioxidant in the lichen *Ramalina lacera* and their response to rehydration. Appl Environ Microbiol 71:6508–6514

Wheeler GL, Jones MA, Smirnoff N (1998) The biosynthetic pathway of vitamin C in higher plants. Nature 393: 365–369

Wildi B, Lütz C (1996) Antioxidant composition of selected high alpine plant species from different altitudes. Plant Cell Environ 19:138–146

Yonekura-Sakakibara K, Saito K (2009) Functional genomics for plant natural product biosynthesis. Nat Prod Rep 26:1466–1487

Zidorn C (2009) Altitudinal variation of secondary metabolites in flowering heads of *Asteraceae*: trends and causes. Phytochem Rev 9:197–203, 7 p

# 9 Interaction of Carbon and Nitrogen Metabolisms in Alpine Plants

F. Baptist and I. Aranjuelo

## 9.1 Introduction

The importance of nitrogen (N) for plant growth has been well understood since the pioneering work of von Liebig (1840), which described the effect of individual nutrients on crops. Since this work, many studies have addressed plant nutrients in general and N in particular, leading to greater understanding of the coupling between N availability, carbon (C) and N fluxes, and whole plant growth.

Optimized whole plant growth requires a close relationship between C and N metabolisms. Although carbon fixation takes place in the leaves, its rate depends on leaf N content and thus on root N uptake, because chemical reactions are catalyzed by enzymes whose activities are based on the energetic substrate provided by photosynthetic C. Hence, the chemical composition of a plant in the different organs must be maintained in a narrow range, which therefore implies a balance between (1) N assimilation from the roots, (2) N allocation to the leaves and (3) C uptake of the leaves. This coupling is largely dependent on environmental constraints.

In alpine areas, the landscape-scale distribution of snow (which is closely related to the mesotopography) is a main driver of plant community composition and functioning. Through its effect on the length of the growing season, snow provides a complex ecological gradient affecting the seasonal course of temperature, light, wind exposure, soil water content and nitrogen availability (Jones et al. 2000). Therefore, all temperature-dependent processes in alpine ecosystems are under the ultimate control of snow cover because it determines growing season length (i.e. the length of the favourable period), soil temperature and water and nitrogen availability (see Fig. 9.1). Typically, between late and early snowmelt locations, the delayed onset of the growing season can be particularly important, up to 40 days in the internal French Alps (Fig. 9.1). Besides, N availability in late snowmelt meadows is generally greater than in dry meadows, at least at the time of snowmelt (Baptist and Choler 2008; May and Webber 1982) and for some locations at all times of the year (Miller et al. 2009). This difference is reflected in all inorganic N pools, with gross and net N mineralization rates being systematically lower in early snowmelt locations compared to late snowmelt locations (Fisk et al. 1998; Miller et al. 2009). Given these constraints, whole plant growth and nutrient acquisition strategies vary greatly over very short distances within alpine areas.

According to the Intergovernmental Panel on Climate Change (IPCC 2007) and the European Environment Agency (EEA 2009), alpine plants will in future be exposed to enhanced $CO_2$ and temperature conditions. As described in more detail below, predicted temperature increases (IPCC 2007) are expected to have direct impact on plant performance. However, it should be also considered that temperature increases will have indirect effects in plants as a consequence of possible modifications of snow cover periods as well as shortages of N and water availability. In order to consider the effects of climate change on alpine plant growth and to identify potential synergistic and

F. Baptist (✉) • I. Aranjuelo
Departamento de Biologia Vegetal, Universitat de Barcelona, Avenida Diagonal, Barcelona, Spain
e-mail: fbaptist@biotope.fr

C. Lütz (ed.), *Plants in Alpine Regions*,
DOI 10.1007/978-3-7091-0136-0_9, © Springer-Verlag/Wien 2012

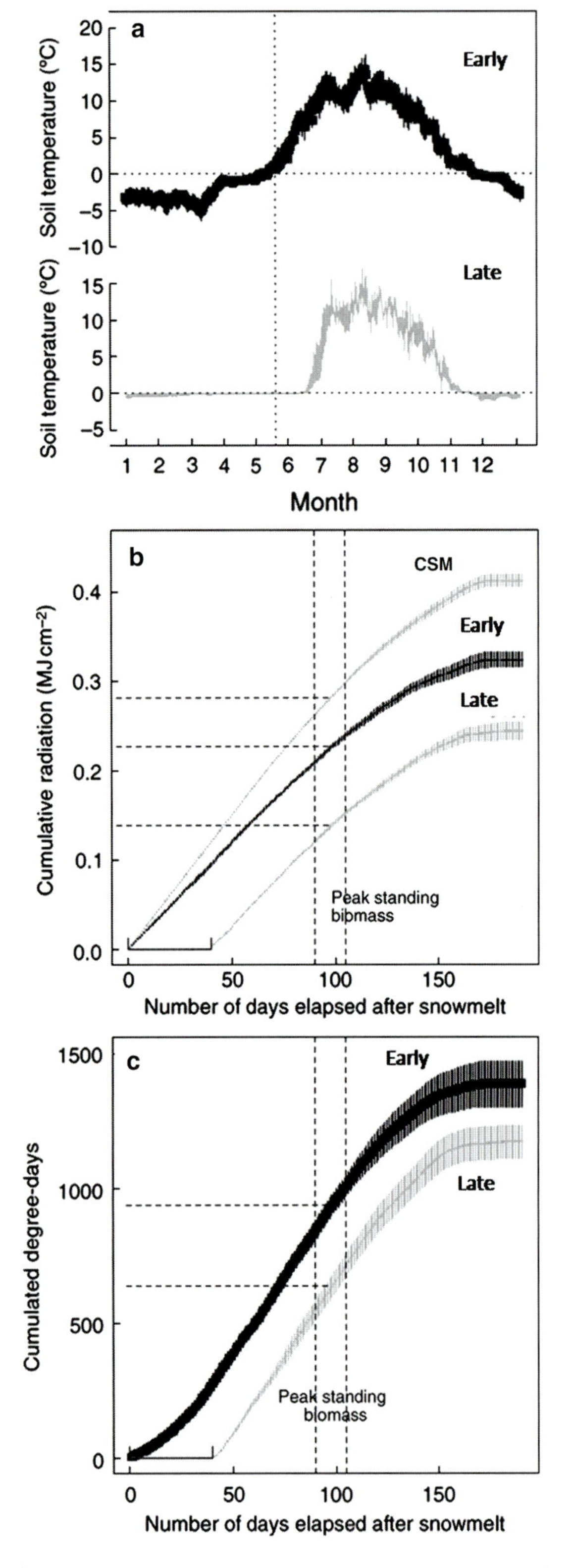


**Fig. 9.1** (a) Time course of daily mean (± standard error) soil temperature 5 cm below the ground surface and (**b**) cumulative radiation during the growing season in the late (*grey*) and early (*black*) snowmelt communities. In (b), the Clear Sky Model (CSM) is shown for comparison, and means (± standard error) were calculated over the period 1999–2004. Data were recorded at Briançon (1,300 m a.s.l.), 30 km from the study site. The Clear Sky Model developed in the framework of the European Solar Radiation Atlas (Rigollier et al. 1999) was used to model the incoming global irradiance on a horizontal plane under cloudless sky cumulative degree-days in early and late-snowmelt locations. (**c**) Cumulative degree-days in early and late-snowmelt locations. Data were averaged over the period 1999–2005 and were recorded at two or three different sites, depending on the year, close to the Galibier pass (France). Modified from Baptist and Choler (2008)

antagonistic phenomena (Valladares and Pearcy 1997), it is crucial to fully understand the coupling between N and C fluxes and pools taking into account variations in plant metabolism along environmental gradients.

In this chapter, we will begin by reviewing the coupling of C and N cycles at the whole plant level along the environmental snow cover gradient with a functional trait approach, before we move on to a more detailed understanding of C and N coupling for C/N uptake and storage in alpine plants. Finally, we will address the potential effect of climatic change (changes in air temperature, snow cover and $CO_2$ concentration) on C and N metabolisms and as a consequence on whole alpine plant growth.

## 9.2 Relation Between Resource Availability, Carbon and Nitrogen Assimilation, Tissue Composition and Whole Plant Growth

### 9.2.1 Carbon and Nitrogen Metabolisms for Resource Acquisition

Plants have several contrasting and complementary strategies for optimizing C and N uptake. However, these strategies do not occur randomly across terrestrial biomes, but partly follow soil N supply. Indeed, a growing body of theoretical (Grime 1977; Chapin 1980; Westoby 1998) and empirical work (Reich et al. 1999; Wright et al. 2004) points to the existence of a fundamental trade-off between rapid acquisition of resources and conservation of resources in plant species along fertility gradients. Cumulative radiation during the growing season in the C140 and C180 communities. The clear sky model is shown for comparison.

Fast-growing species from fertile habitats maximize resource acquisition, whereas slow-growing species in infertile habitats maximize resource conservation. A series of quantitative traits have been associated with this fundamental trade-off in plant function (Reich et al. 1992; Grime 1997; Garnier et al. 1999). Fast-growing species usually have a combination of high Specific Leaf Area (SLA), high tissue N concentration, low tissue density, high rates of C and nutrient uptake, and short-lived leaves. These traits are basically associated with (1) higher rates of maximum photosynthesis because the large amount of leaf organic N is mostly allocated to the photosynthetic machinery (Evans and Seemann 1989) and (2) with short-lived leaves due to low C/N ratios and physically soft foliage. The opposite traits characterize species from nutrient-poor habitats in which the mean residence time of nutrients tends to be maximized through greater organ longevity (in particular leaves) and/or higher resorption of nutrients from senescing organs. Because these traits are easy to measure for a large number of species and sites, they can be considered as precious tools to understand whole plant functioning and to relate it to ecosystem processes.

Within alpine regions, recent studies have demonstrated that consistent shifts in specific richness (e.g. Komarkova and Webber 1978; Kudo and Ito 1992; Theurillat et al. 1994) and in plant functional diversity occur along mesotopographical gradients (Kudo et al. 1999; Choler 2005). A greater Leaf Nitrogen Content (LNC), a higher SLA and a predominance of horizontal leaves (i.e. trait values generally associated with a high capacity for resource acquisition) are common features of species from late snowmelt sites (Choler 2005). These traits ensure efficient carbon fixation at the leaf and canopy levels and are generally associated with higher root respiration rates and higher N uptake (Craine et al. 2002, 2005; Tjoelker et al. 2005). Typically, Bliss (1956) reported at the community level that values of above-ground net primary productivity varied from 0.4 g $m^{-2}$ $d^{-1}$ on a dry windy ridge to 1.9 g $m^{-2}$ $d^{-1}$ in a wet meadow, while above-ground production totalled 280 g $m^{-2}$ in dry meadows, 410 g $m^{-2}$ in moist meadows and 600 g $m^{-2}$ in wet meadows (Fisk et al. 1998). These results underlie a clear reduction of above-ground production from early to late snowmelt locations. In parallel, below-ground productivity (g $m^{-2}$ $d^{-1}$) and total N uptake in semi-controlled conditions was more than twice as high for the late snowmelt species compared to the early snowmelt species (Baptist et al. 2009a). Total N accumulation for production averaged 3.9 g $m^{-2}$ in dry meadows, 5.4 g $m^{-2}$ in moist meadows, and 6.8 g $m^{-2}$ in wet meadows (Fisk et al. 1998) suggesting a higher total N uptake in late snowmelt locations. This tight coupling of C fixation and N uptake in late snowmelt locations allows plants to counterbalance the snow-induced reduction in the carbon uptake period and thus to complete their vegetative life cycle quickly.

Conversely, species from early snowmelt sites are characterized by upright and thick leaves and low SLA, i.e. trait values generally associated with nutrient conservation strategies (Wright et al. 2004). Increased leaf thickness along with reduced surfaces of alpine plants might protect tissues from being abraded by wind-transported particles and ensure structural photoprotection against photoinhibition (Valladares and Pugnaire 1999; Germino and Smith 2000).

Overall, these studies demonstrate the existence of a fundamental trade-off between rapid acquisition of resources and conservation of resources in plant species along snow cover alpine gradients that are regarded as adaptations for overcoming nutrient limitations and other environmental constraints (especially soil water content and fertility). Typically, they reveal a shift from energy-limited plants (i.e. light, because of short growing season) to nutrient and water-limited-plants.

### 9.2.2 Whole Plant Distribution of N-related Function and Compounds in Relation to Root/Shoot C Balance

Although still relatively unknown (Garnier 1991; Osone et al. 2008), growth and carbon allocation patterns are largely affected by the distribution of N-related functions and compounds in plants. Several mechanisms exist including various patterns of biomass allocation and differences in the uptake, assimilation and/or storage of N mineral forms ($NO_3^-$/$NH_4^+$) and organic forms (amino acids) as well as the efficiency of using the assimilated nitrogen to produce new biomass. In order to be assimilated into organic nitrogen, $NO_3^-$ must be reduced by nitrate reductase (NR) to $NO_2^-$ and then by nitrite reductase (NiR) into $NH_4^+$. Both reactions require electrons from photosynthetic electron transport. At the leaf level, $NH_3$ is converted into amino acids by the GS/GOGAT enzyme reaction and carbon skeletons provided by the organic acids derived from the tricarboxylic acid cycle in the mitochondria. These C-skeletons used for amino acid biosynthesis are derived from glycolysis, photosynthetic carbon reduction, the oxidative pentose phosphate pathway and the citric acid cycle, and ATP for the GS/GOGAT reaction is generated by photosynthesis and respiration (Lawlor 2002). Accordingly, the site of $NO_3^-$ reduction (roots and shoots) may have a substantial impact on the C demand for $NO_3^-$ assimilation. For example, species that reduce $NO_3^-$ predominantly in their shoots may have the advantage of being able to use the excess reductant produced in photosynthesis (Pate 1983). By contrast, species that reduce $NO_3^-$ mainly in the roots must obtain their reductants from glycolysis and the oxidative pentose phosphate pathway (Oaks and Hirel 1986). Depending on the plant species and growing conditions, nitrate reduction may be predominantly in the shoot or the root, or there may be some intermediate strategy (Pate 1983; Andrews 1986).

According to Scheurwater et al. (2002) and Andrews (1986), under optimum nutrient availability conditions, shoots are the main site of whole plant $NO_3^-$ reduction in both fast- and slow-growing grasses. However, the mechanisms underlying such a shoot versus root pattern of N reduction may display large interspecific variability in non-optimal environmental conditions. For example, a recent study conducted by Baptist et al. (2009a) with fast (*Carex foetidea*) and slow (*Kobresia myosuroides*) growing alpine plants showed that fast-growing species displayed improved photosynthetic capacity and decreased N reduction capacity in leaves, which was compensated by the preferential C allocation to root growth and/or storage. These plants increased the translocation of reduced N to above-ground organs so as to compensate for the lower N assimilation capacity. The high C flux allocated to the below-ground compartment in *C. foetida* promoted significant levels of $NO_3^-$ reduction in the roots (Pate 1980). Hence, although these results cannot be generalized, they suggest that an allocation-based balance between root N reduction and leaf $CO_2$ assimilation is involved in growth strategies of alpine species growing under short, energy-limited vegetation periods. This coupling between C and N fluxes was less apparent in the case of slow growing species that experienced higher N reduction in the leaves.

Depending on nutrient acquisition strategies, preferential N uptake (mineral, organic) might be expected in order to adapt to N form availability and to optimize uptake and reduction costs especially in the alpine areas. Indeed, approximately 81% of the energy required to synthesize protein with nitrate as the source of nitrogen is used in reducing nitrate and synthesizing amino acids, however when nitrogen is supplied as ammonium, only 2% of the energy involved in protein synthesis is required to synthesize the constituent amino acids. Hence, the conversion of nitrate into amino acids in the cell is a process that consumes large amounts of energy and carbon. Plant roots also display high-affinity uptake systems for amino acids and therefore amino acids can be readily catabolized or used without additional costs.

Numerous studies have addressed the functional significance of the uptake of various forms of N by alpine plants and have generally indicated that all species were capable of taking up organic nitrogen (Chapin et al. 1993; Kielland 1994; Raab et al. 1999; Miller and Bowman 2003), some even equalling or exceeding inorganic N uptake (Raab et al. 1999; Xu et al. 2006). However, relative concentrations of N at the sites of plant sampling did not correspond to patterns of N uptake among species. Instead, species from the same community varied widely in their capacity to take up $NH_4^+$, $NO_3^-$ and glycine, suggesting the potential for differentiation among species in resource (N) use and also during the growing season. For example, while *Festuca eskia* mainly used $NH_4^+$ early and $NO_3^-$ late in the growing season, the reverse was observed for *Nardus stricta* (Pornon et al. 2007). Besides, according to Miller and Bowman (2003), soils from late snowmelt locations and from mid-gradient locations characterized by high amino acid concentration do not support species that exhibit a high capacity for glycine uptake. By contrast, Baptist et al. (2009a) indicated that *Carex foetida*, which grows in late snowmelt locations, displayed higher amino acid uptake compared to species from early snowmelt locations (i.e. *Kobresia myosuroides*). Hence, along snow cover gradients, contrasting and complementary strategies exist for increasing N uptake efficiency, and/or for broadening the options of N uptake from resources of different chemical composition. Although soil N fertility plays an important role in species' distribution and abundance, it is therefore a relatively poor predictor of plant N preferences. Rather, these patterns may favour the potential for species coexistence in a given habitat and might allow plants to adapt to N-form seasonal variations.

## 9.3 Carbon and Nitrogen Storage in Alpine Plants

The acquisition and the allocation of resources are dependent on the build-up of stored carbohydrate and nutrients. This function is particularly important in mountain environments characterized by harsh and constraining conditions (i.e. long, cold winters and short growing seasons) as it gives plants: (1) the support of vegetative regrowth following dormancy (Menke and Trlica 1981), (2) the ability to bridge temporal gaps that exist between resource availability and resource demand (Chapin et al. 1990), (3) the support for sexual or vegetative reproduction during the absence of photosynthesis and (4) the ability to survive calamities such as defoliation, shading or frost. Moreover, the mobilization of stored nitrogen and carbon reserves facilitates competing sinks and permits successful completion of reproduction before the onset of the winter season. Finally, the occurrence of a large concentration of soluble proteins may be important in the low temperature conditions of alpine regions (Öncel et al. 2004). It was typically the case of the leguminous alpine herb, *Oxytropis sericera*, or of members of the Caryophyllaceae that vegetative and reproductive growth was partly supported from stored reserves at least in its earlier stages (Wyka 1999).

Although starch and sucrose are considered as the major storage compounds for cereals and grasses (grains filled with starch) other carbohydrate stores can be metabolized, e.g. fructans in *F. paniculata*, a subalpine species, or cyclitol in some Caryophyllaceae. While *B. bistortoides* relies almost entirely on glucose, fructose and sucrose, *Castilleja puberula* produces high concentrations of mannitol, and *Trifolium nanum* contains high concentrations of cyclitols. According to Monson et al. (2006), in alpine fellfields two groups emerged: the first constituted by *Trifolium sp.* and *Artemisa scopulorum*, which maintain a majority of the soluble carbohydrate as cyclitols, whereas the monocots *Carex* and *Luzula* exhibited little cyclitol and maintained a majority of soluble carbohydrate as sucrose. The selective advantage of fructan or cyclitol as storage carbohydrates is commonly based on the idea that the utilization of the vacuole as a storage compartment would allow plants to exploit constraining environments where periods of positive carbon balance are short and net mobilization of reserves is required to sustain growth (Pollock and Cairns 1991; Monson et al. 2006). Indeed, by maintaining supplies of fructose and sucrose and other ready-to-use C-compounds in vacuoles, these species obviate the need for transport of carbohydrate over distance as in starch storing species (Bloom et al. 1985; Hendry 1987). Besides, in subalpine and alpine ecosystems, plant growth starts, depending on species, either 10 days before and after snowmelt, which largely sensitizes plants to freezing events leading to the loss of tissue (Körner 1999; Inouye 2000, 2008). The presence of such C stores in the stem base and leaf vacuole might offset possible damaging effects associated with frost events (Bloom et al. 1985).

Several studies in controlled environments and field conditions have demonstrated that the availability of N reserves, and particularly the concentration in vegetative storage proteins, is closely related to shoot growth potential (Avice et al. 1996; Justes et al. 2002; Meuriot et al. 2005). However, to our knowledge, no studies have identified patterns of N storage along alpine gradients. The species that possess the highest protein concentration and level of vegetative storage protein (VSP) accumulation consequently exhibit the fastest bud growth, the greatest rate of expansion in leaf area index and the highest shoot production in spring (Justes et al. 2002). This increase in N reserves in perennial organs (taproot), especially in the form of VSPs, can be an important adaptive trait towards tolerance to unpredictable events in alpine environments, to sustain growth at the beginning of the growing season or to withstand processes of cold hardening, which is fundamental in alpine habitats. For example, *Eriophorum vaginatum* reached its maximum growth rate early in the season supported entirely by N stored in the stem at a time when the roots were still frozen in the soil (Shaver et al. 1986). By contrast, according to Jaeger and Monson (1992) and Lipson et al. (1996), rather than using N stores to start its growth earlier in the season, *Bistorta bistortoides* used them to support the high demand for resources encountered during the growing season when leaves competed for substrate that could not be supplied adequately by soil uptake alone. Hence, although the storage organ of *B. bistortoides* accommodates luxury uptake of N, the major function of the rhizome seems to retain a pool of mobile nitrogenous compounds to accommodate predictable seasonal variation in N supply and demand rather than to capitalize on unpredictable events. Similarly, Kleijn et al. (2005) demonstrated that reserves allow *Veratrum album* to complete the above-ground growing cycle as fast as possible and thus reduce the exposure to stochastic events such as frosts. Hence, higher nutrient concentrations and important N stores permit the refilling of carbohydrate stores used to promote growth during the growing season, and to support maintenance metabolism during the winter season. However, according to the literature, most of the studied plants rely primarily on soil N-resources for their seasonal growth rather than on N stores. Hence, a reduction in soil N availability due to lower snow precipitations may result in a reduction of C stores and therefore (1) decreased allocation to growth and reproduction and (2) a lower ability to cope with hazardous events.

## 9.4 Effect of Climatic Changes on Carbon–Nitrogen Interactions in Alpine Plants

### 9.4.1 Global Climate Change in Alpine Areas

The EU White Paper on Adaptation (EEA 2009) names the Alps as among the areas most vulnerable to climate change in Europe, although much uncertainty still exists as to the possible effects of such changes on vegetation communities and ecosystem properties (Theurillat and Guisan 2001). After the industrial revolution, as a consequence of human activity, atmospheric [$CO_2$] has steadily increased from an estimated 280–379 μmol mol$^{-1}$ in 2005 with a current average increase of 1.9 μmol mol$^{-1}$ per year (Alley et al. 2007). According to the predictions of the IPCC (2007), at the end of the present century this concentration may be around 700 μmol mol$^{-1}$, i.e. 2.5 times the preindustrial value. As a consequence, the global mean surface temperature has risen by 0.74ºC ± 0.18ºC over the last 100 years (1906–2005), and Europe has become warmer than the global average especially in the south-west, the north-east and mountain areas (IPCC 2007). Besides, according to Dye and Tucker (2003), between 1972 and 2000 the duration of the snow-free period in northern hemisphere land areas increased by 5–6 days per decade, and earlier snow cover disappearance in spring has been observed to have increased by 3–5 days per decade. Moderate future climate scenarios predict a temperature rise of 3.9ºC up to the end of the twenty-first century for the Alps, with stronger warming during the second half of the century (European Environment Agency 2009; Beniston 2003; Nogués-Bravo et al. 2007) associated with an alpine-wide decline of snow-covered days (see Stewart 2009 for a review). As a consequence, the observed and projected impacts include changes in the hydrological cycle of mountain regions and changing water availability in elevated and surrounding regions, a decline in glacier cover, and a reduction in permafrost, increasing hazards and damage to high-mountain infrastructure and northward and uphill distribution shifts of many European plant species (60% of mountain plant species may face extinction by 2100) (Auer et al. 2007; Beniston 2003; Nogués-Bravo et al. 2007). As a result of the changing precipitation patterns, there will also be a change in the incidence of dry periods. The enhancement of temperature will lead to increased rainfall, less snowfall during the winter, and consequently water availability limitation during the growing season (EEA 2009).

### 9.4.2 Effect of Increasing Temperature and Decreasing Snow Cover Duration on Plant Performance

As mentioned above, warmer conditions and decreased snow cover are expected in the Alps (IPCC 2007, Figs. 9.2 and 9.3), which has the potential to alter individual plant performance by influencing the growing season length and soil microclimate. Warmer conditions are very likely to increase productivity and the biomass of alpine plant communities (Theurillat and Guisan 2001), however, this question has mainly been addressed through modelling approaches. Typically, Riedo et al. (1997) showed that a seasonally uniform temperature increase by 2°C raised net primary production by 50% in an alpine landscape by modelling the productivity of managed grasslands in the Swiss Alps. Similarly, Baptist and

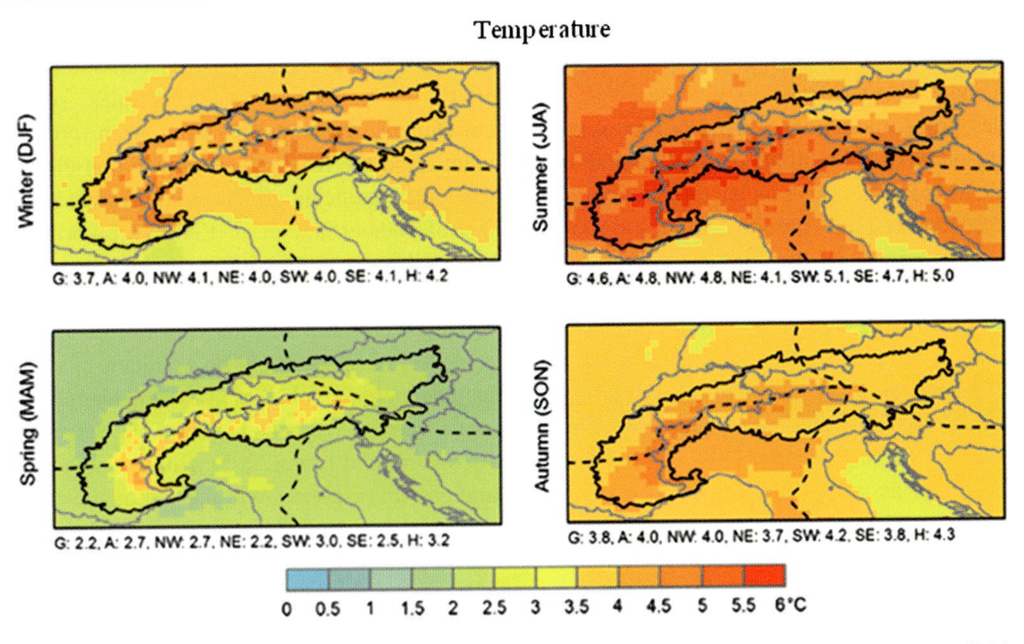


**Fig. 9.2** Seasonal changes in precipitation and temperature up to the end of the twenty-first century, according to CLM scenario A1B. Figure adapted from the European Environment Agency, Regional Climate Change and Adaptation (2009)

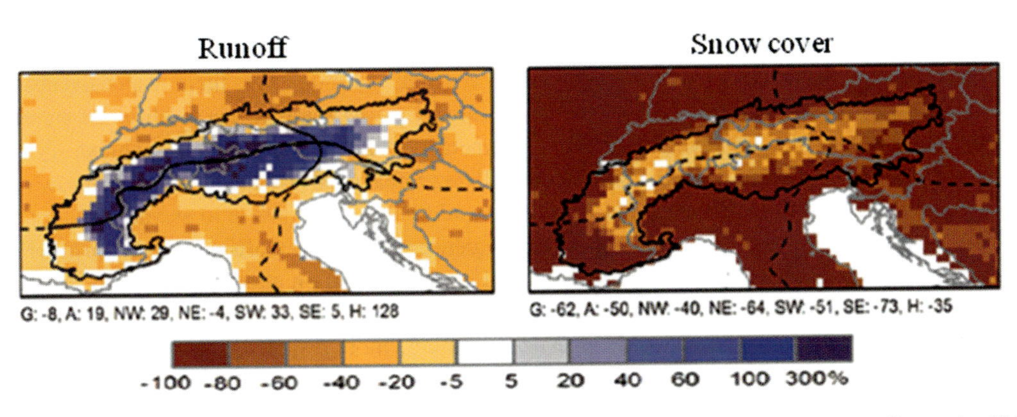


**Fig. 9.3** Run-off and snow cover change up until the end of the twenty-first century in the winter, according to the CLM A1B scenario. Figure adapted from European Environment Agency, Regional Climate Change and adaptation (2009)

Choler (2008) demonstrated that the snow-induced changes in the length of the growing season might have a great impact on the seasonal gross primary productivity of alpine plant communities.

However, in real field conditions, one might predict a small impact on carbon uptake if plant communities are dominated by periodic species, i.e. species with a fixed, genetically controlled growing period (Sørensen 1941). Conversely, the short-term consequences for ecosystem productivity would be stronger if aperiodic species, i.e. species able to extend their vegetative growth, were dominant. Besides, there is a realistic possibility of an increased frequency of freezing events during periods when plants are active in mountain environments, because a warmer climate might advance phenology in many alpine plants with a current risk of freezing temperatures (Inouye 2000, 2008). Indeed, during winter, plants are usually protected from low temperature by snow cover or by specific features (e.g. higher concentration of sugars or supercooling, see previous section). The situation is different once snow disappears because plants generally display rapid dehardening. Hence, if snowmelt is advanced, the plants will be exposed to freezing temperatures making them much more sensitive to frosts (Körner 1999; Baptist et al. 2009b; Sierra-Almeida and Cavieres 2010).

Besides, since microbiota-derived N represents an important source of N for alpine plants (Schmidt et al. 2007), the temperature effect on bacterial activity will strongly condition plant growth. Studies on plant-microbe interactions in the alpine N cycle have revealed a seasonal separation of N use, with plants absorbing N primarily during the summer months and microbes immobilizing N primarily during the autumn months (Jaeger et al. 1999; Lipson et al. 1999; Schmidt et al. 2007). The peak of nutrients at snowmelt, coupled with the mineral release from snowmelt water, is crucial as it partly supports seasonal plant growth, representing up to 7–12% of total N uptake during the growing season (Bilbrough et al. 2000). Unlike early snowmelt species, late snowmelt alpine species strongly rely on this flush of mineral nitrogen (e.g. Kleijn et al. 2005; Monson et al. 2006), which allows rapid expansion of photosynthetic tissues and ensures efficient light capture and carbon fixation (Bryant et al. 1998; Baptist et al. 2010). In the context of global climate change we can expect that, despite a potential plant growth-stimulating increase in temperature, reduced snow cover and advanced snowmelt might dramatically impact the growth of species from late snowmelt locations (Björk and Molau 2007). This might occur due to (1) the growth period not matching the period with large soil N availability and (2) because the soil may dry out to a great degree during summer. Further studies addressing climate change effects and the overlap between environmental conditions advantageous for plant growth and N release from microorganism collapse will be crucial in order to understand the response of alpine plants to future climate conditions and their redistribution within alpine environments.

### 9.4.3 Elevated $CO_2$ Effect on C and N Interactions in Alpine Plants

Changes in C and N metabolism, implying changes to overall plant growth, cannot be dissociated from the 'fertilization effect' of increasing atmospheric $CO_2$. Understanding how plants will respond to the rapid $CO_2$ increase and developing knowledge about their capacity to adapt is an essential initial step in understanding the full impact that global climate change will have on terrestrial ecosystems (Leakey et al. 2009). During recent decades, several $CO_2$ enrichment experiments have been conducted in plants, and the results have shown that elevated $CO_2$ can have significant effects on the growth and physiology of plants (Saxe et al. 1998; Ainsworth and Long 2005; Körner et al. 2005; Hovenden et al. 2008; Aranjuelo et al. 2008). Primary effects of $CO_2$ enhancement on plants are well documented (Nowak et al. 2004; Long et al. 2004; Aranjuelo et al. 2009) and include increased plant biomass and leaf net photosynthetic rates (Long et al. 2004; Nowak et al. 2004; Ainsworth and Long 2005). However, photosynthetic and growth responses will depend on their genetically determined potential (Long et al. 2004; Nowak et al. 2004; Aranjuelo et al. 2009). Although most of the research on climate change effects in plants has been conducted in fast-growing plants, it should be considered that slow-growing plants account for a large proportion of species. Because these plants usually grow in extreme environmental conditions, such as alpine environments, it is not clear to what extent their growth rate and, consequently, their responsiveness to elevated $CO_2$ is going to be conditioned by their own metabolic limitations. In this context, Poorter and Pérez-Soba (2001) stated that slow-growing plants would be less responsive to elevated $CO_2$ as a consequence of their metabolism and the construction costs. However, based on photosynthetic models, Lloyd and Farquhar (2000) suggested that slow-growing plants would be more responsive to enhanced $CO_2$ than fast-growing plants. In a first report, Körner and Diemer (1987) compared the $CO_2$ response curves of 12 plants, later extended to 20 species (Körner and Pelaez Menendez-Riedl 1989) grown at different altitudes and $CO_2$ levels and observed that high altitude plants with lower ambient $CO_2$ increased their efficiency of $CO_2$ utilization. In a later study conducted by Körner et al. (1997), where alpine plants were exposed to elevated $CO_2$ conditions (355 versus 680 $\mu$mol mol$^{-1}$) in open top chambers (OTC), the authors observed that plant biomass was not affected by the $CO_2$ level. The study also indicated that ecosystem gas exchange was increased in plants grown under elevated $CO_2$ conditions. This increase in $CO_2$ concentration may enhance the potential net photosynthesis for $C_3$ plants, because ribulose-1,5-bisphophate carboxylase/oxygenase (rubisco) is not $CO_2$ saturated at the current concentration (Drake et al.,1997). This enzyme catalyses the photosynthesis and photorespiration reactions, but the current atmospheric $CO_2$ concentration is insufficient to saturate Rubisco in $C_3$ plants. Thus, an increase in ambient $CO_2$ increases the leaf internal $CO_2$ concentration and the $CO_2/O_2$ ratio at the Rubisco site, which favours carboxylation rather than oxygenation of ribulose-1,5-bisphosphate (RuBP) (Andrews and Lorimer 1987). In this context, a study conducted in alpine plants exposed to $CO_2$ conditions by Körner et al. (1997) revealed that although photosynthetic activity increased under elevated $CO_2$ conditions, no statistical differences were observed in biomass production. The excess C associated with the enhanced $CO_2$ fixation in these plants was accumulated in the leaves (Körner et al. 1997). This study suggested that with some exceptions, total non-structural carbohydrates (TNC) increased (mainly due to sucrose, glucose and fructose enhancement) under elevated $CO_2$ conditions. As observed by the authors, these results are in agreement with previous studies (Sakai and Larcher 1987; Guy 1990) which support the idea that plants from cold habitats accumulate soluble sugars so as to maintain a high degree of frost resistance throughout the vegetation period (see above).

Many studies have shown that photosynthesis acclimates to elevated $CO_2$ over long-term experiments, a process often referred to as 'down-regulation' (Long et al. 2004). A study conducted on alpine plants described that although exposure to elevated $CO_2$ conditions increased $CO_2$ utilization efficiency, depending upon the extent and duration of $CO_2$ enhancement, a tendency to downward adjustment of photosynthesis was observed (Körner and Diemer 1994; Körner et al. 1997). Reduced or acclimated stimulation of photosynthesis is attributed to stomatal (Naumburg et al. 2004) and non-stomatal limitations (Aranjuelo et al. 2009). Non-stomatal limitations reduce photosynthesis due to reduced light capture (Aranjuelo et al. 2008) or decreased Rubisco carboxylation of RuBP (Stitt and Krapp 1999; Long et al. 2004; Aranjuelo et al. 2005). There are two basic mechanisms by which Rubisco down-regulation occurs. The first mechanism hypothesizes that the reduction in Rubisco content occurs as a consequence of the leaf C build-up (Moore et al. 1999; Aranjuelo et al. 2009). According to the second mechanism, decreases in Rubisco content may reflect a general decrease in leaf N availability (Ainsworth and Rogers 2007).

Based on the C build-up theory, enhancement of leaf carbon content caused by the greater photosynthetic rates of plants exposed to elevated $CO_2$ induces suppression of gene encoding for proteins belonging to the photosynthetic apparatus, resulting in decreased photosynthetic capacity (Moore et al. 1999; Jifon and Wolfe 2002). When plants exposed to elevated $CO_2$ are limited in their ability to increase C sink strength, they decrease their photosynthetic activity to balance C source activity and sink capacity (Thomas and Strain 1991). Although the C source capacity increases as a consequence of photosynthesis and carbohydrate synthesis during the early stages of elevated $CO_2$ exposure (Drake et al. 1997), the capacity to make use of such an increased C supply will condition responsiveness of the photosynthetic apparatus in the long term (Aranjuelo et al. 2009). The ability to "use" such an increase in C depends on the capacity of the actual sinks or development of new sinks (Stitt and Krapp 1999; Lewis et al. 2002; Aranjuelo et al. 2009). As explained by Körner et al. (1997), there are four pathways through which excess C could be diverted: (1) respiratory losses during the dormant season, (2) C accumulation in soil microorganisms, (3) accretion of soil organic matter, and (4) export from the system in the form of dissolved organic matter. In the case of alpine plants, the previously mentioned study conducted by Körner et al. (1997) did not detect significant differences in respiratory losses. Likewise, elevated $CO_2$ had no effect on soil respiration, microbial biomass or on soil C and N. These results suggest that the inability of such plants to "use" the extra C caused the carbohydrate build-up that led to adjustments in photosynthetic activity in alpine plants exposed to elevated $CO_2$ conditions. Furthermore, it should be considered that, since N availability is a key factor conditioning responsiveness of the photosynthetic apparatus, there is evidence that the carbohydrate-mediated repression of photosynthetic genes is more severe in nitrogen-deficient plants than in nitrogen-depleted plants (Stitt and Krapp 1999).

The second hypothesis states that Rubisco activity decreases due to the relocation of N within the plant. Recent studies indicated that, under elevated $CO_2$ conditions, plants increased their N use efficiency (NUE) through the redistribution of the excess N invested in Rubisco (Ainsworth and Rogers 2007). Low leaf N could lead to either a proportional (Geiger et al. 1999) or a selective (Reviere-Rolland et al. 1996) reduction in Rubisco. This hypothesis suggests that there is N limitation where N uptake from soils fails to keep pace with photosynthesis and C acquisition. According to the description of Körner et al. (1997), in alpine plants, the exposure to elevated $CO_2$ has no measurable effect on sugar and amino acids exuded by the roots into the soil. This study also showed that, at elevated $CO_2$ concentration, a low soil N supply could limit photosynthesis, leading to diminished plant N availability in the long term. Limited C supply to soil microbiota together with the low soil N availability of those plants might have limited N availability. N availability is a critical factor, limiting plant growth and increasing the response to elevated $CO_2$ conditions. Since low N availability often strongly limits biomass production in alpine plants (Bowman et al. 1993; Haselwandter et al. 1983), changes in plant N availability induced by exposure to elevated $CO_2$ could modify rates of N cycling and cause shifts in plant species dominance. However, according to the observations of Arnone (1997, 1999) on different high-elevation native Swiss alpine plants, elevated $CO_2$ does not modify soil N content.

These studies highlighted the fact that the $CO_2$ effect on photosynthetic activity and consequently on plant growth will be strongly mediated by N availability. Although little is known about the $CO_2 \times N$ effect in alpine plants, studies conducted on other plants suggest that the role of N availability will be a key factor conditioning the capacity of such plants to develop or increase C sink strength. More research is therefore needed on this topic.

## 9.5 Conclusions and Perspectives

In alpine areas, the productivity of most communities and species is constrained by the supply of N, but also by the length of the growing season and water availability. These constraints are distributed along topographical gradients and vary from energy-limited species to nutrient- and water-limited plants from late to early snowmelt locations. As a result the acquisitive nutrient strategies of alpine plants shift towards conservative nutrient strategies. Little information is available concerning the coupling between C and N fluxes within the alpine plants as well as the residence time of C and N in different organs. In fertile habitats, species experienced high C fixation coupled with higher root N uptake, although most species reduce nitrogen in the leaves rather than in the roots. By contrast, alpine species do not display a correlation between nutrient acquisition strategies and preferential N-form uptake. This process appears to be highly idiosyncratic and can by be interpreted as an adaptation to seasonal variations in N-form and a way to maintain plant coexistence in the ecosystem. Alpine plants store large amounts of C and N in the different storage organs (roots, bulb etc.). However, even though C stores appear to be crucial for the start of growth at the beginning of the growing season, N reserves seem to support the high demand for resources during the growing season when soil uptake alone cannot provide adequate supplies for whole plant demand.

As discussed above, although very few studies have considered the predicted climate change effect on the performance of alpine plants, it is crucial to improve our knowledge of this topic, because plant growth and species distribution will be strongly affected in the near future. Lengthened growing seasons might lead to an increase in net primary productivity depending on the phenological and frost tolerance features of alpine plants. Also, for the correct performance of alpine plants it will be crucial that the growth period matches the period of large soil N availability, namely at snowmelt. Accordingly, the species growing in late snowmelt locations might be the most sensitive to climatic change, as they generally display a low capacity to recover from frost events and depend greatly on the pulse of nitrogen at snowmelt.

Furthermore, due to synergistic and antagonistic phenomena, future studies should consider the interaction of the elevated $CO_2$ effect and other predicted growth-limiting conditions, such as temperature, and N and water availability. The very few studies conducted up to this point on this topic show that the $CO_2$ effect will be mediated by N and temperature. However, very little is known about the key processes involved in these responses. Knowledge of these topics is imperative to further understand how alpine plants (at the individual and community level) will cope with climatic change in the following decades.

## References

Ainsworth EA, Long SP (2005) What have we learned from 15 years of free-air $CO_2$ enrichment (FACE)? A meta-analytic review of responses of photosynthesis, canopy properties and plant production to rising $CO_2$. New Phytol 165:351–372

Ainsworth EA, Rogers A (2007) The response of photosynthesis and stomatal conductance to rising [$CO_2$]: mechanisms and environmental interactions. Plant Cell Environ 30:258–270

Alley R, Berntsen T, Bindoff NL et al (2007) Climate change 2007: the physical 613 science basis. In: Summary of policy-makers fourth assessment report of working 614 group I, Intergovernmental panel on climate change, Geneva, Switzerland

Andrews M (1986) The partitioning of nitrate assimilation between root and shoot of higher plants. Plant Cell Environ 9:511–519

Andrews JT, Lorimer GH (1987) Rubisco: structure, mechanisms and prospects for improvement. In: Hatch MD, Broadman NK (eds) Biochemistry of plants, vol 10. Academic, New York, pp 132–207

Aranjuelo I, Irigoyen JJ, Pérez P, Martínez-Carrasco R, Sánchez-Díaz M (2005) The use of temperature gradient tunnels for studying the combined effect of $CO_2$, temperature and water availability in $N_2$ fixing alfalfa plants. Ann Appl Biol 146:51–60

Aranjuelo I, Irigoyen JJ, Sánchez-Díaz M, Nogués S (2008) Carbon partitioning in $N_2$ fixing *Medicago sativa* plants exposed to different $CO_2$ and temperature conditions. Funct Plant Biol 35:306–317

Aranjuelo I, Pardo T, Biel C, Savé R, Azcón-Bieto J, Nogués S (2009) Leaf carbon management in slow-growing plants exposed to elevated $CO_2$. Glob Chang Biol 15:97–109

Arnone JA III (1997) Indices of plant N availability in an alpine grassland under elevated atmospheric $CO_2$. Plant Soil 190: 61–66

Arnone JA III (1999) Symbiotic $N_2$ fixation in a high Alpine grassland: effects of four growing seasons of elevated $CO_2$. Funct Ecol 13:383–387

Auer I, Böhm R, Jurkovic A et al (2007) HISTALP – Historical instrumental climatological surface time series of the Greater Alpine Region 1760–2003. Int J Climatol 27:17–46

Avice JC, Ourry A, Lemaire G, Boucaud J (1996) Nitrogen and carbon flows estimated by $^{15}N$ and $^{13}C$ pulse chase labelling during regrowth of alfalfa. Plant Phys 112:281–290

Baptist F, Choler P (2008) A simulation on the importance of growing season length and canopy functional properties on the seasonal gross primary production of temperate alpine meadows. Ann Bot 101:549–559

Baptist F, Tcherkez G, Aubert S, Pontailler JY, Choler P, Noguès S (2009a) $^{13}C$ and $^{15}N$ allocations of two alpine species from early and late snowmelt locations reflect their different growth strategies. J Exp Bot 60:2725–2735

Baptist F, Flahaut C, Streb P, Choler P (2009b) No increase in alpine snowbed productivity in response to experimental lengthening of the growing season. Plant Biol. doi:10.1111/j.1438-8677.2009.00286.x

Baptist F, Yoccoz G, Choler P (2010) Direct and indirect control by snow cover over decomposition in alpine tundra along a snowmelt gradient. Plant soil 328:397–410

Beniston M (2003) Climatic change in mountain regions: a review of possible impacts. Clim Chang 59:5–31

Bilbrough CJ, Welker JM, Bowman WD (2000) Early spring nitrogen uptake by snow-covered plants: a comparison of arctic and alpine plant function under the snowpack. Arct Antarctic Alp Res 32(2):404–411

Björk RG, Molau U (2007) Ecology of alpine snowbed and the impact of global change. Arc Antarctic Alp Res 39:34–43

Bliss L (1956) A comparison of plant development in micro-environments of arctic and alpine tundras. Ecol Monogr 26: 303–307

Bloom AJ, Chapin FS, Mooney HA (1985) Resource limitation in plants – an economic analogy. Ann Rev Ecol Syst 16: 363–392

Bowman WD, Theodose TA, Schardt JC, Conant RT (1993) Constraints of nutrient availability on primary production in two alpine tundra communities. Ecology 74:2085–2097

Bryant DM, Holland EA, Seastedt TR, Walker MD (1998) Analysis of litter decomposition in an alpine tundra. Can J Bot 76:1295–1304

Chapin F (1980) The mineral nutrition of wild plants. Ann Rev Ecol Systematics 11:37–52

Chapin FS, Schulze ED, Mooney HA (1990) The ecology and economics of storage in plants. Ann Rev Ecol Syst 21: 423–447

Chapin F, Moilanen L, Kielland K (1993) Preferential use of organic nitrogen for growth by a non-mycorrhizal arctic sedge. Nature 361:743–751

Choler P (2005) Consistent shifts in alpine plant traits along a mesotopographical gradient. Arct Antarctic Alp Res 37(4): 444–453

Craine JM, Tilman D, Wedin D, Reich P, Tjoelker M, Knops J (2002) Functional traits, productivity and effects on nitrogen cycling of 33 grassland species. Funct Ecol 16:563–574

Craine JM, Lee WG, Bond WJ, Williams RJ, Johnson LC (2005) Environmental constraints on a global relationship among leaf and root traits of grasses. Ecology 86:12–19

Drake BG, González-Meler MA, Long SP (1997) More efficient plants: a consequence of rising atmospheric $CO_2$? Ann Rev Plant Phys Plant Mol Biol 48:609–639

Dye DG, Tucker CJ (2003) Seasonality and trends of snow-cover, vegetation index, and temperature in northern Eurasia. Geophys Res Lett 30:9–12

European Environment Agency (2009) Regional climate change and adaptation. The Alps facing the challenge of changing water resources 8, ISSN 1725–9177

Evans JR, Seemann JR (1989) The allocation of protein nitrogen in the photosynthetic apparatus: costs, consequences, and control. In: Briggs WR (ed) Photosynthesis. Alan R. Liss Press, New York

Fisk MC, Schmidt SK, Seastedt TR (1998) Topographic patterns of above- and belowground production and nitrogen cycling in Alpine tundra. Ecology 79:2253–2266

Garnier E (1991) Resource capture, biomass allocation and growth in herbaceous plants. Trends Ecol Evol 6:126–131

Garnier E, Salager JL, Laurent G, Sonie L (1999) Relationships between photosynthesis, nitrogen and leaf structure in 14 grass species and their dependence on the basis of expression. New Phytol 143:119–129

Geiger M, Haake V, Ludewig F, Sonnewald U, Stitt M (1999) The nitrate and ammonium nitrate supply have a major influence on the response of photosynthesis, carbon metabolism and growth to elevated carbon dioxide in tobacco. Plant Cell Env 22:1177–1199

Germino MJ, Smith WK (2000) High resistance to low temperature photoinhibition in two alpine, snowbank species. Physiol Plant 110:89–95

Grime J (1977) Evidence for the existence of three primary strategies in plants and its relevance to ecological and evolutionary theory. Am Nat 111:1169–1194

Grime J (1997) Biodiversity and ecosystem function: the debate deepens. Science 277:1260–1261

Guy CL (1990) Cold acclimation and freezing stress tolerance: role of protein metabolism. Ann Rev Plant Phys Plant Mol Biol 41:187–223

Haselwandter K, Hofmann A, Holzmann HP, Read DJ (1983) Availability of nitrogen and phosphorus in the nival zone of the Alps. Oecologia 57:266–269

Hendry G (1987) The ecological significance of fructan in a contemporary flora. New Phytol 106:201–216

Hovenden MJ, Karen EW, Vander Schoor JK, Williams AL, Newton PCD (2008) Flowering phenology in a species-rich temperate grassland is sensitive to warming but not to elevated $CO_2$. New Phytol 178(4):815–822

Inouye DW (2000) The ecological and evolutionary significance of frost in the context of climate change. Ecol Lett 3: 457–463

Inouye D (2008) Effects of climate change on phenology, frost damage, and floral abundance of montane wildflowers. Ecol 89:353–362

IPCC (Intergovernmental Panel on Climate Change) (2007) Climatic change 2007: the physical science basis. In: Proceedings of the 10th session of working group I of the IPCC, Paris, February 2007

Jaeger C, Monson R (1992) Adaptive significance of nitrogen storage in *Bistorta bistortoides*, an alpine herb. Oecologia 92:121–131

Jaeger C, Monson RK, Fisk MC, Schmidt S (1999) Seasonal partitioning of nitrogen by plants and soil microorganisms in an alpine ecosystem. Ecol 80:1883–1891

Jifon JL, Wolfe DW (2002) Photosynthetic acclimation to elevated $CO_2$ in *Phaseolus vulgaris L.* is altered by growth response to nitrogen supply. Glob Chang Biol 8:1018–1027

Jones H, Pomeroy J, Walker DA, Hoham R (2000) Snow ecology: an interdisciplinary examination of snow-covered ecosystems. Cambridge University Press, Cambridge

Justes E, Thiébeau P, Avice JC, Lemaire G, Volenec JJ, Ourry A (2002) Influence of sowing dates, N fertilization and irrigation on autumn VSP accumulation and dynamics of spring regrowth in alfalfa (*Medicago sativa L.*). J Exp Bot 53: 111–121

Kielland K (1994) Amino acid absorption by arctic plants: implications for plant nutrition and nitrogen cycling. Ecol 75: 155–181

Kleijn D, Treier UA, Muller Scharer H (2005) The importance of nitrogen and carbohydrate storage for plant growth of the alpine herb *Veratrum album*. New Phytol 166:565–575

Komarkova V, Webber PJ (1978) An alpine vegetation map of Niwot Ridge, Colorado. Arct Alp Res 1:1–29

Körner C (1999) Alpine plant life. Springer Verlag, Berlin/Heidelberg/New York

Körner C, Diemer M (1987) In situ photosynthetic response to light, temperature and carbon dioxide in herbaceous plants from low and high altitude. Funct Ecol 1:179–194

Körner C, Diemer M (1994) Evidence that plants from high altitudes retain their greater photosynthetic efficiency under elevated $CO_2$. Funct Ecol 8:58–68

Körner C, Pelaez Menendez-Riedl S (1989) The significance of developmental aspects in plant growth analysis. In: Lambers H, Cambridge ML, Konings H, Pons TL (eds) Causes and consequences of variation in growth rate productivity of higher plants. Springer Verlag Academic Publishing, The Hague, pp 141–157

Körner C, Diemer M, Scäppi B, Niklaus P, Arnone J III (1997) Response of alpine grassland to four seasons of $CO_2$ enrichment: a synthesis. Acta Ecol 18:165–175

Körner C, Asshoff R, Bignucolo O, Hättenschwiler S, Keel SG, Peláez-Riedl S, Pepin S, Siegwolf RTW, Zotz G (2005) Carbon flux and growth in mature deciduous forest trees exposed to elevated $CO_2$. Science 309:1360–1362

Kudo G, Ito K (1992) Plant distribution in relation to the length of the growing season in a snow-bed in the Taisetsu Mountains, northern Japan. Vegetatio 98:319–328

Kudo G, Nordenhall U, Molau U (1999) Effects of snowmelt timing on leaf traits, leaf production, and shoot growth of alpine plants: comparisons along a snowmelt gradient in northern Sweden. Ecoscience 6:439–450

Lawlor D (2002) Carbon and nitrogen assimilation in relation to yield: mechanisms are the key to understanding production systems. J Exp Bot 53:773–787

Leakey ADB, Ainsworth EA, Bernacchi CJ, Rogers A (2009) Elevated $CO_2$ effects on plant carbon, nitrogen, and water relations: six important lessons from FACE. J Exp Bot 60: 2859–2876

Lewis JD, Wang XZ, Griffin KL, Tissue DT (2002) Effects of age and ontogeny on photosynthetic responses of a determinate annual plant to elevated $CO_2$ concentrations. Plant Cell Env 25:359–368

Liebig J (1840) Die organische chemie in ihrer anwendung auf agrikultur und physiologie. Friedrich Vieweg, Braunschweig

Lipson D, Monson R, Bowman W (1996) Luxury uptake and storage of nitrogen in the rhizomatous alpine herb, *Bistorta bistortoides*. Ecology 77:569–576

Lipson DA, Schmidt SK, Monson RK (1999) Links between microbial population dynamics and nitrogen availability in an alpine ecosystem. Ecology 80:1623–1631

Lloyd J, Farquhar GD (2000) Do slow-growing species and nutrient-stressed plants consistently respond less to elevated $CO_2$? A clarification of some issues raised by Poorter (1998). Glob Chang Biol 6:871–876

Long SP, Ainsworth EA, Rogers A, Ort DR (2004) Rising atmospheric carbon dioxide: plants FACE the future. Ann Rev Plant Biol 55:591–628

May DE, Webber PJ et al (1982) Spatial and temporal variation of vegetation and its productivity on Niwot Ridge, Colorado. In: Ecological studies in the Colorado alpine, a Festschrift for John W. Marr. Occasional Paper Number 37. Institute of

Artic and Alpine Research, University of Colorado, Boulder/ Colorado, pp 35–62

Menke J, Trlica M (1981) Carbohydrate reserve, phenology, and growth cycles of nine Colorado range species. J Range Management 34:269–277

Meuriot F, Simon JC, Decau MP, Prudhomme MP, Morvan-Bertrand A, Gastal F, Volenec JJ, Avice JC (2005) Contribution of initial C and N reserves in *Medicago sativa L.* recovering from defoliation: modulation by the cutting height and the residual leaf area. Funct Plant Biol 32:321–334

Miller AE, Bowman WD (2003) Alpine plants show species-level differences in the uptake of organic and inorganic nitrogen. Plant Soil 250:283–292

Miller A, Schimel J, Sickman J, Skeen K, Meixner T, Melack J (2009) Seasonal variation in nitrogen uptake and turnover in two high-elevation soils: mineralization responses are site-dependent. Biogeochemistry 93:253–270

Monson RK, Rosenstiel TN, Forbis TA, Lipson DA, Jaeger CH (2006) Nitrogen and carbon storage in alpine plants. Integrative Comparative Biol 46:35–48

Moore BD, Cheng SH, Sims D, Seemann JR (1999) The biochemical and molecular basis for photosynthetic acclimation to elevated atmospheric $CO_2$. Plant Cell Environ 22:567–582

Naumburg E, Loik ME, Smith SD (2004) Photosynthetic responses of *Larrea tridentata* to seasonal extreme temperatures under elevated $CO_2$. New Phytol 162:323–330

Noguès-Bravo D, Araujo M, Errea M, Martínez-Rica J (2007) Exposure of global mountain systems to climate warming during the 21st century. Glob Env Chang 17:420–428

Nowak RS, Ellsworth DS, Smith S (2004) Functional responses of plants to elevated atmospheric $CO_2$: do photosynthetic and productivity data from FACE experiments support early prediction? New Phytol 162:253–280

Oaks A, Hirel B (1986) Nitrogen metabolism in roots. Ann Rev Plant Physiol 36:345–365

Öncel I, Yurdakulol E, Keles Y, Kurt L, YIldIz A (2004) Role of antioxidant defense system and biochemical adaptation on stress tolerance of high mountain and steppe plants. Acta Oecol 26:211–218

Osone Y, Ishida A, Tateno M (2008) Correlation between relative growth rate and specific leaf area requires associations of specific leaf area with nitrogen absorption rate of roots. New Phytol 179:417–427

Pate JS (1980) Transport and partitioning of nitrogenous solutes. Ann Rev Plant Physiol 31:313–340

Pate JS (1983) Patterns of nitrogen metabolism in higher plants and their ecological significance. In: Lee JA, McNeill S, Rorison IH (eds) Nitrogen as an ecological factor. Blackwell Scientific Publishing, Oxford, pp 225–255

Pollock CJ, Cairns AJ (1991) Fructan metabolism in grasses and cereals. Ann Rev Plant Physiol Plant Mol Biol 42:77–101

Poorter H, Pérez-Soba M (2001) The growth response of plants to elevated $CO_2$ under non-optimal environmental conditions. Oecologia 129:1–20

Pornon A, Escavarage N, Lamaze T (2007) Complementarity in mineral nitrogen use among dominant plant species in a subalpine community. Am J Bot 11:1778–1785

Raab TK, Lipson DA, Monson RK (1999) Soil amino acid utilization among species of the cyperaceae: plant and soil processes. Ecol 80:2408–2419

Reich P, Walters M, Ellsworth D (1992) Leaf-life span in relation to leaf, plant, and stand characteristics among diverse ecosystems. Ecol Monogr 62:2142–2147

Reich PB, Ellsworth DS, Walters MB, Vose JM, Gresham C, Volin JC, Bowman WD (1999) Generality of leaf trait relationships: a test across six biomes. Ecology 80: 1955–1969

Reviere-Rolland H, Contard P, Betsche T (1996) Adaptation of pea to elevated $CO_2$: rubisco, phosphoenolpyruvate carboxylase and chloroplast phosphate translocator at different levels of nitrogen and phosphorus nutrition. Plant Cell Env 19:109–117

Riedo M, Grub A, Rosset M, Fuhrer J (1997) A pasture simulation model for dry matter production, and fluxes of carbon, nitrogen, and water energy. Ecol Model 105:141–183

Rigollier C, Bauer O, Wald L (1999) On the clear sky model of the 4th European Solar Radiation Atlas with respect to the Heliosat method. Solar Energy 68:33–48

Sakai A, Larcher W (1987) Frost survival of plants. Response and adaptation to freezing stress, Ecological Studies 62. Springer, Berlin

Saxe H, Ellsworth DS, Heath J (1998) Tree and forest functioning in an enriched $CO_2$ atmosphere. New Phytol 139: 395–436

Scheurwater M, Koren M, Lambers H, Atkin OK (2002) The contribution of roots and shoots to whole plant nitrate reduction in fast- and slow-growing grass species. J Exp Bot 53: 1635–1642

Schmidt SK, Costello EK, Nemergut DR, Cleveland CC, Reed SC, Weintraub MN, Meyer AF, Martin AM (2007) Biogeochemical consequences of rapid microbial turnover and seasonal succession in soil. Ecology 88:1379–1385

Shaver G, Chapin F, Gartner B (1986) Factors limiting seasonal growth and peak biomass accumulation in *Eriophorum vaginatum* in Alaskan tussock tundra. J Ecol 74:983–989

Sierra-Almeida A, Cavieres L (2010) Summer freezing resistance decreased in high-elevation plants exposed to experimental warming in the central Chilean Andes. Oecologia 163: 267–276

Sørensen T (1941) Temperature relations and phenology of the northeast Greenland flowering plants. Meddelelser om Grønland 125:1–305

Stewart IT (2009) Changes in snowpack and snowmelt runoff for key mountain regions. Hydrol Process 23:78–94

Stitt M, Krapp A (1999) The interaction between elevated carbon dioxide and nitrogen nutrition: the physiological and molecular background. Plant Cell Env 22:583–621

Theurillat JP, Guisan A (2001) Potential impact of climate change on vegetation in the European Alps: a review. Clim Chang 50:77–109

Theurillat J-P, Aeschimann D, Küpfer P, Spichiger R (1994) The higher vegetation units of the Alps. Colloq Phytosociol 23:189–239

Thomas RB, Strain BR (1991) Root restriction as a factor in photosynthetic acclimation of cotton seedlings grown in elevated carbon dioxide. Plant Phys 96:627–634

Tjoelker MG, Craine JM, Wedin D, Reich PB, Tilman D (2005) Linking leaf and root trait syndromes among 39 grassland and savannah species. New Phytol 167:493–508

Valladares F, Pearcy RW (1997) Interactions between water stresses, sunshade acclimation, heat tolerance and photoinhibition in the sclerophyll *Heteromeles arbutifoliar*. Plant Cell Env 20:25–36

Valladares F, Pugnaire FI (1999) Tradeoffs between irradiance capture and avoidance in semi-arid environments assessed with a crown architecture model. Ann Bot 83:459–469

Westoby M (1998) A leaf-height-seed (LHS) plant ecology strategy scheme. Plant Soil 199:213–227

Wright IJ, Reich PB, Westoby M, Ackerly DD, Baruch Z, Bongers F, Cavender Bares J, Chapin T, Cornelissen JHC, Diemer M, Flexas J, Garnier E, Groom PK, Gulias J, Hikosaka K, Lamont BB, Lee T, Lee W, Lusk C, Midgley JJ, Navas ML, Niinemets U, Oleksyn J, Osada N, Poorter H, Poot P, Prior L, Pyankov VI, Roumet C, Thomas SC, Tjoelker MG, Veneklaas EJ, Villar R (2004) The world-wide leaf economics spectrum. Nature 428:821–827

Wyka T (1999) Carbohydrate storage and use in an alpine population of the perennial herb, *Oxytropis sericea*. Oecologia 120(2):198–208

Xu XL, Ouyang H, Kuzyakov Y, Richter A, Wanek W (2006) Significance of organic nitrogen acquisition for dominant plant species in an alpine meadow on the Tibet plateau, China. Plant Soil 285:221–231

# From the Flower Bud to the Mature Seed: Timing and Dynamics of Flower and Seed Development in High-Mountain Plants

# 10

Johanna Wagner, Ursula Ladinig, Gerlinde Steinacher, and Ilse Larl

## 10.1 Introduction

High mountains are climatically extreme environments. Short growing seasons and low temperatures are the most important factors limiting plant life at higher altitudes. In the mountains of temperate and cold climates, the period available for growth, flowering and seed production varies with relief and snow accumulation in winter (e.g. Crawford 2008; Galen and Stanton 1991; Kudo 1991, 1992; Galen and Stanton 1995; Kudo and Suzuki 1999; Inouye et al. 2002, 2003; Körner 2003; Ladinig and Wagner 2005; Molau et al. 2005; Kudo and Hirao 2006; Ladinig and Wagner 2007; Inouye 2008). In the European Alps, the growing season lasts 3–5 months in the alpine belt and 1–3 months in the ice-free areas of the nival belt (Larcher 1980; Larcher and Wagner 2009; Wagner et al. 2010). Not only short snow-free periods but also large temperature fluctuations and sudden cold spells with fresh snow, which can occur at any time during the growing season, are typical of mountain habitats. This produces a stop–start situation, additionally shortening the time available for growth and development. The plant species differ in how well they have adapted to such climatic extremes which increase with elevation. Accordingly, in the Alps, species richness decreases from more than 200 species in the upper alpine zone to about 30 species in the nival zone (Grabherr et al. 1995). Only a dozen specialists still occur above 4,000 m a.s.l. (Ozenda 1988; Grabherr et al. 1995; Körner 2003). To be successful in such a harsh environment, plants need to cope with temperature extremes while actively growing (Larcher and Wagner 1976; Neuner et al. 1999; Taschler and Neuner 2004; Larcher et al. 2010), to maintain metabolism over a broad temperature range (Larcher and Wagner 1976; Körner and Diemer 1987) and to complete vegetative and reproductive development within a short period of time.

Reproductive development, which is particularly susceptible to disturbances, requires the precisely coordinated timing of different processes from floral induction to seed maturation. During floral induction the shoot apex shifts from vegetative to reproductive, forming an inflorescence or a single flower. In most mountain plants, floral development is initiated 1 year prior to maturation (Billings and Mooney 1968; Mark 1970; Nakhutsrishvili 1999; Larl and Wagner 2006; Ladinig and Wagner 2009), or even earlier (Diggle 1997). Overwintering flower buds are also the rule in most arctic plants (Sørensen 1941). The earlier floral development starts and the further developed flower buds enter winter, the earlier they bloom in the following growing season (Molau et al. 2005). Thus, the timing of reproductive phases in the year of anthesis is determined by the course of floral development in the preceding year.

Anthesis is the functional phase of a flower. The length of time a flower is functional depends on a variety of factors. On the one hand, species-specific properties such as type of gender sequence (adichogamous or dichogamous) and pollination mechanisms are decisive. On the other hand environmental factors such as temperature and, in insect

J. Wagner (✉) • U. Ladinig • G. Steinacher • I. Larl
Institute of Botany, University of Innsbruck, Innsbruck, Austria
e-mail: Johanna.Wagner@uibk.ac.at

C. Lütz (ed.), *Plants in Alpine Regions*,
DOI 10.1007/978-3-7091-0136-0_10, © Springer-Verlag/Wien 2012

pollinated flowers, pollinator frequency affect the course of anthesis. Fertilization marks the onset of seed development which is comprised of histogenesis (formation of seed tissues and early embryogenesis) and maturation. With the release of mature seeds the reproductive cycle is terminated. Mature seeds of many alpine plants exhibit relative dormancy (Amen 1966; Billings and Mooney 1968; Giménez-Benavides et al. 2005; Shimono and Kudo 2005) and can persist for a variable length of time in seed banks. This persistence is pivotal in renewing populations (Stöcklin and Bäumler 1996; Erschbamer et al. 2001; Marcante et al. 2009).

Timing and dynamics of all these reproductive processes depend on both the species-specific developmental pattern and on environmental factors – in particular on temperature and photoperiod. In the literature, the time course of reproductive development in high-mountain plants is mainly documented by phenological observations at the population level (e.g. Arroyo et al. 1981; Bahn and Körner 1987; Prock 1990) or at the plot level (e.g. Kudo 1991; Theurillat and Schlüssel 2000; Kudo and Hirao 2006; Inouye 2008; Makrodimos et al. 2008). Such records show the length of different phenophases such as prefloration (time span between snowmelt and first flowering), anthesis, seed development and seed maturity for a cohort of individuals, however, they provide little information about the species-specific developmental dynamics in individual flowers.

To get a more in-depth view of reproductive processes in high-mountain plants, the reproductive timing and the development dynamics were analysed on the basis of single flowers in a multi-year study. Eleven abundant herbaceous plant species with different altitudinal distributions in the European Alps were studied (Table 10.1). Species are common either in the alpine zone (*Gentianella germanica*, *Ranunculus alpestris*, *Saxifraga androsacea*, *S. caesia*) or from the subnival to the nival zone (*Androsace alpina*, *Cerastium uniflorum*, *R. glacialis*, *S. biflora*, *S. bryoides*). *Saxifraga moschata* and *S. oppositifolia* cover a particularly wide altitudinal range and occur from the alpine to the nival zone. Some of the nival species have even been recorded above 4,000 m at climatically favourable microsites.

In this chapter we give an overview of the species-specific patterns and strategies of reproductive development. Special focus is given to developmental dynamics and to the influence of the environmental factors, temperature and day length. We further address the question of whether reproductive strategies differ with respect to the altitudinal distribution and what impact prolongation of the growing season might have on the reproductive performance in the investigated species. Some of the results have already been presented in individual publications (Ladinig and Wagner 2005, 2007, 2009; Larl and Wagner 2006; Steinacher and Wagner 2010; Wagner et al. 2010; Steinacher and Wagner 2011). Here we draw general conclusions from the comparative analyses.

## 10.2 Study Sites and Methods

Most investigations took place between 2001 and 2008. The investigations were carried out at different elevations at four localities in the Tyrolean Alps (alpine zone: Hafelekar 2,320 m a.s.l., Northern Calcareous Alps, 47°18′N, 11°23′E; subnival zone: forelands of the Tux Ferner 2,650 m a.s.l., Zillertal Alps, 47°04′N, 11°40′E and the Schaufelferner 2,850 m a.s.l., Stubai Alps, 46°59′N, 11°07′E). At each alpine and subnival locality, early and late-thawing sites were chosen. Plant temperatures (boundary layer temperatures) were recorded at all sites at hourly intervals throughout the investigation period, using small data loggers (Tidbit, Onset, Bourne, MA, USA). To follow the developmental dynamics exactly, all investigations were conducted on individually labelled plants and flowers. Structural changes to reproductive tissues were quantitatively recorded using different microscopic methods (DIC, SEM, fluorescence microscopy) and image analysis software (Optimas 6.5, Optimas Corp., Seattle, WA, USA). For more details see e.g. Ladinig and Wagner (2007, 2009), Steinacher and Wagner (2010), and Wagner et al. (2010).

## 10.3 Timing of Flower Development

The majority of the investigated species extend the reproductive cycle over two growing seasons, but there were marked differences among species in the extent of flower preformation and the timing of the different reproductive phases (Fig. 10.1). Most

**Table 10.1** Characteristics of study species

| Species | Abbr. | Geographical distribution | Mountain belt[a] | Vertical distribution in the European Alps (m a.s.l.)[b] | Sampling sites[c] | Flowering time | Gender sequence |
|---|---|---|---|---|---|---|---|
| *Androsace alpina* L. | A. alp | European Alps | Subnival – nival | 2400–4000 [4200] | TxG, StG | July–August | Protandrous |
| *Cerastium uniflorum* (Clairv.) | C. uni | European Alps | Subnival – nival | 2000–3400 | StG | July–August | Protandrous |
| *Gentianella germanica* (Willd.) subsp. *germanica* | G. ger | Alpine grasslands in Western and Central Europe | Subalpine – alpine | 500–2400 [2700] | PK | Sept–Oct | Adichogamous |
| *Ranunculus alpestris* L. | R. alp | European mountains | Alpine | 1700–2800 [2940] | HK | June | Adichogamous |
| *Ranunculus glacialis* L. | R. gla | Arctic, European mountains | Subnival – nival | 2300–4000 [4275] | StG | June–July | Adichogamous |
| *Saxifraga androsacea* L. | S. and | Eurasic mountains | Alpine | 1800–3000 | HK | June–July | Protogynous |
| *Saxifraga biflora* All. | S. bif | European Alps | Subnival – nival | 2200–4000 [4450] | TxG | July–August | Protogynous |
| *Saxifraga bryoides* L. | S. bry | European mountains | Subnival – nival | 2000–4000 [4200] | StG | July–August | Protandrous |
| *Saxifraga caesia* L. | S. cae | European mountains | Alpine | 1600–3000 | HK | July–August | Protandrous |
| *Saxifraga moschata* Wulfen | S. mos | Eurasic mountains | Alpine – nival | 1600–4000 [4200] | HK | June–July | Protandrous |
| *Saxifraga oppositifolia* | S. opp | Arctic; mountains in Europe, Asia, N-America | Alpine – nival | 1800–3800 [4500] | HK, TxG | May–July | Protogynous |

[a]Mountain belt: subnival = alpine-nival ecotone (Pauli et al. 1999), nival = glacier zone (above the permafrost limit; Grabherr et al. 2003)

[b]Vertical distribution according to Anchisi (1985), Hegi (1975), Kaplan (1995), Körner (2011), Landolt (1992) and Zimmermann (1975); numbers in square brackets give the highest localities in the Swiss Alps reported up to date

[c]Sampling sites: *HK* Mt. Hafelekar, *PK* Mt. Patscherkofel, *TxG* Tux Glacier foreland, *StG* Stubai glacier foreland

investigated species show a two-season strategy, i.e. flower bud initiation occurs in the first year and flowering and fruiting in the second year. Only two species (*C. uniflorum*, *S. caesia*) follow the one-season strategy and develop the flower buds completely in the year of anthesis.

In *S. oppositifolia*, single terminal flowers develop on short-stem shoots. The preformed flower buds overwinter in a nearly fully differentiated pre-meiotic state. Meiosis is passed immediately after snowmelt; often female gametogenesis is still ongoing during anthesis (Wagner and Tengg 1993; Ladinig 2005; Larl and Wagner 2006). Anthesis starts about 1 week after snowmelt regardless of the date of snowmelt. This differs from flower bud initiation. In our investigations we found flower bud initials only in June and July. This means when anthesis occurs in May – which is the case in earlier melting sites in the alpine zone – flower bud formation starts about 1 month later. At later melting sites in the subnival zone, anthesis and flower bud formation started at the same time. This has led to the assumption that flower initiation of *S. oppositifolia* is day-length dependent and occurs under long-day conditions only (Larl and Wagner 2006). For arctic ecotypes it is possible that the long-day requirement for floral induction is particularly marked, as the plants experience 24-h days during the period of active growth.

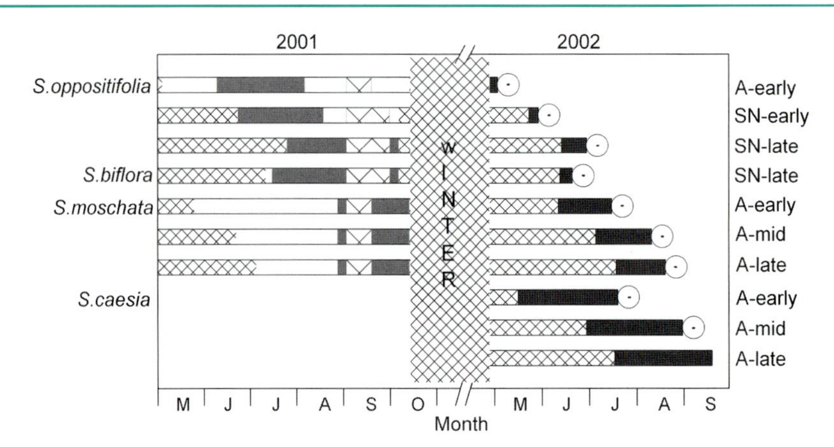


**Fig. 10.1** Timing of flower development in saxifrages in the investigation period 2001–2002 at climatically different sites. A alpine site; SN subnival site; Early, Mid, Late refer to the melting dates of the winter snow cover. *Columns* show different events during the year; *narrow hatched*: winter snow cover; *wide hatched*: temporary snow cover; *white*: periods without active flower development; *grey*: period of flower development in the first year; *black*: period of flower development in the year of anthesis (corresponds to the prefloration period); *round symbol*: anthesis

*S. biflora* shows a similar developmental pattern to the closely related *S. oppositifolia* (Hörandl and Gutermann 1994; Gugerli 1997), however development proceeds faster. Unlike *S. oppositifolia*, *S. biflora* inflorescences bear 1–12 flowers. As *S. biflora* colonizes late melting sites in the subnival and nival zone, reproductive and vegetative development starts under the thinning snow cover, a phenomenon regularly observed when radiation reaches the ground (Kimball and Salisbury 1974; Salisbury 1985). The terminal flower starts anthesis about 1 week after snowmelt, lateral flowers open at intervals over the following week. During anthesis of the current year, flower buds for the following year show all flower whorls in a primordial state (Larl 2007). Flower preformation is terminated at the time of fruit maturity of the current year flowers, which occurs about 8 weeks after the plants have become snow-free. Terminal flowers reach a well-differentiated pre-meiotic state, lateral flowers lag somewhat behind.

*S. moschata* flowers about 1 month after snowmelt, which in our study on an alpine population was at the end of June in earlier melting sites and at the end of July for later sites. Irrespective of the flowering date, flower development did not start before late August. As a consequence, flower buds entered winter in an early primordial state and had to pass through most floral development stages in the year of anthesis (Larl 2007).

Similarly, flowers of *S. bryoides* develop largely or even completely in the year of anthesis (Ladinig and Wagner 2009). New floral apices appear as day-length decreases from August on. Flower buds attain only primordial stages before winter and form three cohorts of flowers in the second year. The most developed buds immediately resume floral development after winter and bloom about 7 weeks later. A second cohort of buds flowers about 10 weeks after snowmelt, whereas a third cohort does not develop beyond a middle stage. At the end of the growing season, flower buds of different stages are present, but only primordial stages survive winter.

*S. caesia* follows the one-season strategy (Larl 2007). The transition from the vegetative to reproductive apex (Fig. 10.2) possibly occurs during snowmelt in spring. In individuals becoming snow-free in early May it took 3 weeks until the floral apex of the terminal flower bud became visible. Bolting began 6 weeks and anthesis about 2 months after snowmelt. In late melting individuals (at the end of June–early July), development was clearly accelerated: early stages of flower development were passed within 1 week and anthesis set in within 7 weeks. At the latest melting site (mid-July) floral development was markedly retarded again and flower buds did not enter anthesis before winter.

Among the non-saxifrages, the *Ranunculus* species show the most advanced flower preformation (Widmann and Wagner, unpublished). In *R. glacialis*, inflorescences develop at the end of lateral branches of the below-ground sympodial rhizome system. In plants emerging from the winter snow, the shoot apex is already floral when it appears, which suggests

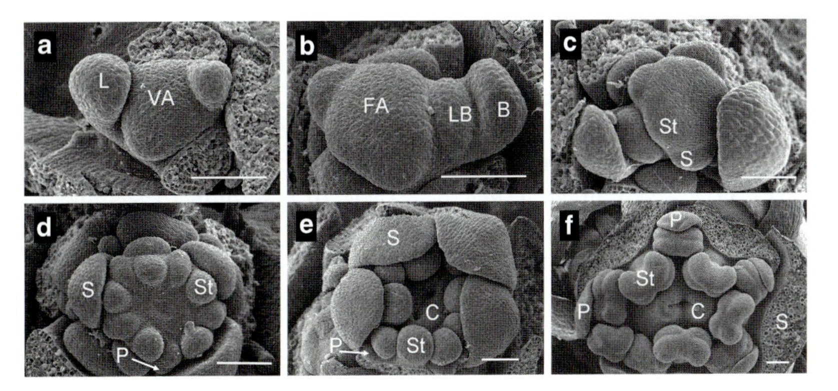


**Fig. 10.2** Stages of floral development in *Saxifraga caesia*. (**a**) Stage 0: vegetative shoot apex (*VA*) with alternately arranged leaf primordia (*L*). (**b**) Stage 1: floral apex (*FA*) of the terminal flower forms; bract (*B*) with lateral flower bud (*LB*) visible. (**c**) Stage 2: sepal primordia (*S*) arise, stamen primordia (*St*) weakly visible. (**d**) Stage 3: stamen primordia clearly visible, petal primordia (*P*) appear. (**e**) Stage 4: carpels (*C*) emerge. (**f**) Early stage 5: stamen primordia differentiate into filaments and anthers, carpels begin to elongate. During the remaining course of stage 5 floral organs further elongate and differentiate (not shown). Carpels become cone-shaped and ovule primordia emerge. Meiosis occurs shortly before anthesis

that the transition from the vegetative to reproductive apex occurs at the end of the previous growing season. Flower bud preformation goes on below ground during flowering and fruiting of the current year and stops when the above-ground parts of the plants senesce. By this time sepals fully cover the flower bud, petals are still short, stamens begin to differentiate into filaments and anthers, and in the still poorly developed carpels ovule primordia emerge. After winter, flower buds need 2–3 more weeks before entering anthesis. During this period stamens and carpels further differentiate and sporogenesis and gametogenesis take place.

Taking all species together, the length of the prefloration period was negatively correlated with the degree of flower bud preformation at snowmelt in spring (Fig. 10.3). This signifies that the timing of flower development and the state of flower bud preformation in the first year had a clear impact on the length of the prefloration period (snowmelt to first flowering) in the second year. Or in other words, the differences in flowering phenology among different species at the same site to a large extent reflect the species-specific pattern of flower preformation. This is in accordance with what Molau et al. (2005) report for tundra plants in northern Swedish Lapland, when relating prefloration periods to the winter bud stages documented by Sørensen (1941) in northeast Greenland.

Among the species investigated in our study, *S. oppositifolia* had the shortest prefloration period (6–10 days, Larl and Wagner 2006) which is in the range reported for arctic and alpine genotypes (Bliss

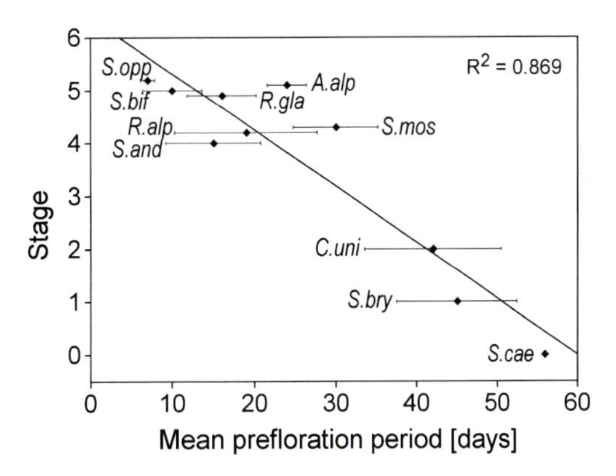


**Fig. 10.3** Correlation between the stage of flower preformation at snowmelt in spring and the length of the prefloration period ($r = 0.94$, Pearson, $p < 0.001$). For each species, the median of the maximum stage of flower development in $n = 10$ individuals emerging from the snow was determined and plotted against the mean prefloration period of individuals with the same melting date at the same site; *error bars* indicate the minimum and maximum of the first individual flowering; for staging see Fig. 10.2. For abbreviations of species names see Table 10.1. Data: S. Widmann, unpublished

1971; Stenström and Molau 1992; Stenström et al. 1997). The prefloration time in *S. biflora* is similarly short. In order of increasing length, it is followed by *R. alpestris* (7–14 days), *S. androsacea* (9–20 days), *R. glacialis* (14–21 days, up to 30 days in the nival zone, Wagner et al. 2010) and *A. alpina* (21–28 days). *S. moschata* needed about 1 month (Ladinig and Wagner 2005). The longest prefloration periods were observed in *C. uniflorum* (6 weeks), *S. bryoides*

(6–7 weeks, Ladinig and Wagner 2009) and *S. caesia* (8 weeks), in which flower preformation in the year before was limited (*S. bryoides*) or completely absent (*C. uniflorum*, *S. caesia*).

The total time taken from flower bud initiation to anthesis was longest in *S. oppositifolia*, *S. biflora*, *R. glacialis*, *R. alpestris* (about 1 year) and was shortest in *C. uniflorum* and *S. caesia* (6 and 8 weeks, respectively). Active flower development (i.e. the time between flower initiation and anthesis excluding the periods of winter dormancy and summer snow cover, which are not effectively used for development), however, did not differ much among species and generally amounted to 6–8 weeks. This signifies that there is little difference in the time needed to construct reproductive tissues among species, and it is the species-specific timing which causes developmental diversity.

## 10.4 Anthesis – The Functional Phase of the Flower

At corolla opening, the flower enters the functional phase, which in hermaphroditic flowers comprises the male phase (pollen dissemination) and the female phase (pollen deposition on the stigma and fertilization). The length of time a flower is functional may be an important determinant of male and female reproductive success (Evanhoe and Galloway 2002). Floral longevity is basically species-specific and depends on heritable traits such as gender sequence, breeding system and flower morphology (Primack 1985). However, flower longevity can be optimized by natural selection in response to the pollination environment (Ashman and Schoen 1994). Several studies have shown that flower longevity generally increases with altitude (Arroyo et al. 1981; Primack 1985; Bingham and Orthner 1998; Blionis and Vokou 2002), which is seen as compensation for the variability in pollinator visitation rates in the stochastic high mountain climate (Primack 1978; Arroyo et al. 1985; Muñoz and Arroyo 2006). Within a plant species, flower longevity is plastic and not a fixed trait. It may be extended or shortened in response to short-term environmental variations (Evanhoe and Galloway 2002; Clark and Husband 2007; Lundemo and Totland 2007). For 26 species tested in the European Alps, a mean flower longevity of 8.7 days was found (Fabbro and Körner 2004; Steinacher and Wagner 2010). However, there is a high variation among species, and within a species among different investigation periods, ranging from a few days (e.g. *G. germanica*, *R. alpestris*, *C. uniflorum*) to more than 2 weeks (*R. glacialis*, saxifrages); (Table 10.2). For the species listed in Table 10.2 we further tested the potential flower longevity, i.e. the capacity to prolong flower functions (corolla life-time; duration of stigma, style and ovule receptivity) in the case when pollinators are absent or rare (Steinacher and Wagner 2010). Unpollinated flowers generally increased longevity, but the plasticity of single floral functions was quite different. Among the female functions, stigma receptivity could be maintained longest (maximum stigma life-times were 29 days in *R. glacialis* and 24 days in *G. germanica*). Ovule receptivity, however, ceased between 16 and 20 days after onset of anthesis in most species. In some species, corolla life-time was even less plastic. Thus, the maximum longevities of individual flowers with fresh corolla and receptive pistils were around 20 days in saxifrages but only 8 days in *R. alpestris* and *C. uniflorum*.

As soon as compatible pollen is deposited on the stigma, the progamic phase, i.e. the period between pollination and fertilization, starts. Pollen germination and pollen tube growth are strongly temperature-dependent. Mountain plants show a wide optimum temperature range for progamic processes (Steinacher and Wagner 2011), which is consistent with the high temporal variability as a result of large diurnal variations in site temperatures (see Larcher, chap. 3, this book; Neuner and Buchner, chap. 6, this book). In the studied species (listed in Table 10.2), most progamic processes were still functioning at near freezing temperatures, which can be seen as an adaptation to the generally low night temperatures in high mountains (about 5°C in the alpine zone, 3–5°C in the subnival zone and $\leq$0–3°C in the nival zone; Larcher and Wagner 2009). At the other extreme, sexual functions were still intact at 25–30°C, which corresponds to the flower temperatures on clear summer days (Luzar and Gottsberger 2001; Steinacher and Wagner 2011). The length of the progamic phase strongly depends on the speed of pollen tube growth and on the species-specific lengths of stigma and style, which is the distance the pollen tubes have to cover. Highest speeds were attained at 30°C (mean growth rates 3,100 $\mu m\ h^{-1}$ in *G. germanica*, 1,550 $\mu m\ h^{-1}$ in *C. uniflorum*, 418 $\mu m\ h^{-1}$ in *R. glacialis*, and 250 $\mu m\ h^{-1}$ in

**Table 10.2** Actual longevity (natural pollination) and potential longevity (emasculated, pollinators excluded) of corollas, and lengths of gender phases in climatically different years and periods

| Species | Year | Site | Mean temperatures [°C]/days with snow | Actual longevity [days] | | | Potential longevity [days] | | Source |
|---|---|---|---|---|---|---|---|---|---|
| | | | | Corolla longevity | Male phase | Female phase | Corolla longevity | Female phase | |
| *C. uniflorum* | 2001 | TxG | 11.4/0 | Up to 4 | Up to 2 | Up to 2 | – | – | E |
| | 2007 | StG | 10.1/1 | 4–8 (11) | 1–4 (6) | 2–7 | 4–7 | 6–8 | B |
| *G. germanica* | 1998 | PK | 12.1/0 | Up to 5 | Up to 2 (3) | Up to 5 | – | – | F |
| | 2005 | PK | 6.3/2 | 3–5 | 2–5 | 3–5 | 6–15 | 14–24 | B |
| *R. alpestris* | 2003 | HK | 13.0/0 | 6–8 | 4–5 (6) | Up to 8 | 9–10 | 10–11 | A |
| | 2005 | HK | 10.2/3 | 4–9 | 4–7 | 7–11 | 4–9 | 10–18 | B |
| *R. glacialis* | 2001 | TxG | 9.5/0 | 6–7 | 8 | 5–8 | – | – | E |
| | 2003 | StG | 11.3/0 | 6–7 | 4–7 | 4–7 | 8–9 | >20 | A |
| | 2007 | TxG | 9.5/6 | 8–18 | 7–17 | 8–18 | 4–19 | 16–29 | B |
| *S. bryoides* | 2002 | StG | 6.5/3 | 12 – (14) | Up to 8 | 3–5 | – | – | C |
| | 2003 | StG | 9.3/1 | Up to 10 | Up to 5 | 3–4 | – | – | C |
| | 2007 | StG | 10.1/1 | 8–11 | 2–4 | 3–7 | 8–19 | 10–16 | B |
| *S. caesia* | 2005 | HK | 7.0/7 | 17–19 | 6–11 | 6–8 | 13–24 | 7–13 | B |
| *S. moschata* | 2001 | HK | 9.6/1 | – | 7–8 | Up to 6 | – | – | D |
| | 2005 | HK | 7.9/7 | 11–13 | 4–7 | 4–7 | 13–22 | 6–13 | B |

Temperatures are mean temperatures during the investigation periods. The duration of respective flower functions are ranges or maxima in days. Numbers in *brackets* refer to single flowers. Male phase: first anther dehiscing until all pollen sacs empty; female phase in individual flowers: stigma expanded and turgid
Investigation sites: *HK* Mt. Hafelekar, *PK* Mt. Patscherkofel, *TxG* Tux Glacier, *StG* Stubai glacier
*A* present authors, unpublished results; *B* Steinacher and Wagner (2010); *C* Ladinig and Wagner (2007); *D* Ladinig and Wagner (2005); *E* S. Erler, unpublished; *F* Wagner and Mitterhofer (1998). – no observations

*R. alpestris*) and at 25°C (saxifrages, mean maximal growth rate about 600 μm h$^{-1}$). These speeds are within the range reported for lowland plants: e.g. *Lilium longiflorum* 2,400 μm h$^{-1}$ at 17–20°C (Janson et al. 1994), *Prunus avium* 300 μm h$^{-1}$ at 25°C and *Primula obconica* 290 μm h$^{-1}$ at 30°C (Lewis 1942). Due to efficient pollen tube growth in most mountain species, the progamic phase lasted only a few hours at 20–30°C and between 12 and 30 h at 5°C (Steinacher and Wagner 2011). In comparison, lowland plants need 12–72 h at 20–30°C (Dafni et al. 2005) but mostly show a drastically reduced pollen performance below 10°C (e.g. Lewis 1942; Pasonen et al. 2000).

The full period from onset of anthesis until fertilization (Figs. 10.4 and 10.5) primarily depends on whether pistils are receptive from the very beginning (adichogamous and protogynous flowers), or whether the female phase follows the male phase (protrandrous flowers).

Additionally, weather conditions (temperature, precipitation) and related pollinator frequency, and the speed of pollen tube growth (see above) affect this time span. Fertilization occurs fastest in the adichogamous species *G. germanica*, whose stigma is already receptive on the first day of anthesis (Steinacher and Wagner 2010). Seed development starts only a few hours after pollination (Steinacher and Wagner 2011). In the adichogamous to weakly protandrous species *R. glacialis*, stigmas lie close together at the onset of anthesis, however stigma tips are already papillous and receptive (Steinacher and Wagner 2010). Nevertheless, first fertilizations were observable only 2–3 days after onset of anthesis (DAA). This is because only about 60% of ovules contain mature embryo sacs when flowers open. In the remainder of ovules, gametogenesis was still going on. Similar holds true for the protogynous saxifrages *S. androsacea*, *S. oppositifolia* and *S. biflora*. Fertilizations occurred from two DAA in *S. androsacea* and after four to five DAA in *S. oppositifolia* and *S. biflora*. In the two latter species most ovules were still in an early stage of gametogenesis at the beginning of flowering (Wagner and Tengg 1993; Ladinig 2005). Thus pistils were ready for pollination but not for fertilization which can be seen as a mechanism to make self-fertilization more difficult.

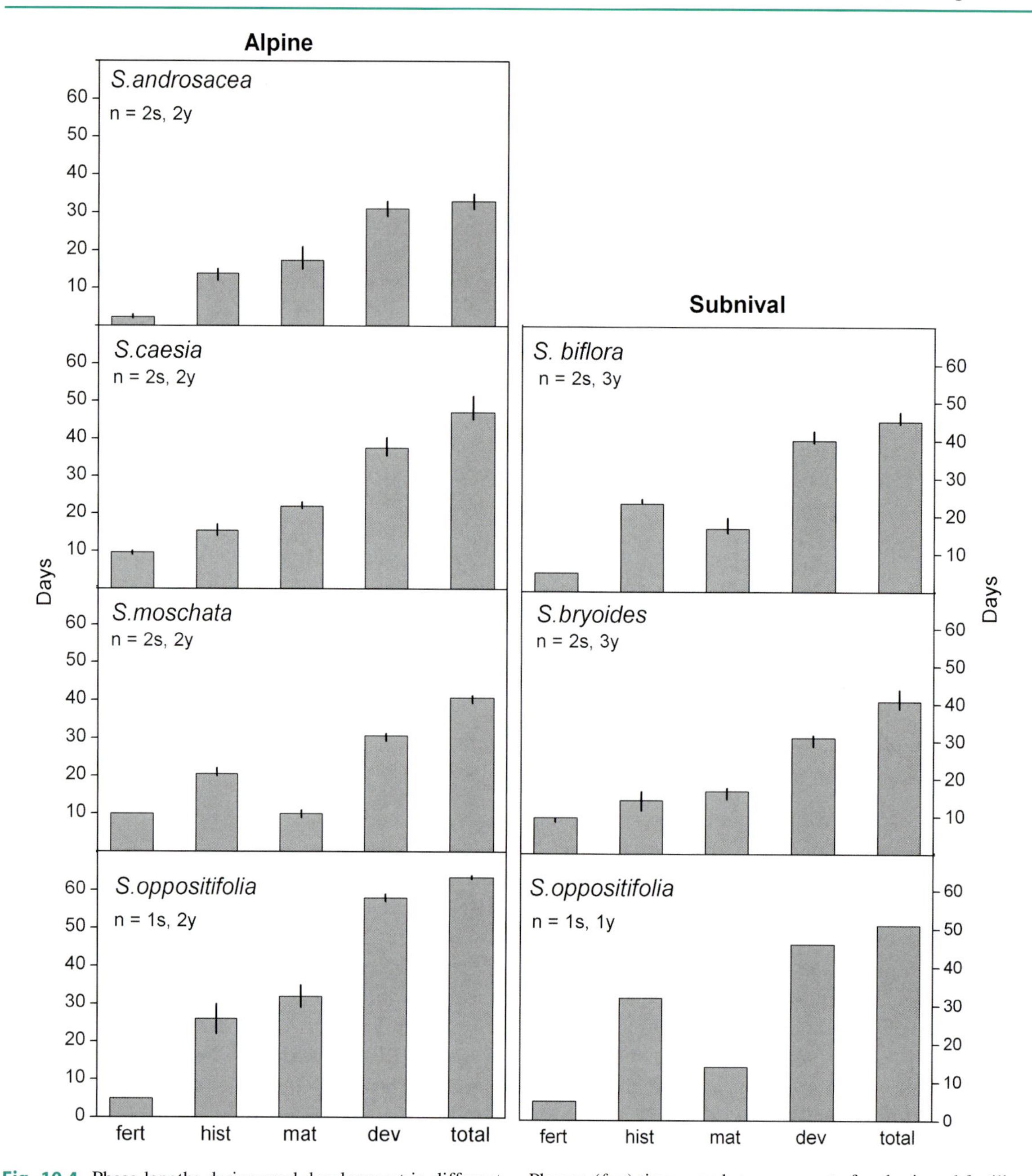


**Fig. 10.4** Phase lengths during seed development in different *Saxifraga* species at alpine and subnival sites. *Columns* give the mean duration ±SD for n sites (*s*) and years (*y*). Data based on about 100 investigated flowers and 300 seeds per site and year. Phases: (*fert*) time-span between onset of anthesis and fertilization, (*hist*) histogenesis, (*mat*) maturation phase, (*dev*) seed development from fertilization to seed maturity, and (*total*) total phase from onset of anthesis until seed maturity

The moderately protrandrous flowers of *C. uniflorum* and *A. alpina* were fertilized four to six DAA, whereas the markedly protrandrous flowers of *S. caesia*, *S. moschata* and *S. bryoides* needed up to 10 days for fertilization to take place (Ladinig 2005; Ladinig and Wagner 2005, 2007).

## 10.5 Dynamics of Seed Development

Fertilization marks the onset of seed development which, depending on the species, started 2–10 days after onset of anthesis (DAA) (cf. Figs. 10.4 and 10.5).

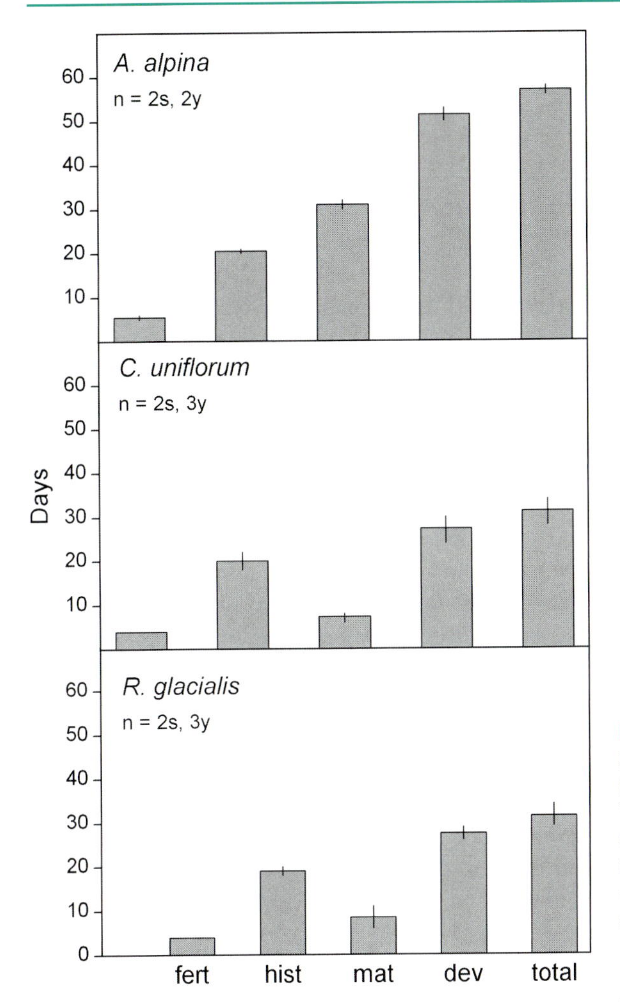


**Fig. 10.5** Phase lengths during seed development in the nival plant species *A. alpina*, *C. uniflorum* and *R. glacialis*. Columns give the mean duration ±SD for n sites (*s*) and years (*y*). Data based on about 100 investigated flowers and 300 seeds per site and year. For phases see Fig. 10.4

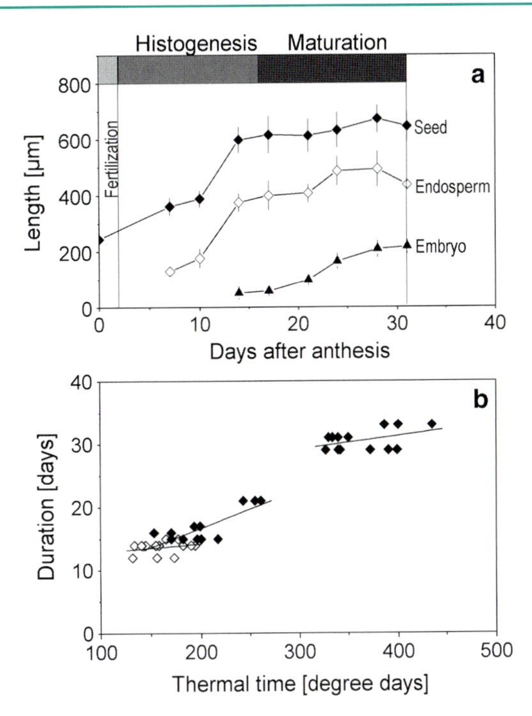


**Fig. 10.6** Seed development in *Saxifraga androsacea*. (**a**) Dynamics of seed development expressed as the increase in the length of the entire seed, the endosperm, and the embryo. Values are means ±SD of 100 seeds and 30 embryos on average. (**b**) Duration of histogenesis (*open symbols*), maturation period (*grey symbols*) and total period for seed development (*black symbols*) plotted against thermal time. *Trend lines*: linear regression for each period

Two main phases of seed development can be distinguished: (1) histogenesis during which the seed coat, the nutrient tissue (endosperm or perisperm) and a globular embryo form and (2) seed maturation which comprises seed filling (reserve deposition), further embryo growth and maturation drying to acquire desiccation tolerance (Fig. 10.6a).

The length of time taken to complete histogenesis differed markedly among species, even within the same genus. Within the saxifrages (Fig. 10.4), the mean time taken to form the seed tissues was relatively short in *S. androsacea* (15 days), *S. bryoides* (15 days) and *S. caesia* (16 days). The longest time taken to complete histogenesis was observed in the subnival population of *S. oppositifolia* (32 days), whereas the alpine population completed this phase in only 26 days. The length of histogenesis in saxifrages appears to be linked to seed size, as there is a positive correlation (Pearson $r = 0.93$, $p = 0.007$) between seed size and the length of histogenesis (Fig. 10.7). *R. glacialis*, *C. uniflorum* and *A. alpina* are exceptions with histogenesis taking only around 20 days despite seeds being comparatively large. The mode of resource allocation and the developmental pattern in these species might play a role. *C. uniflorum* and *A. alpina* partition most of their dry matter in above-ground tissues, and obviously invest carbon in offspring rather than in filling large below-ground reserve pools (Körner and Renhardt 1987). Furthermore, in the seeds of *C. uniflorum*, perisperm, instead of endosperm, evolves rather quickly from the existing nucellus tissue. In *R. glacialis*, the above-/below-ground dry matter ratio is comparatively small (Körner and Renhardt 1987; Prock and

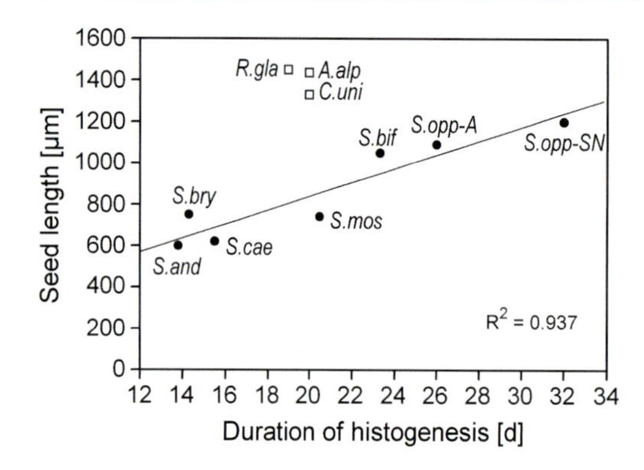


**Fig. 10.7** Relationship between seed size and duration of histogenesis, the period of seed growth. There is a good correlation (Pearson r = 0.93, p = 0.007) between seed size and the length of histogenesis in saxifrages, but not for the remainder of species. For abbreviations of species names see Table 10.1

Körner 1996). However, the comparatively large leaves already show a highly positive leaf carbon balance 3–4 weeks after snowmelt (Diemer and Körner 1996), at a time when anthesis is largely over and young seeds start to develop. Thus, an abundant supply of carbohydrates by the leaves can be assumed.

In most species investigated here, the zygote divides in an early stage of nuclear endosperm development. Firstly, a proembryo consisting of suspensor cells and an apical cell is formed. From the apical cell the embryo proper arises, which as a rule attains the globular stage at the end of histogenesis (Akhalkatsi and Wagner 1997; Ladinig 2005; Wagner et al. 2010). During seed maturation, the embryo further enlarges within the nutrient tissue. The final stage of embryo development depends both on the species and on the climatic site conditions. In a long growing season under favourable weather conditions, embryos of saxifrages mostly reach the early to late torpedo stage. In cool and short seasons, however, embryo growth often does not go beyond the heart stage, and seeds with underdeveloped embryos are shed (Ladinig and Wagner 2005, 2007). Mature seeds of *R. glacialis* generally contain a morphologically undifferentiated embryo in the late globular or early heart stage with only one poorly developed cotyledon (Wagner et al. 2010). By contrast, embryos of *C. uniflorum* are usually highly developed: the hypocotyl and cotyledons elongate markedly and due to spatial restrictions within the seed become curved (Wagner and Tengg 1993). During maturation, seeds of most investigated species become dormant. The mechanisms range from cold-stratification requirements (*C. uniflorum*, *G. germanica*) to complex still unknown dormancy mechanisms (*R. glacialis*, most saxifrages).

Depending on the plant species, the maturation phase lasted from 1 week (*C. uniflorum*, *R. glacialis*) to 1 month (*A. alpina*). No relationships between the length of the maturation phase on the one hand and the length of histogenesis, seed size and embryo size on the other hand could be found. The maturation process obviously follows a species-specific autonomous programme and moreover seems to be scarcely affected by temperature (Wagner and Reichegger 1997; Wagner and Mitterhofer 1998).

The period for seed development (histogenesis plus maturation) in a single flower is shortest in *R. glacialis* and *C. uniflorum* (28 days on average; Fig. 10.5). *S. androsacea*, *S. bryoides* und *S. moschata* require about 31 days (Fig. 10.4). Seed development lasts longest in *S. oppositifolia* (subnival site 46 days, alpine site 58 days), and *A. alpina* (52 days) which represents a doubling of time compared with the fastest group of species. Adding the time for seed development to the time between onset of anthesis and fertilization results in a total postfloration period of about 1 month for an individual flower in the fastest group of species (Fig. 10.8a).

Within the saxifrages development is only as fast in *S. androsacea*, mainly because flowers are protogynous and histogenesis is particularly short. Most saxifrages need between 40 and 50 days. However, the alpine population of *S. oppositifolia* needs more than 60 days and the postfloration period in *A. alpina* is similarly long (57 days on average). Remarkably, seeds of *S. oppositifolia* mature more rapidly in the subnival than in the alpine population. It is possible that subnival genotypes – which have adapted to the shorter growing season by enhancing development including floral development and leaf turnover (Larl and Wagner 2006) – have evolved. Evidence for adaptive variation within *S. oppositifolia* comes from early and late flowering genotypes in the high Arctic which differ markedly in morphology, growth speed, and various ecophysiological characteristics (Crawford et al. 1995; Brysting et al. 1996; Kume et al. 1999).

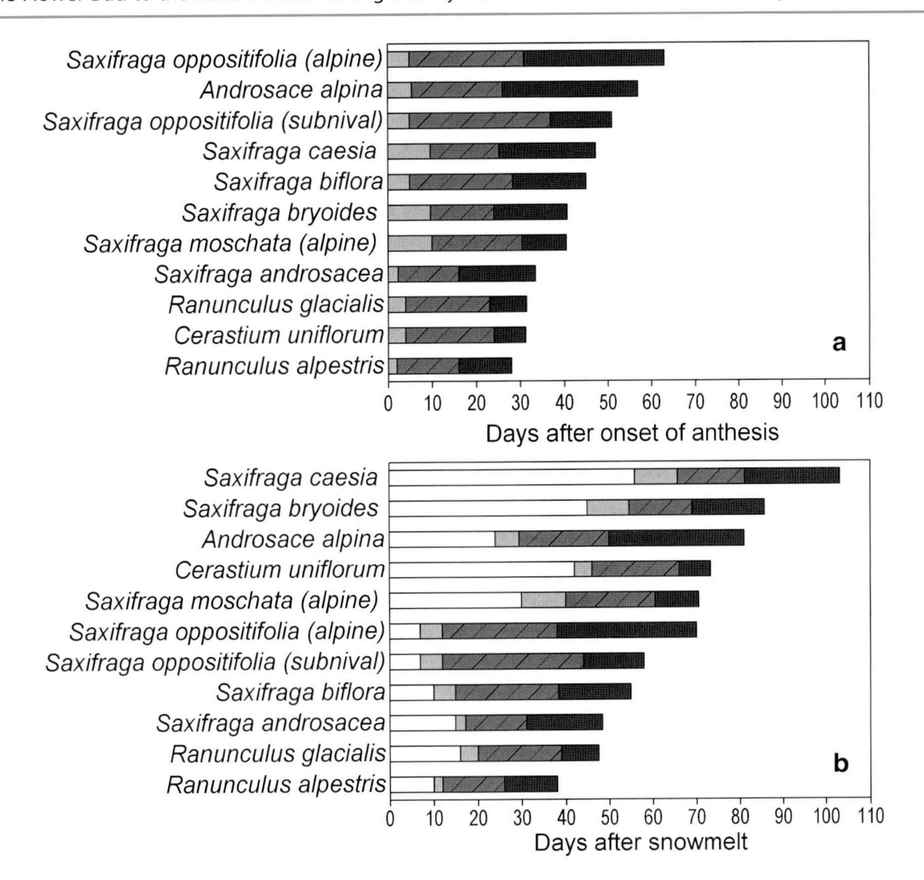


Fig. 10.8 (**a**) Species order according to the average length of the postfloration period (i.e. the time-span between onset of anthesis and seed maturity) of individual flowers presented by the total length of each column. Within the column the time spans from onset of anthesis until fertilization (*grey filling*), for histogenesis (*hatched*) and for maturation (*dark grey*) are indicated. (**b**) Species order after adding the prefloration period (i.e. the time span between snowmelt and first flowering; *white filling*) to the columns in A. The total column length gives the time from snowmelt to first seed maturity

## 10.6 Time Lapse from Snowmelt to Seed Maturity

The species-specific period required for reproductive development within a growing season comprises the prefloration period (snowmelt until onset of anthesis) and the postfloration period (onset of anthesis until seed maturity); (Fig. 10.8b). The minimum total period to produce at least some mature seeds is shortest in *R. alpestris*, *R. glacialis* and *S. androsacea* (40–44 days), somewhat longer in *S. biflora* (55 days), followed by *A. alpina*, *C. uniflorum*, *S. oppositifolia* and *S. moschata* (about 70 days). The longest reproductive periods were found in *S. bryoides* (90 days) and *S. caesia* (110 days). This period can be seen as decisive for the colonization potential of a species in high mountain habitats. Only species that regularly produce mature seeds have the chance to establish at a site. To prevent seed production from being reduced to a level below that necessary for recruitment, at least some individuals within a population have to produce mature seeds. To achieve this, more time than the minimal periods indicated above is necessary. For the investigated species, the period needed to complete reproductive development in most individuals at a site extends by a further 2–3 weeks (Fig. 10.9).

Comparing the time required from snowmelt to first seed maturity with the length of the growing season shows that in the alpine zone only later flowering individuals such as *S. caesia* are at risk of not maturing seeds in time, particularly at mid and late melting sites.

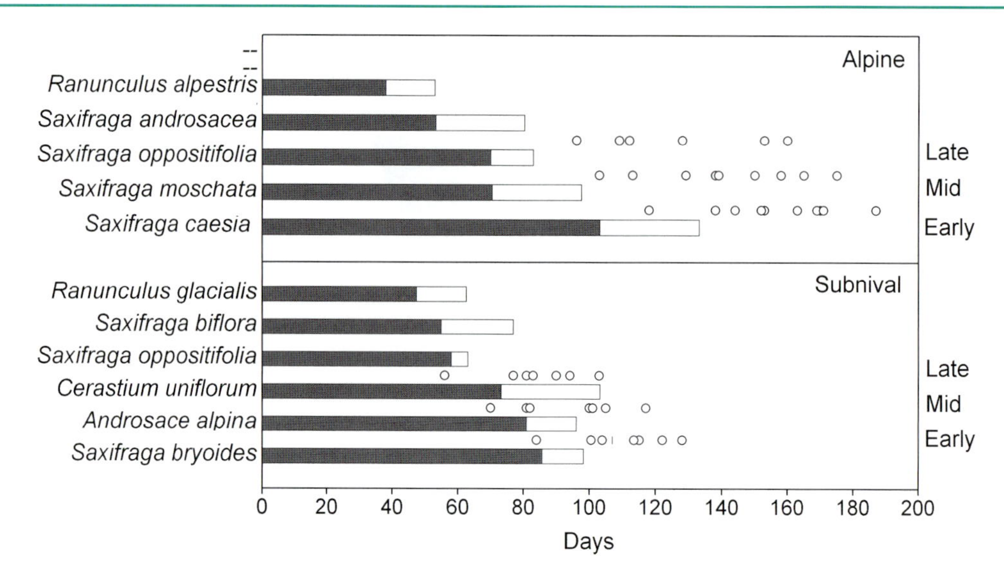


**Fig. 10.9** Time span between snowmelt and first seed maturity (*dark bar*) and seed maturity in most individuals (*right end of the white bar*) compared with the lengths of the growing seasons 2003–2009 at early, mid and late-melting alpine and subnival sites (*open circles*). In the alpine zone only the late flowering species *S. caesia* was at risk of not maturing seeds. In the subnival zone, *R. glacialis* was never at risk, whereas the remainder of species, depending on the site, failed to mature seeds in 1–4 years. Mean melting dates at early/mid/late sites were day numbers 125/140/150 for the alpine zone and day numbers 155/170/180 for the subnival zone

The situation is different in the subnival zone. During the investigation period 2003–2009 the seed crop of *R. glacialis* was never at risk. Only in 1 year did *S. biflora* and *S. oppositifolia* fail to produce any mature seeds and then only at a later melting site. *C. uniflorum* failed to produce mature seeds in the same year at both a mid and late melting site. For *A. alpina* and *S. bryoides* reproductive success was ensured only at early melting sites, but failed at mid and late melting sites in 3 and 4 years out of 7, respectively.

## 10.7 Reproductive Development and Temperature

Temperature generally influences the length of reproductive phases as reported for a number of arctic and alpine plant species (Sandvik and Totland 2000; Inouye et al. 2002; Molau et al. 2005 and citations therein; Huelber et al. 2006). This in principle also applies to the species investigated here, but our analyses have shown that the temperature effect is not linear over the full temperature range. We tested the hypothesis that variation in the speed of development was affected by site temperatures, by calculating cumulative degree-days (thermal time) as a measure for total heat accumulation during different developmental phases. Contrary to our expectations, there was frequently no relationship or even a positive one between the heat sum and the length of the different developmental phases, e.g. in *S. moschata* (Ladinig and Wagner 2005) and *S. oppositifolia* (Larl and Wagner 2006). A further example is shown in Fig. 10.6b for the phases of seed development in *S. androsacea*. In different periods of investigation, histogenesis remained constant at around 13 days, while the temperature sum varied between 30 and 200 degree days. Maturation and the whole period for seed development turned out to be even longer the higher the temperature sum was. A missing or a positive relationship between developmental time and thermal time indicates that temperature was not generally limiting reproductive development, on the contrary, there were warmer periods during which tissue differentiation could not be further accelerated. However, in some cases the heat sum did matter. So, the length of seed development and its sub-phases were negatively correlated with thermal time in *S. bryoides* (Ladinig and Wagner 2007) and *G. germanica* (Wagner and Mitterhofer 1998). In both cases seed development occurred relatively late in the growing season (August and September–October, respectively), when a heat surplus is less likely. Thus, thermal time is useful only to a degree for explaining development times in mountain plants and mostly fails during warmer periods in high summer and at thermally favoured, sun-lit sites where plants can be up to 20 K warmer than the free air temperature (Larcher and Wagner 2009; Larcher, chap. 3, this book; Neuner and Buchner, chap. 6, this book). The relationships between the length of a developmental phase and the frequency of hours which can not be used for growth was more consistent (Ladinig and Wagner 2007; Wagner et al. 2010). For cold-adapted plant species the thermal limit for growth usually lies around 2–3°C (Körner 2003, 2006). Thus in mountain plants, the length of a developmental phase is not necessarily a function of the sum of warm hours but to a large extent depends on the frequency of hours with low temperatures when development slows down or even stops.

## 10.8 Reproductive Development and Day Length

Photoperiod can affect both flower initiation (primary induction) and flowering (secondary induction; Heide 1994). Particularly in cold habitats with marked seasonal variations in climate, photoperiod control plays a crucial role in the correct timing of developmental processes. But this also means that the time available for certain developmental processes to be initiated and completed is minimized. Most knowledge about the effect of photoperiod on primary and secondary induction stems from laboratory investigations on cold-adapted plants of northern origin (e.g. Heide 1989, 1992a, b; Heide et al. 1990). In phytotron experiments most species show a short day (SD) requirement for flower induction over a wide temperature range, which is remarkable, as plants usually do not experience SD in their natural environment at high latitudes during the growing season. However, most species also have an alternative long day (LD) pathway for floral initiation at low temperatures (12°C and lower) and possibly initiate floral primordia in the late arctic summer in response to low temperatures (Heide et al. 1990). Flowering occurs under LD conditions only, whereby a 24 h LD (i.e. continuous light) is most promoting. European ecotypes of *Oxyria digyna* from different latitudes (45–78°N) show a short-long-day response as well. However, the critical day lengths vary among plants of different origins, which clearly indicates adaptation to the respective environment (Heide 2005). The SD response for flower initiation was greatest in provenances from Central Europe and decreased with increasing latitude; conversely, the critical day length for LD secondary induction of flowering increased from the southern to the northern populations. Keller and Körner (2003) investigated the role of photoperiodism and temperature on flowering (second induction) in 23 high-elevation species of the European Central Alps in the laboratory. About half of the species were found to be sensitive to photoperiod and flowered under LD conditions only; the rest were either insensitive to photoperiod or needed decreasing day length for flowering (three species). Most species were insensitive to temperature under LD-conditions, whereas an increase in temperature enhanced flowering under SD conditions. Interestingly, typical high-elevation species such as *C. uniflorum*, *Elyna myosuroides*, *R. glacialis*, *S. oppositifolia* and *S. seguieri* were insensitive to both photoperiod and temperature.

In our *in situ* study, *S. oppositifolia* was found to initiate flower buds (primary induction) only under long-day conditions (above 15 h) in June and July (Larl and Wagner 2006). Individuals thawing in early May did not develop floral shoot apices until the beginning of June. By contrast, individuals that became snow-free in July started floral development immediately after thawing. Similarly, in *S. biflora* early floral stages were observed immediately after thawing in mid-July. Flower development quickly proceeded during August, but at this time no more flowers were initiated (Larl 2007). Thus, *S. oppositifolia* and *S. biflora* follow the strategy of long-day plants combined with a low temperature requirement during winter as a precondition for flowering in the following growing season. Proleptic flowers in *S. oppositifolia* without winter vernalization were not functional, showing malformed anthers and pistils. Unlike *S. oppositifolia* and *S. biflora*, *S. moschata* did not set flowers before the end of August (day length 12–13 h), irrespective of the date of thawing in spring (cf. Fig. 10.1). This species obviously needs a decreasing day length to shift from the vegetative to the reproductive state and thus can be classified as short-day plant with a vernalization requirement for flowering in the next growing season. *S. caesia* is different from the other saxifrages as this species sets flowers in the year of anthesis, irrespectively of the day-length. When cultivated under lowland conditions, floral initiation occurs already in March (day length 11–12 h) and flowering starts in mid-May (A. Seiwald and J. Wagner, unpublished). At the alpine sites, floral initiation can be observed soon after snowmelt from late May onwards.

Thus, day length seems to affect flower initiation in at least three of the investigated saxifrages, however the date of anthesis is not affected. First flowering was primarily dependent on the date of snowmelt and set in after a species-specific prefloration period which is needed to complete floral development after winter dormancy (cf. Fig. 10.3). However, it has to be added that a possible LD requirement for flowering would have remained undetected in our field studies, as most of the investigated species had already experienced photoperiods of 15 h (passed in mid-May in Central Europe) and longer when thawing. According to Keller and Körner (2003) the critical photoperiod is 15 h, below which photoperiod-sensitive species show a response. In this context the observation that *S. biflora* individuals, transplanted to the alpine site, did not start flowering before mid-June though plants had become snow-free 2–3 weeks earlier is noteworthy. This would point to a distinct LD requirement for flowering, which is met at the later melting sites where this species usually occurs (J. Wagner, pers. obs.).

## 10.9 Differences in Reproductive Strategies Between Alpine and Nival Plant Species

One of our objectives was to find out whether plant species colonizing the nival zone employ special reproductive strategies enabling quick and effective seed production. Our investigations have shown that there is more than one reproductive strategy which is suited to the particular climatic requirements in the nival zone. There are species restricted to the alpine zone that reproduce quickly and effectively (e.g. *R. alpestris*, *S. androsacea*) and typical nival plant species that require a surprisingly long period for reproduction (*S. bryoides*, *A. alpina*, *C. uniflorum*).

*R. glacialis* appears best adapted for a life at high altitudes combining a short developmental period with a high phenological plasticity, and a relatively high reproductive success even at nival sites (Wagner et al. 2010). *S. biflora*, though requiring about 55 days for seed production, can use even shorter snow-free periods by maturing seeds below the snow (Ladinig 2005). In contrast to *S. oppositifolia*, *S. biflora* does not show reduced reproductive fitness when thawing late and flowering and fruiting in August. The sexual reproduction of *S. bryoides* is amazingly vulnerable to climatic extremes. Summer snow fall and frost from −2°C and lower regularly injure a large number of flower buds, flowers and young fruits (Ladinig et al. in prep.). In addition, the species is a typical seed-risker in the sense of Molau (1993), failing to mature seeds when winter conditions set in too early. But a high seed number in a single matured capsule might compensate for the regular losses of reproductive structures (Ladinig and Wagner 2007). Equally, *C. uniflorum* and *A. alpina* are at high risk of losing the seed crop when the growing season is too short. Accordingly, fruit set varies between 0% and 100% (*C. uniflorum*) and between 0% and 60% (*A. alpina*) among sites and years (G. Steinacher, S. Erler unpublished). These results show that quick and efficient reproduction, though advantageous, does not seem to be a prerequisite to colonize the nival zone. More important than a yearly seed crop might be the individual lifetime-reproductive success. Compact alpine cushion plants can live for several decades (Morris and Doak 1998) and contribute to seed banks in climatically favourable growing seasons (Molau and Larsson 2000).

Germination and seedling establishment is the most critical phase in the life cycle of a plant. Though climatic and mechanistic constraints increase with altitude, recent studies did not show significant relationships between establishment and altitude (Venn and Morgan 2009; Cavieres et al. 2007). Rather, the combination of various microsite factors such as shelter, soil moisture, substrate type, and extreme low and high substrate temperatures seems to be more decisive for seedling survival (Giménez-Benavides et al. 2007a; Wenk and Dawson 2007; Venn and Morgan 2009). Seed size may play a positive role in seedling emergence, particularly at higher altitudes. When compared to related lowland species, alpine species tend to have larger seeds as was shown for 29 species pairs in the Swiss Alps (Pluess et al. 2005). As seedlings of large-seeded species have higher survivorship than those of small-seeded species (Westoby et al. 1997), there might be a selection pressure for species with heavier seeds at higher altitude (Pluess et al. 2005). The species investigated in our study largely match this thesis (cf. Fig. 10.7). Within the saxifrages, the pure alpine species *S. androsacea* and *S. caesia* form the smallest seeds, *S. oppositifolia* the largest at both sites. When comparing all species, the typical nival species *A. alpina*, *R. glacialis*, and *C. uniflorum* are characterised by particularly large seeds.

## 10.10 Reproduction in a Changing Climate

The extraordinarily warm year of 2003 can serve as a model for how climate warming, together with a longer growing season, could affect the reproductive performance of high mountain plants. The growing season started about 1 month earlier at all sites and warm and dry weather conditions prevailed until the beginning of September. During the growing season, mean boundary layer temperatures were about 3 K higher at all altitudinal levels than in climatically normal years, and precipitation was 30–40% less than usual (Ladinig and Wagner 2005, 2007, 2009). Under these climatic conditions reproductive development was expected to be enhanced, the reproductive success to be particularly high and more shoot apices were expected to shift from vegetative to reproductive, increasing the flower frequency in the following year. As already stated earlier, above a certain temperature threshold, more warmth could not accelerate reproductive development. However, the long growing season was beneficial for later flowering species in that fruits became ripe long before onset of winter conditions. Nevertheless, reproductive success was not increased in all species. For *S. moschata*, *S. bryoides* and *S. caesia* the period of seed development coincided with the warmest but also driest months June and July. Enduring drought led to substantial losses during seed development (Ladinig and Wagner 2005, 2007). On the other hand, species with quick, early seed development such as *R. glacialis* and *R. alpestris* were not affected by summer drought, because they were supplied with sufficient water from snowmelt during their active phase. In these species, seed set was significantly higher than in a standard year. A check of the flowering frequency in the 2004 growing season did not show a significant increase in flowering shoots. On the contrary, *S. oppositifolia* sets even less flowers than in the preceding year, and in *S. moschata* a high percentage of shoots had died off.

These examples show that the effect of a change in site climate on different plant species would strongly depend on the seasonal timing of their development. Climate warming and a longer growing season might be beneficial for middle to late flowering species that need longer for reproductive development, but would increase the risk of heat and drought damage (Buchner and Neuner 2003; Giménez-Benavides et al. 2007b). When the growing season starts earlier, early flowering species are more at risk of damage from late spring frost events. An increase in the frequency of frost damage because of earlier snowmelt has been reported for both high altitudes (Bannister et al. 2005; Inouye 2008) and high latitudes (Molau 1996, 1997). Hence, the impacts of a changing climate differ among species according to the phenological response. The persistence of each species will essentially depend on how often and to what extent sensitive developmental phases are impaired by changing climatic forces.

**Acknowledgements** This work was supported by funding from the Austrian Science Foundation (FWF-projects P15595-B3 "Diversity of sexual reproduction in high-mountain plants" and P18398-BO3 "Pollen tube growth and pistil receptivity of high-mountain plants under extreme climatic conditions") to J. Wagner. We thank S. Erler and S. Widmann for providing data, and W. Sakai for SEM preparation. We further thank the Patscherkofelbahn and the Stubai Gletscherbahn for free transportation by cable-car.

## References

Akhalkatsi M, Wagner J (1997) Comparative embryology of three Gentianaceae species from the central Caucasus and the European Alps. Plant Syst Evol 204:39–48

Amen RD (1966) The extent and role of seed dormancy in alpine plants. Quart Rev Biol 4:271–281

Anchisi E (1985) Quatrieme contribution à l´étude de la flore valaisanne. Bull Murithienne 102:115–126

Arroyo MTK, Armesto JJ, Villagran C (1981) Plant phenological patterns in the high Andean Cordillera of central Chile. J Ecol 69:205–223

Arroyo MTK, Armesto J, Primack R (1985) Community studies in pollination ecology in the high temperate Andes of central Chile. II. Effect of temperature on visitation rates and pollination possibilities. Plant Syst Evol 149:187–203

Ashman TL, Schoen DJ (1994) How long should flowers live? Nature 371:788–791

Bahn M, Körner C (1987) Vegetation und Phänologie der hochalpinen Gipfelflur des Glungezer in Tirol. Ber Nat med Verein Innsbruck 74:61–80

Bannister P, Maegli T, Dickinson KJM, Halloy STP, Knight A, Lord JM, Mark AF, Spencer KL (2005) Will loss of snow cover during climatic warming expose New Zealand alpine plants to increased frost damage? Oecologia 144:245–256

Billings WD, Mooney HA (1968) The ecology of arctic and alpine plants. Biol Rev 43:481–529
Bingham RA, Orthner AR (1998) Efficient pollination of alpine plants. Nature 391:238–239
Blionis GJ, Vokou D (2002) Structural and functional divergence of *Campanula spatula* subspecies on Mt Olympos (Greece). Plant Syst Evol 232:89–105
Bliss LC (1971) Arctic and alpine plant life cycles. Annu Rev Ecol Syst 2:405–438
Brysting AK, Gabrielsen TM, Sørlibråten O, Ytrehorn O, Brochmann C (1996) The purple saxifrage, *Saxifraga oppositifolia*, in Svalbard: two taxa or one? Polar Res 15:93–105
Buchner O, Neuner G (2003) Variability of heat tolerance in alpine plant species measured at different altitudes. Arct Antarct Alp Res 35:411–420
Cavieres L, Badano EI, Sierra-Almeida A, Molina-Montenegro MA (2007) Microclimatic modifications of cushion plants and their consequences for seedling survival of native and non-native herbaceous species in the high Andes of central Chile. Arct Antarct Alp Res 39:229–236
Clark MJ, Husband BC (2007) Plasticity and timing of flower closure in response to pollination in *Chamerion angustifolium* (Onagraceae). Int J Plant Sci 168:619–625
Crawford RMM (2008) Plants at the margin. Ecological limits and climate change. Cambridge University Press, Cambridge/New York/Melbourne
Crawford RMM, Chapman HM, Smith LC (1995) Adaptation to variation in growing season length in arctic populations of *Saxifraga oppositifolia* L. Bot J Scotland 41:177–192
Dafni A, Kevan PG, Husband BC (2005) Practical pollination biology. Enviroquest Ltd., Cambridge
Diemer M, Körner C (1996) Lifetime leaf carbon balances of herbaceous perennial plants from low and high altitudes in the central Alps. Funct Ecol 10:33–43
Diggle PK (1997) Extreme preformation in alpine *Polygonum viviparum*: an architectural and developmental analysis. Am J Bot 84:154–169
Erschbamer B, Kneringer E, Niederfriniger-Schlag R (2001) Seed rain, soil seed bank, seedling recruitment, and survival of seedlings on a glacier foreland in the central Alps. Flora 196:304–312
Evanhoe L, Galloway LF (2002) Floral longevity in *Campanula americana* (Campanulaceae): a comparison of morphological and functional gender phases. Am J Bot 89:587–591
Fabbro T, Körner C (2004) Altitudinal differences in flower traits and reproductive allocation. Flora 199:70–81
Galen C, Stanton M (1991) Consequences of emergence phenology for reproductive success in *Ranunculus adoneus* (Ranunculaceae). Am J Bot 78:978–988
Galen C, Stanton M (1995) Responses of snowbed plant species to changes in growing-season length. Ecology 76:1546–1557
Giménez-Benavides L, Escudero A, Pérez-García F (2005) Seed germination of high mountain mediterranean species: altitudinal, interpopulation and interannual variability. Ecol Res 20:433–444
Giménez-Benavides L, Escudero A, Iriondo JM (2007a) Local adaption enhances seedling recruitment along an altitudinal gradient in a high mountain mediterranean plant. Ann Bot 99:723–734
Giménez-Benavides L, Escudero A, Iriondo JM (2007b) Reproductive limits of a late-flowering high-mountain mediterranean plant along an elevational climate gradient. New Phytol 173:367–382
Grabherr G, Gottfried M, Gruber A, Pauli H (1995) Patterns and current changes in alpine plant diversity. In: Chapin FS III, Körner C (eds) Arctic and alpine biodiversity, Ecological studies 113. Springer, Berlin, pp 167–181
Grabherr G, Nagy L, Thompson DBA (2003) An outline of Europe's alpine areas. In: Nagy L, Grabherr G, Körner C, Thompson DBA (eds) Alpine biodiversity in Europe, Ecological studies 167. Springer, Berlin/Heidelberg, pp 3–12
Gugerli F (1997) Hybridization of *Saxifraga oppositifolia* and *S. biflora* (Saxifragaceae) in a mixed alpine population. Plant Syst Evol 207:255–272
Hegi G (1975) Illustrierte Flora von Mitteleuropa, vol V/3. Paul Parey, Berlin
Heide OM (1989) Environmental control of flowering and viviparous proliferation in seminiferous and viviparous Arctic populations of two *Poa* species. Arct Alp Res 21:305–315
Heide OM (1992a) Flowering strategies of the high-arctic and high-alpine snow bed grass species *Phippsia algida*. Physiol Plant 85:606–610
Heide OM (1992b) Experimental control of flowering in *Carex bigelowii*. Oikos 65:371–376
Heide OM (1994) Control of flowering and reproduction in temperate grasses. New Phytol 128:347–362
Heide OM (2005) Ecotypic variation among European arctic and alpine populations of *Oxyria digyna*. Arct Antarc Alp Res 37:233–238
Heide OM, Pedersen K, Dahl E (1990) Environmental control of flowering and morphology in the high-arctic *Cerastium regelii*, and the taxonomic status of *C. jenisejense*. Nord J Bot 10:141–147
Hörandl E, Gutermann W (1994) Populationsstudien an Sippen von *Saxifraga* sect. Porphyrion (Saxifragaceae) in den Alpen: I. Hybriden von *S. biflora* und *S. oppositifolia*. Phyton (Horn) 34:143–167
Huelber K, Gottfried M, Pauli H, Reiter K, Winkler M, Grabherr G (2006) Phenological responses of snowbed species to snow removal dates in the central Alps: implications for climate warming. Arct Antarc Alp Res 38:99–103
Inouye DW (2008) Effects of climate change on phenology, frost damage, and floral abundance of montane wildflowers. Ecology 89:353–362
Inouye D, Morales MA, Dodge GJ (2002) Variation in timing and abundance of flowering by *Delphinium barbeyi* Huth (Ranunculaceae): the roles of snowpack, frost, and La Niña, in the context of climate change. Oecologia 130: 543–550
Inouye DW, Saavedra F, Lee-Yang W (2003) Environmental influences on the phenology and abundance of flowering by *Androsace septentrionalis* (Primulaceae). Am J Bot 90:905–910
Janson J, Reinders MC, Valkering AGM, Vantuyl JM, Keijzer CJ (1994) Pistil exudate production and pollen-tube growth in *Lilium longiflorum* Thunb. Ann Bot 73:437–446
Kaplan K (1995) Saxifragaceae. In: Weber HE (ed) Gustav Hegi – Illustrierte Flora von Mitteleuropa, vol 4/2A. Blackwell, Berlin, pp 130–229
Keller F, Körner C (2003) The role of photoperiodism in alpine plant development. Arct Antarc Alp Res 35:361–368
Kimball SL, Salisbury FB (1974) Plant development under snow. Bot Gaz 135:147–149

Körner C (2003) Alpine plant life, 2nd edn. Springer, Berlin
Körner C (2006) Significance of temperature in plant life. In: Morison JIL, Morecroft MD (eds) Plant growth and climate change. Blackwell, Oxford, pp 48–69
Körner C (2011) Coldest places on earth with angiosperm plant life. Alp Botany 121:11–22
Körner C, Diemer M (1987) In situ photosynthetic responses to light, temperature and carbon dioxide in herbaceous plants from low and high altitude. Funct Ecol 1:179–194
Körner C, Renhardt U (1987) Dry matter partitioning and root length/leaf area rations in herbaceous perennial plants with diverse altitudinal distribution. Oecologia 74:411–418
Kudo G (1991) Effects of snow-free period on the phenology of alpine plants inhabiting snow patches. Arct Alp Res 23:436–443
Kudo G (1992) Performance and phenology of alpine herbs along a snow-melting gradient. Ecol Res 7:297–304
Kudo G, Hirao AS (2006) Habitat-specific responses in the flowering phenology and seed set of alpine plants to climate variation: implications for global-change impacts. Popul Ecol 48:49–58
Kudo G, Suzuki S (1999) Flowering phenology of alpine plant communities along a gradient of snowmelt timing. Polar Biosci 12:100–113
Kume A, Nakatsubo T, Bekku Y, Masuzawa T (1999) Ecological significance of different growth forms of purple saxifrage, *Saxifraga oppositifolia* L., in the High Arctic, Ny-Ålesund, Svalbard. Arct Antarct Alp Res 31:27–33
Ladinig U (2005) Reproductive development of Saxifraga-species in a high mountain climate. Doctoral thesis, University Innsbruck
Ladinig U, Wagner J (2005) Sexual reproduction of the high mountain plant *Saxifraga moschata* Wulfen at varying lengths of the growing season. Flora 200:502–515
Ladinig U, Wagner J (2007) Timing of sexual reproduction and reproductive success in the high-mountain plant *Saxifraga bryoides* L. Plant Biol 9:683–693
Ladinig U, Wagner J (2009) Dynamics of flower development and vegetative shoot growth in the high mountain plant *Saxifraga bryoides* L. Flora 204:63–73
Landolt E (1992) Unsere Alpenflora. Fischer, Stuttgart
Larcher W (1980) Klimastreß im Gebirge – Adaptationstraining und Selektionsfilter für Pflanzen. Rheinisch-Westfälische Akad Wiss 291:49–88
Larcher W, Wagner J (1976) Temperaturgrenzen der $CO_2$-Aufnahme und Temperaturresistenz der Blätter von Gebirgspflanzen im vegetationsaktiven Zustand. Oecol Plant 11:361–374
Larcher W, Wagner J (2009) High mountain bioclimate: temperatures near the ground recorded from the timberline to the nival zone in the Central Alps. Contrib Nat Hist Berne 12:765–782
Larcher W, Kainmüller C, Wagner J (2010) Survival types of high mountain plants under extreme temperatures. Flora 205:3–18
Larl I (2007) Flower development in high mountain Saxifrages. Doctoral thesis, University Innsbruck
Larl I, Wagner J (2006) Timing of reproductive and vegetative development in *Saxifraga oppositifolia* in an alpine and a subnival climate. Plant Biol 8:155–166
Lewis D (1942) The physiology of incompatibility in plants I. The effect of temperature. Proc R Soc Lond B Biol Sci 131:13–26
Lundemo S, Totland Ø (2007) Within-population spatial variation in pollinator visitation rates, pollen limitation on seed set, and flower longevity in an alpine species. Acta Oecol 32:262–268
Luzar N, Gottsberger G (2001) Flower heliotropism and floral heating of five alpine plant species and the effect on flower visiting in *Ranunculus montanus* in the Austrian Alps. Arct Antarct Alp Res 33:93–99
Makrodimos N, Blionis GJ, Krigas N, Vokou D (2008) Flower morphology, phenology and visitor patterns in an alpine community on Mt Olympos, Greece. Flora 203: 449–468
Marcante S, Schwienbacher E, Erschbamer B (2009) Genesis of a soil seed bank on a primary succession in the central Alps (Ötztal, Austria). Flora 204:434–444
Mark AF (1970) Floral initiation and development in New Zealand alpine plants. N Z J Bot 8:67–75
Molau U (1993) Relationships between flowering phenology and life history strategies in tundra plants. Arct Alp Res 25:391–402
Molau U (1996) Climatic impacts on flowering, growth and vigour in an arctic–alpine cushion plant, *Diapensia lapponica*, under different snow cover regimes. Ecol Bull 45:210–219
Molau U (1997) Phenology and reproductive success in arctic plants: susceptibility to climate change. In: Oechel WC, Callaghan T, Gilmanov T, Holten JI, Maxwell B, Molau U, Sveinbjörnsson B (eds) Global change and arctic terrestrial ecosystems, Ecological studies 124. Springer, Berlin/Heidelberg, pp 153–170
Molau U, Larsson E-L (2000) Seed rain and seed bank along an alpine altitudinal gradient in Swedish Lapland. Can J Bot 78:728–747
Molau U, Nordenhäll U, Eriksen B (2005) Onset of flowering and climate variability in an alpine landscape: a 10-year study from Swedish Lapland. Am J Bot 92:422–431
Morris WF, Doak DF (1998) Life history of the long-lived gynodioecious cushion plant *Silene acaulis* (Caryophyllaceae), inferred from size-based population projection matrices. Am J Bot 85:784–793
Muñoz A, Arroyo MTK (2006) Pollen limitation and spatial variation of reproductive success in the insect-pollinated shrub *Chuquiraga oppositifolia* (Asteraceae) in the Chilean Andes. Arct Antarct Alp Res 38:608–613
Nakhutsrishvili G (1999) The vegetation of Georgia (Caucasus). Braun Blanquetia 15:5–74
Neuner G, Braun V, Buchner O, Taschler D (1999) Leaf rosette closure in the alpine rock species *Saxifraga paniculata* Mill.: significance for survival of drought and heat under high irradiation. Plant Cell Environ 22:1539–1548
Ozenda P (1988) Die Vegetation der Alpen. Elsevier, München
Pasonen HL, Kapyla M, Pulkkinen P (2000) Effects of temperature and pollination site on pollen performance in *Betula pendula* Roth – evidence for genotype-environment interactions. Theor Appl Genet 100:1108–1112
Pauli H, Gottfried M, Grabherr G (1999) Vascular plant distribution patterns at the low-temperature limits of plant life – the alpine-nival ecotone of Mount Schrankogel (Tyrol, Austria). Phytocoenologia 29:297–325
Pluess A, Schütz W, Stöcklin J (2005) Seed weight increases with altitude in the Swiss Alps between related species but not among populations of individual species. Oecologia 144:55–61

Primack R (1978) Variability in New Zealand montane and alpine pollinator assemblages. N Z J Ecol 1:66–73
Primack RB (1985) Longevity of individual flowers. Annu Rev Ecol Syst 16:15–37
Prock S (1990) Symphänologie der Pflanzen eines kalkalpinen Rasens mit besonderer Berücksichtigung der Wachstumsdynamik und Reservestoffspeicherung charakteristischer Arten. Ber Nat-med Verein Innsbruck 77:31–56
Prock S, Körner C (1996) A cross-continental comparison of phenology, leaf dynamics and dry matter allocation in arctic and temperate zone herbaceous plants from contrasting altitudes. Ecol Bull 45:93–103
Salisbury FB (1985) Plant adaptations to the light environment. In: Kaurin A, Junttila O, Nilsen J (eds) Plant production in the north. Norwegian University Press, Tromsø, pp 43–61
Sandvik S, Totland O (2000) Short-term effects of simulated environmental changes on phenology, reproduction, and growth in the late-flowering snowbed herb *Saxifraga stellaris* L. Ecoscience 7:201–213
Shimono Y, Kudo G (2005) Comparisons of germination traits of alpine plants between fellfield and snowbed habitats. Ecol Res 20:189–197
Sørensen T (1941) Temperature relations and phenology of the northeast Greenland flowering plants. Medd Grønland 125:1–307
Steinacher G, Wagner J (2010) Flower longevity and duration of pistil receptivity in high mountain plants. Flora 205:376–387
Steinacher G, Wagner J (2011) Effect of temperature on the progamic phase in high-mountain plants. Plant Biology, in press
Stenström M, Molau U (1992) Reproductive ecology of *Saxifraga oppositifolia*: phenology, mating system, and reproductive success. Arct Alp Res 24:337–343
Stenström M, Gugerli F, Henry GHR (1997) Response of *Saxifraga oppositifolia* L. to simulated climate change at three contrasting latitudes. Global Change Biol 3:44–54
Stöcklin J, Bäumler E (1996) Seed rain, seedling establishment and clonal growth strategies on a glacier foreland. J Veg Sci 7:45–56
Taschler D, Neuner G (2004) Summer frost resistance and freezing patterns measured *in situ* in leaves of major alpine plant growth forms in relation to their upper distribution boundary. Plant Cell Environ 27:737–746
Theurillat J-P, Schlüssel P (2000) Phenology and distribution strategy of key plant species within the subalpine-alpine ecocline in the Valaisan Alps (Switzerland). Phytocoenologia 30:439–456
Venn SE, Morgan JW (2009) Patterns in alpine seedling emergence and establishment across a stress gradient of mountain summits in south-eastern Australia. Plant Ecol Div 1:5–16
Wagner J, Mitterhofer E (1998) Phenology, seed development, and reproductive success of an alpine population of *Gentianella germanica* in climatically varying years. Bot Acta 111:159–166
Wagner J, Reichegger B (1997) Phenology and seed development of the alpine sedges *Carex curvula* and *Carex firma* in response to contrasting topoclimates. Arct Alp Res 29:291–299
Wagner J, Tengg G (1993) Phänoembryologie der Hochgebirgspflanzen *Saxifraga oppositifolia* und *Cerastium uniflorum*. Flora 188:203–212
Wagner J, Steinacher G, Ladinig U (2010) *Ranunculus glacialis* L.: successful reproduction at the altitudinal limits of higher plant life. Protoplasma 243:117–128
Wenk EH, Dawson TE (2007) Interspecific differences in seed germination, establishment, and early growth in relation to preferred soil type in an alpine community. Arct Antarct Alp Res 39:165–176
Westoby M, Leishman M, Lord J (1997) Comparative ecology of seed size and dispersal. In: Silvertown J, Franco M, Harper JL (eds) Plant life histories – ecology, phylogeny and evolution. Cambridge University Press, Cambridge, pp 143–162
Zimmermann W (1975) Ranunculaceae. In: Rechinger KH, Damboldt J (eds) Gustav Hegi – Illustrierte Flora von Mitteleuropa, vol 3/3. Paul Parey, Berlin, pp 53–341

# Plant Water Relations in Alpine Winter

# 11

Stefan Mayr, Peter Schmid, and Barbara Beikircher

## 11.1 Introduction

Plants in alpine ecosystems are trans exposed not only to overall extreme climatic conditions but also to an enormous spatial and temporal variability in environmental factors. Water relations, as a central physiological aspect of plant life, are adjusted to this environment and mirror its extraordinary heterogeneity. In temperate mountains, plants also have to overcome pronounced contrasts between the hydraulic situation in summer and winter.

In summer, hydraulic conditions for plants at higher elevation are often favorable: while evaporative forces decrease with altitude due to lower air temperatures, annual precipitation increases in most mountains of the temperate zone (e.g. Körner 2003, also see Körner 2007). In consequence, soil moisture is relatively high (e.g. Neuwinger-Raschendorfer 1963) and in most cases does not limit plant life in alpine areas (although indirect effects on mineral nutrition might play an important role; Körner 2003). However, there are local or temporal dry conditions related to spatial patterns in topography (e.g. in the lee of high mountains; Tranquillini 1979), geology, exposure, debris content and soil structure, vegetation cover, etc. Wind exposure is also a major factor influencing desiccation of soils (Neuwinger 1980). At extreme sites, like rock faces, where soil volumes are too small to enable sufficient water storage and where high temperatures force desiccation, plants show adaptations to drought stress (e.g. succulents, Körner 2003). Even trees of the timberline ecotone can suffer from some drought stress in summer, as shown by Anfodillo et al. (1998), Badalotti et al. (2000) or Giger and Leuschner (2004).

In winter, the hydraulic situation remarkably differs from summer conditions: for months, precipitation due to low air temperatures takes place only in the form of snow and thus does not contribute to soil moisture (unless snow melts). Soil temperatures usually fall below zero, whereby duration and depth of freezing periods in the soil mainly depend on the snow cover, as demonstrated by Aulitzky (1961) or Platter (1976). Early snowfall and a permanent snow cover during winter can prevent soils from reaching sub-zero temperatures, whereas wind-exposed sites were found to freeze permanently and down to a depth of 1 m. For plants, the access to soil water is blocked completely when the soil or the root system is frozen. Even when only upper soil layers are frozen, the respective xylem sections are ice-filled and do not allow water transport (see 11.6). Above the snow cover, transpirational forces are high because of overheating effects. There are many sunny periods in winter, radiation is high and amplified by reflection from the snow cover (Turner 1961, Tranquillini 1979). This leads to an increase in temperature of objects above the snow cover and, in consequence, of humidity gradients between the overheated object and the cold, dry air. This is of relevance for all plants which are higher than the snow cover as well as for plants growing on sites without snow cover. For instance, overheating of conifer needles above air temperature can reach more than 20 K in winter (Mayr et al. 2006b), so that evaporative forces reach values

S. Mayr (✉) • P. Schmid • B. Beikircher
Institute of Botany, University of Innsbruck, Innsbruck, Austria
e-mail: stefan.mayr@uibk.ac.at

C. Lütz (ed.), *Plants in Alpine Regions*,
DOI 10.1007/978-3-7091-0136-0_11, © Springer-Verlag/Wien 2012

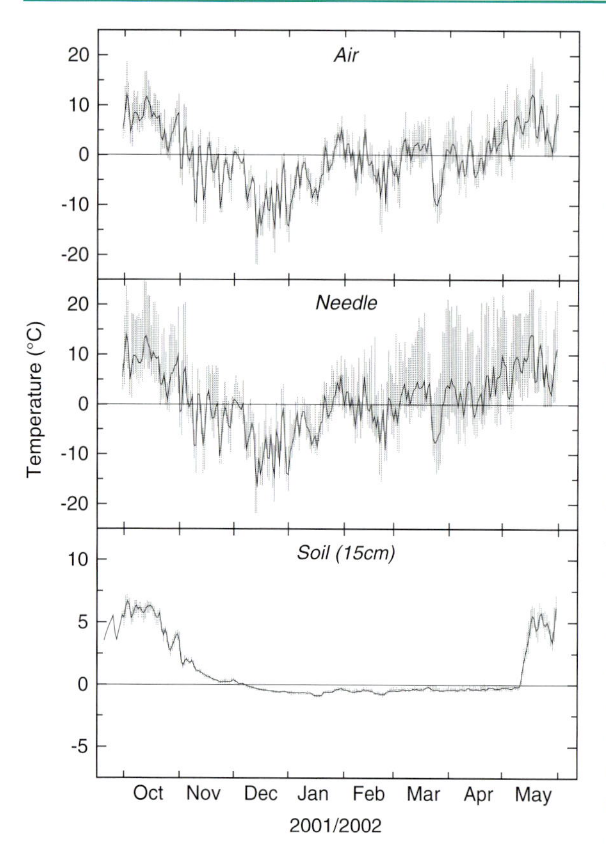


**Fig. 11.1** Temperature courses at the European Central Alps timberline during winter 2001/2002. Mean (*solid line*), maximum and minimum (*grey*) temperatures of air, sun-exposed needles and soil in 15 cm depth are given. Temperatures were monitored in Praxmar (2,010 m) on *Picea abies*. Unpublished data of the study by Mayr et al. (2006b)

like in summer (Mayr 2007). Figure 11.1 shows courses of air, soil and needle temperatures in Norway spruce growing at the Central Alps timberline. Please note that maximum needle temperatures exceed air temperatures especially in late winter, when soil temperatures are still subzero.

Regarding plant hydraulics, temperatures during winter are not only relevant because ice in the soil or in the axis system blocks water shifts. It is known that temperature fluctuations in plant parts above the snow cover can be extreme, regarding both temperature differences and velocity of temperature changes. Mayr et al. (2006b) reported daily temperature amplitudes of up to 30 K and maximum freezing and thawing rates of 5.4 and 7.0 K $h^{-1}$ in twigs of *Picea abies*. This again is due to strong overheating during sunny days and negative net radiation in clear nights. Frequent freeze-thaw events can impair xylem function, when embolism is induced by the formation and expansion of air bubbles formed in the conduits (see 11.6). The temperature regime at high altitudes also shortens the vegetation period and limits plant growth, which may affect the formation of hydraulically relevant structures such as xylem, bark or cuticles.

This article focuses on winter water relations, although winter and summer hydraulics cannot be dealt with completely independently from each other. For instance, plant water status at the beginning of winter, which is determined by water supply in autumn, can substantially influence the hydraulic situation during winter months. Boyce and Saunders (2000) demonstrated for *Picea engelmannii* and *Abies lasiocarpa* that low summer temperatures inhibit xylem formation and affect the water transport in the following months. On the other hand, impairments of the hydraulic system caused by winter stress can affect plants during the consecutive spring and vegetation period. In the following, we will first deal with water uptake from the soil, focus on the role of snow covers for plant hydraulics, discuss water loss, transport and storage during winter, and finish with a chapter on the dynamics of plant water relations during winter. As mentioned, winter is especially critical for plants not covered by snow. For this reason, the main parts of the following chapters focus on trees and shrubs of the timberline ecotone.

## 11.2 Access to Soil Water

Root systems of alpine plants are adapted to soils varying in structure and humidity on a small spatial scale. As stated by Körner (2003), "alpine plants almost always have a small fraction of deeper roots, not uncommonly reaching to 1 m in depth. These roots secure some water supply even under conditions when surface soils desiccate." A remarkable example of an extended root system provides *Pinus cembra*: when specimens, derived from seeds hidden by nutcrackers (*Nucifraga caryocatactes*; Mattes 1982), grow at the top of rocks, their roots can span meters to reach the soil and soil water. However, these observations do not allow a prediction on functional traits of an alpine plant's root system, as the efficiency for water uptake may be strongly influenced on the overall area of the fine root's absorption zone and transport resistances in

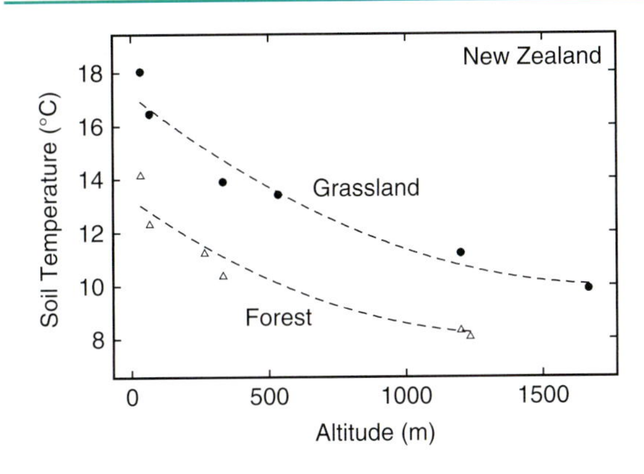


**Fig. 11.2** Altitudinal change in soil temperatures. Soils below grassland (*closed circles*) and forest (*open triangles*) were measured at several altitudes. Modified according to Körner et al. (1986)

the root system. Low soil temperatures in general reduce root growth (e.g. Tranquillini 1973, 1979, Häsler et al. 1999, Alvarez-Uria and Körner 2007), which may even cause a sink limitation and, according to Körner's theory on the formation of the timberline (Körner 1998), limit tree growth in higher altitudes (Fig. 11.2).

Low temperatures also directly affect water uptake by the roots, e.g. as shown for *Picea glauca* (Goldstein et al. 1985): at the treeline in Alaska, a reduction in water uptake took place when soil temperatures dropped below 9°C. In alpine plants, this may be of relevance in late autumn as well as in early spring and impair possibilities to fill up internal water reservoirs before the onset of winter and possibilities to improve a plant's water status at the beginning of the vegetation period. Thus, even when the soil stays unfrozen below the snow cover (e.g. Aulitzky 1961, Platter 1976), water uptake may be reduced due to low soil temperatures. This will be problematic for plants which are higher than the snow cover but not for plants buried by the snow (see below).

A freezing of the soil in late autumn or early winter causes a complete blockage of water supply from the soil, and in alpine areas this blockage often lasts for several months. Tranquillini (1979) estimated that water uptake is usually blocked from December to April at European timberline locations with little or no snow cover. Sap flow measurements on *Pinus cembra* demonstrated that water uptake is not possible unless all parts of the basal water transport way are ice-free (Mayr et al. 2003c). So even when soil layers have already melted, the frozen stem base situated in the snow cover prevents any water flow to higher crown parts.

## 11.3 Snow Cover

From a hydraulic point of view, snow covers are an excellent shield against unfavorable conditions during winter. First, high humidity in the snow pack reduces transpirational water losses of plants. Melting snow may provide water not only for the soil but even for water uptake over axes or leaves (Katz et al. 1989). Second, snow protects from wind. This reduces transpirational forces on the one hand and mechanical damage (ice blast) on the other. Third, temperatures within snow layers are moderate (around 0°C) and radiation is reduced, so that overheating effects do not play a role. Snow also reduces temperature fluctuations, which is of importance as freeze-thaw cycles can affect the water transport system of plants (see 11.6).

However, there might be indirectly negative effects on water relations, as soil development and thus water storage capacity is poor at sites with a long lasting snow cover. Many plants can only survive under the humid conditions below the snow cover. When the snow cover is lacking or removed (e.g. by wind, avalanches, or human activities), plants are exposed to intense drought stress (besides other stresses such as high radiation) and, in consequence, can dehydrate lethally.

Plants and even plant parts below snow covers thus dramatically differ in water relations compared to plants or plant parts above the snow cover (Hadley and Smith 1986, Boyce and Lucero 1999, Mayr et al. 2003b). This can easily be demonstrated by measurements of water potentials in mid-winter, as shown in Fig. 11.3. At a rocky site with only partial snow cover, *Calluna*, *Vacccinium* and *Rhododendron* specimens below the snow exhibited only moderately negative potentials (−0.5 to −1.5 MPa). In plants growing at snow-free sites, water potentials decreased below −6 MPa. The uncovered *Rhododendron* probably did not survive winter as, at a potential of −6.8 MPa, its water transport system is already completely damaged (Mayr et al. 2010a). In contrast, *Calluna*, adapted to wind-blown sites, is known to withstand winter desiccation. Similarly, within tree

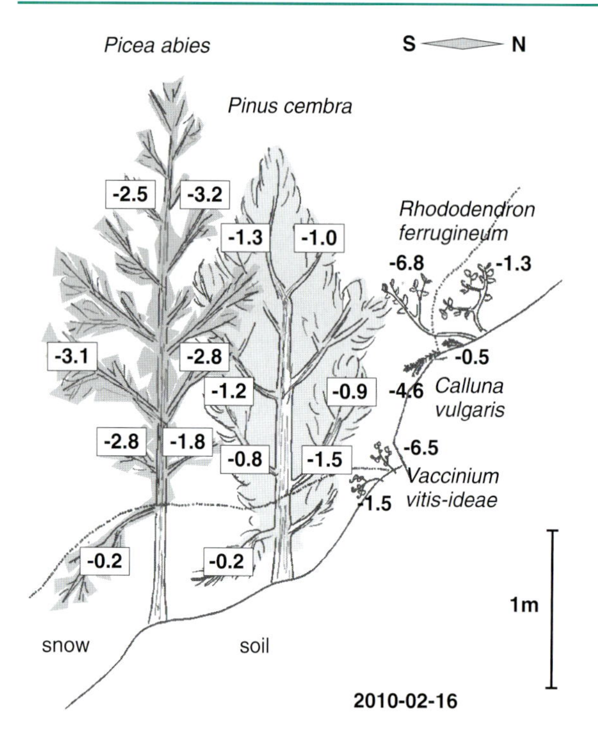


**Fig. 11.3** Water potentials at the European Central Alps timberline in winter. Water potentials (MPa) of two conifer trees and three shrubs were measured with a Scholander apparatus on twigs cut above and below the snow cover in February 2010

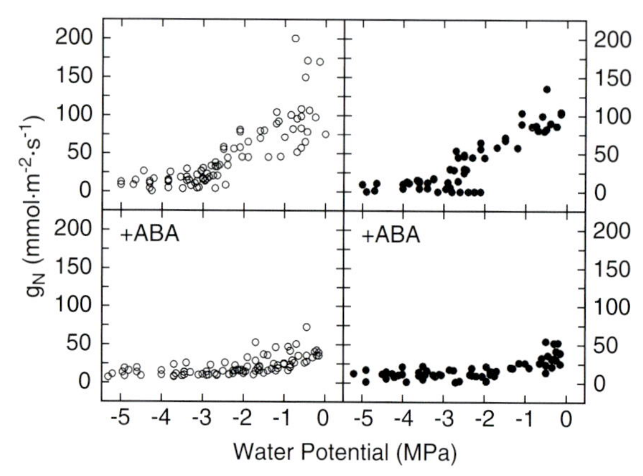


**Fig. 11.4** Conductance ($g_N$) of needles at different water potentials. Measurements were made on sun-exposed (*open symbols*) or shaded (*closed symbols*) needles of *Picea abies*. End twigs were watered with abscisic acid solution (100 mM) or tap water, and dehydrated to various extents under dark conditions. Transpiration was measured gravimetrically and water potentials with a Scholander apparatus. Modified according to Schwienbacher (2008)

crowns, branches buried by the snow showed potentials near zero (−0.2 MPa) while upper, exposed crown parts exhibited potentials down to -3.2 MPa (also see Mayr and Charra-Vaskou 2007).

## 11.4 Water Loss

While plants below the snow cover are situated in a comfortably humid environment, plant parts above the snow cover are closely coupled to atmospheric conditions: low air humidity, wind exposure and, very important, overheating effects (see 11.1) lead to steep transpirational gradients, whereas restrictions in water supply force these plants to reduce water losses. Larger trees at the timberline have to cope with these conditions every winter. Small trees, shrubs or other small plants are affected when their height exceeds the height of the snow cover or when the snow cover is lacking (see 11.3).

Deciduous plants have a reduced transpiring surface during winter and are thus generally less prone to transpirational water losses. However, the axes system of trees is still exposed to the atmosphere so that, amplified by overheating, peridermal transpiration can be increased. Accordingly, considerable desiccation stress was observed in some deciduous trees (Richards and Bliss 1986, Tranquillini and Platter 1983). Evergreen species have to keep stomata closed during winter and require a cuticular protection which sufficiently minimizes water losses as long as water supply is blocked. According to Michaelis (1934a) and Tranquillini (1976, 1979), evergreen trees of the timberline close their stomata in autumn, with low soil and air temperatures triggering this long-term closure (e.g. Christersson 1972, Havranek 1972, Larcher 1972, Smith et al. 1984, Goldstein et al. 1985) and abscisic acid probably playing a central role in signal transduction (Christmann et al. 1999). According to Bauer et al. (1994) the reduced stomatal conductance is a response to the reduced photosynthetic capacity in frost-hardened conifers.

However, conifer species still exhibit some transpiration, which may partly be related to incomplete stomata closure. Figure 11.4 demonstrates that application of abscisic acid leads to more than 50% reduction in leaf conductance in a water potential range between 0 and -3 MPa. This indicates that stomata may not be completely closed during winter and explains the high values of needle conductance observed in some species, such as *Picea abies* (e.g. Baig and Tranquillini 1976, Anfodillo et al. 2002, Mayr et al. 2003c). According to Michaelis (1934a, b), the vegetation period at higher altitudes may limit the maturation of needles and their cuticular layers (Günthardt and Wanner 1982, Tranquillini and Platter 1983). Baig and Tranquillini (1976) showed that the thickness of the cuticle of *Picea abies* and *Pinus cembra* decreases with altitude, and cuticular transpiration rates of several species were found to increase (Baig et al. 1974, Sowell et al. 1982, Tranquillini 1974, Platter 1976). In addition, ice-blast caused by strong winds may damage the cuticle and, in consequence, increase needle conductance and water losses (e.g. Holzer 1959, Hadley and Smith 1983, 1986, 1989, Van Gradingen et al. 1991). Biotic interactions may also amplify water losses, as demonstrated in Mayr et al. (2010b).

There are species well-adapted to high transpirational forces. *Pinus cembra* showed a three-fold lower needle conductance than *Picea abies* and, accordingly, water potentials during winter reached only −2.3 MPa in *Pinus cembra*, but −3.5 MPa in *Picea abies* (Mayr et al. 2003c, also see Wieser 2000). A closure of flushes might help *Pinus cembra* to further reduce winter transpiration (Mayr et al. 2003c).

## 11.5 Water Storage and Ice Blockages

In woody plants, water is stored in several organs, including leaves, roots and the axes system (Holbrook 1995, Larcher 1963; also see Larcher 1957). Again, during winter this is of relevance for all plants higher than the snow cover, as stored water can buffer water deficits: a study on *Picea abies* (Mayr and Charra-Vaskou 2007) revealed that water reservoirs in large trees can help the plant to avoid critical water potentials during winter. It should be noted that the hydraulic situation in a tree crown above the snow cover is complex and highly dynamic due to changing ice blockages within the tree. As long as parts of the axes system stay frozen, ice blockages can avoid water shifts within the stem, within branches or between stem, branches and leaves. The extent of these blockages depends on air temperatures and exposure of crown parts (Mayr et al. 2006b). Ice blockages can even be the reason for strange water potential patterns within trees: north-exposed branches may stay frozen for a longer time and thus attached needles may become disconnected from water stored in axes. In consequence, these needles can reach lower water potentials than sun-exposed needles with higher transpirational losses. Mayr et al. (2003c) demonstrated that the water potential of *Picea abies* end twigs was −2.7 MPa in shaded, but only −2 MPa in the south exposed crown parts in January 2002.

**Table. 11.1** Maximum water content of crown sections in timberline conifers

| | Stem | Lower branches | Upper branches |
|---|---|---|---|
| *Picea abies* | | | |
| Length (m) | 1.230 ± 0.180 | 0.301 ± 0.020 | 0.306 ± 0.015 |
| Water content (mL) | 438 ± 139 | 17 ± 2 | 18 ± 2 |
| *Pinus cembra* | | | |
| Length (m) | 1.312 ± 0.300 | 0.367 ± 0.029 | 0.242 ± 0.022 |
| Water content (mL) | 821 ± 524 | 43 ± 6 | 26 ± 5 |

The length and water content (axes xylem, periderm/bark, needles) of the stem (n = 10) and lower (n = 40) or upper branches (n = 60) are given for *Picea abies* and *Pinus cembra*. Mean ± SE. Schmid and Mayr, unpublished

The ice bodies within axes on the other hand represent water sources when they melt. It is suggested that basal stem parts, which are situated in the snow cover during several winter months, store a lot of water until these sections melt the first time in late winter and water potentials equilibrate within connected parts of the crown (Mayr and Charra-Vaskou 2007). Table 11.1 gives an overview of stored water volumes in different parts of trees growing at the alpine timberline. Like above-ground axes, roots may also serve as water reservoirs, as shown by Boyce and Lucero (1999) on *Picea engelmannii*. According to Tranquillini (1957), alpine winter can lead to more than 50% reduction in water reserves of trees.

## 11.6 Water Transport

The plant water transport system is responsible for the supply of tissues in leaves and other organs with water taken up by the roots. As long as massive blockages (e.g. frozen stem base or soil layers) prevent mass transport of water, the transport system is not relevant, but any damage during winter may affect water transport during the consecutive vegetation period. The transport function of the xylem can be affected by drought- or freeze-thaw induced formation of embolism (e.g. Zimmermann 1983, Cochard and Tyree 1990, Sperry and Sullivan 1992, Tyree et al. 1994, Nardini et al. 2000, Hacke and Sperry 2001), and these stress factors can amplify each other (Mayr et al. 2003c). Excessive drought stress in combination with numerous freeze-thaw events (more than 100 per winter; Groß et al. 1991, Mayr et al. 2006b) can cause up to 100% loss of conductivity in conifers growing at the alpine timberline (Sperry and Sullivan 1992, Sperry et al. 1994, Sparks and Black 1998, Sparks et al. 2001, Mayr et al. 2002, 2003a, b, c). This is remarkable as conifers were known to be very resistant to freeze-thaw induced embolism (e.g. Sperry and Sullivan 1992, Sperry et al. 1994), and it has so far not been possible to explain the effect of repeated freeze-thaw cycles by the classical "bubble expansion hypothesis": it is assumed that air bubbles form in the ice during freezing. During thawing, these bubbles can either re-dissolve or, when the bubble is too big or the tension in the surrounding sap is too high, expand and form an embolism. There is no reason why bubble expansion should not take place during an early cycle when it happens in a later one. Furthermore, ultrasonic emission analysis indicated processes occurring during freezing and not only on thawing, when bubble expansion is expected (Fig. 11.5; Kikuta and Richter 2003, Mayr et al. 2007, Mayr and Zublasing 2010, Mayr and Sperry 2010; also see Zweifel and Häsler 2000). The "bubble expansion hypothesis" was validated by centrifuge (Mayr and Sperry 2010) and other experiments (Pittermann and Sperry 2003, 2006), but the origin of ultrasonic emissions during freezing is still not understood. It should be noted that freezing and thawing may also be of relevance for roots. Freeze-thaw events in the soil during spring and autumn may affect root xylem which is known to be very sensitive to this kind of stress (Sperry and Ikeda 1997, Kavanagh et al. 1999, Hacke et al. 2000).

But how do alpine trees and shrubs survive despite the risk of water transport failure? The hydraulic safety of alpine conifer species was found to be in the range of low altitude species (e.g. Mayr et al. 2003c, Martinez-Vilalta et al. 2004; also see Fig. 11.8) although some adaptation was observed: vulnerability to drought induced embolism decreased with altitude in *Picea abies* twigs (Mayr et al. 2002). Figure 11.6 shows the bordered pits in earlywood, which obviously have smaller apertures at higher altitude. These smaller pores might enable a more stable sealing of the torus and thus more efficiently avoid air seeding (Zimmermann 1983). In addition, tracheids at higher altitude are narrow, exhibit thick cell walls and a high lignin content (Gindl et al. 2001, Mayr et al. 2002). The needle xylem of alpine *Pinus* species (*Pinus cembra, Pinus mugo*) was found to be significantly more resistant to xylem deformation as well as embolism when compared with low altitude pines (Cochard et al. 2004). Avoidance of critical water loss or sufficient water storage are other strategies to reduce embolism formation (see above). Most fascinating is the ability of several species to completely refill their xylem during late winter and spring (Sperry and Sullivan 1992, Sperry et al. 1994, Mayr et al. 2003c). The underlying

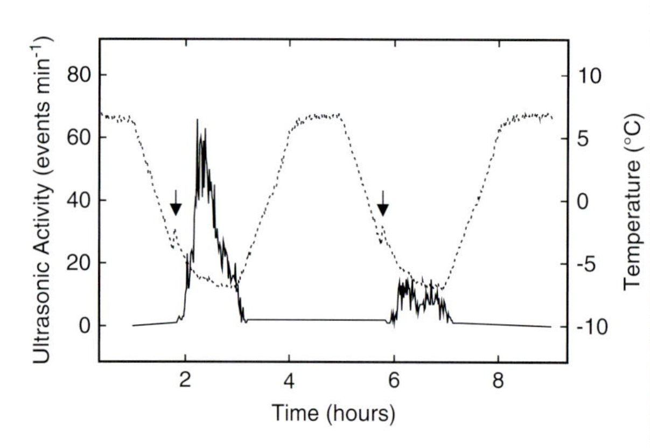


**Fig. 11.5** Ultrasonic activity during freeze-thaw events in the xylem of *Larix decidua*. The emission of ultrasonic events was counted in axis sections dehydrated to −1.8 MPa and exposed to two freeze–thaw cycles in a temperature chamber. The *solid line* shows the ultrasound activity and the *dotted line* shows the temperature course. The exotherm (*arrows*) indicates the onset of freezing. Modified according to Mayr and Zublasing (2010)

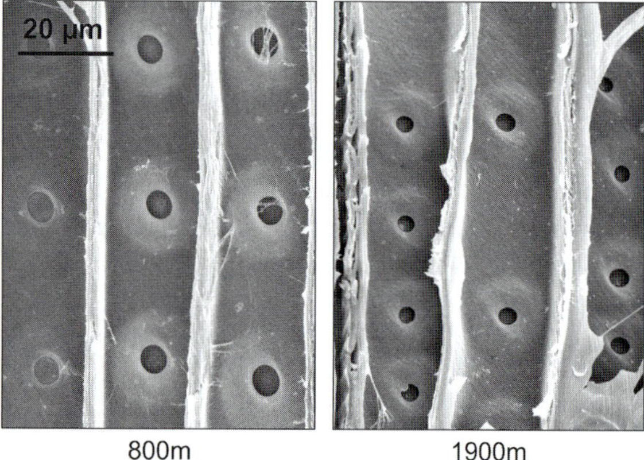


**Fig. 11.6** Pit sizes in earlywood of *Picea abies*. Pictures show radial longitudinal sections of xylem samples harvested at 800 and 1,900 m and analysed with a scanning electron microscope

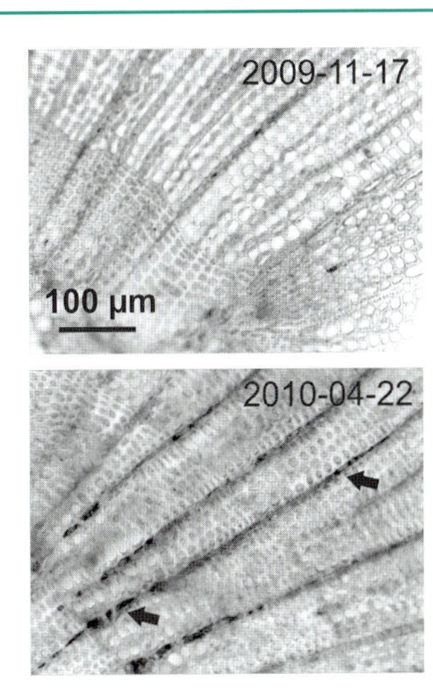


**Fig. 11.7** Starch content of ray parenchyma in branch wood. Iodine staining of starch in parenchyma cells in cross-sections of *Picea abies* branch wood. Branches were harvested in November and April at the Central Alps timberline (Birgitz Köpfl, 2030 m). Arrows indicate examples of regions with high starch content

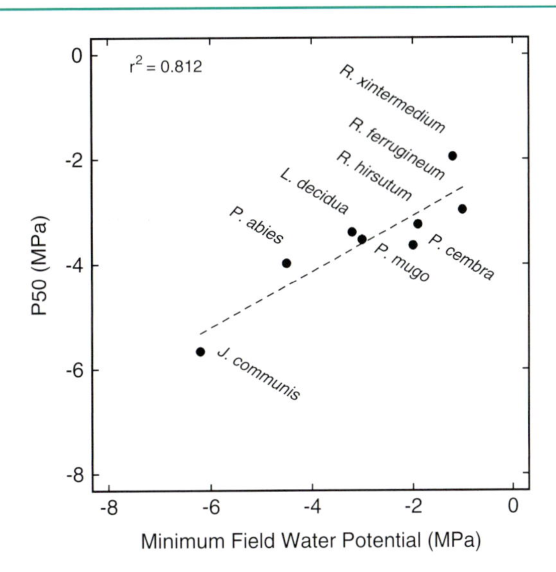


**Fig. 11.8** Correlation of thresholds for drought-induced embolism and minimum field water potentials. The water potential at 50% loss of conductivity (P50) is plotted *versus* the minimum water potentials observed in winter (*Juniperus communis*, *Picea abies*, *Larix decidua*, *Pinus mugo* and *Pinus cembra*) or in summer (*Rhododendron* species). Data are from Mayr et al. (2006a) and Mayr et al. (2010a)

mechanism (as for other species showing refilling; e.g. Zwieniecki and Holbrook 2009) is not yet understood. Katz et al. (1989) and Sparks et al. (2001) suggested water uptake via branches or leaves. Changes in starch content in the bark and ray parenchyma cells might indicate active processes of living cells within the wood (Fig. 11.7).

## 11.7 Frost Drought Dynamics

Temporal patterns of plant hydraulics during winter are very different from summer dynamics. During the vegetation period, transpirational water losses are compensated by water uptake and balanced on a small time scale by stomata regulation. In winter, when water uptake is blocked, plants are a kind of closed system but still with dynamic changes at several levels: in the long term, the water content of the system decreases due to transpirational water losses until access to soil water or water uptake over the surface is possible. The mean water content is thus lowest at the end of winter, but this is superimposed by short term effects: ice blockages within the crown can vary due to changing temperatures, resulting in a complex and permanently changing pattern of hydraulic subsystems (Mayr and Charra-Vaskou 2007). But, as mentioned, ice within the xylem may not only separate crown parts from each other but also serve as water reservoir when it thaws. In consequence, not all parts of the crown exhibit identical water potentials and reach their lowest water potentials at the same time. Embolism formed during winter contributes to separation of crown sections, and refilling processes in late winter further complicate the spatial and temporal patterns. For plants under the snow, conditions are rather constant unless the snow height decreases and plant parts are uncovered. Branches of flexible shrubs (e.g. Körner 2003) can be above and below the snow several times within one winter and, in consequence, their water potentials may be the result of a complex hydraulic history.

It is fascinating that, despite this spatial and temporal variability, thresholds for drought induced vulnerability correlate with minimum water potentials occurring at the respective timberline sites (Fig. 11.8). It is suggested that vulnerability thresholds are a good estimate for drought stress intensities plants are exposed to. Figure 11.8 demonstrates that the xylem of alpine trees and shrubs, which are higher than the snow cover, is obviously adapted to the low water potentials occurring during winter. By contrast,

**Fig. 11.9** Needle damage after frost drought at Birgitz Köpfl, Tirol, 2,030 m. Needles damaged by frost drought during winter became brown at the beginning of May 2005

plants covered by snow during winter are adapted to summer extremes in potential.

Damage from drought stress during winter can usually not be observed before spring. As alpine plants are well-adapted to their environment, massive drought damage of woody species is a rather local (single plants or distinct parts of plants, see Fig. 11.9) and occasional event (Tranquillini 1979). Water deficits in living tissues become obvious at the beginning of the vegetation period as dieback of plant parts or growth deficits. Sometimes one can observe interesting patterns of drought damage on end twigs of *Picea abies*: on the underside of these twigs, significantly more needles show drought damage than on the upper side, where sun exposure was highest. This can only be explained by ice blockages in the lower, shaded part of the twig, which remained frozen for a longer period in winter (see 11.6) and thus caused water potentials to reach critical levels.

In summary, the hydraulic situation of alpine woody plants during winter is extremely complex and depends on numerous interacting factors. Further research activities at a cellular and whole plant level are required to understand the underlying mechanisms and ecophysiological consequences. For some aspects, such as refilling processes, alpine plants can even serve as model systems. Improved, comprehensive knowledge on plant hydraulics will be a prerequisite for models and strategies for future developments of the sensitive alpine ecosystems, especially in times of global and local changes.

## References


Alvarez-Uria P, Körner C (2007) Low temperature limits of root growth in deciduous and evergreen temperate species. Funct Ecol 21:211–218

Anfodillo T, Rento S, Carraro V, Furlanetto L, Urbinati C, Carrer M (1998) Tree water relations and climatic variations at the alpine timberline: seasonal changes of sap flux and xylem water potential in *Larix decidua* Miller, *Picea abies* (L.) Karst. and *Pinus cembra* L. Ann For Sci 55:159–172

Anfodillo T, DiBisceglie DP, Urso T (2002) Minimum cuticular conductance and cuticle features of *Picea abies* and *Pinus cembra* needles along an altitudinal gradient in the Dolomites (NE Italian Alps). Tree Physiol 22:479–487

Aulitzky H (1961) Die Bodentemperaturen in der Kampfzone oberhalb der Waldgrenze und im subalpinen Zirben-Lärchenwald, vol 59. Mitteilungen der Forstlichen Bundesversuchsanstalt, Mariabrunn, pp 153–208

Badalotti A, Anfodillo T, Grace J (2000) Evidence of osmoregulation in *Larix decidua* at Alpine treeline and comparative responses to water availability of two co-occurring evergreen species. Ann For Sci 57:623–633

Baig MN, Tranquillini W (1976) Studies on upper timberline: morphology and anatomy of Norway spruce (*Picea abies*) and stone pine (*Pinus cembra*) needles from various habitat conditions. Can J Bot 54:1622–1632

Baig MN, Tranquillini W, Havranek WM (1974) Cuticuläre Transpiration von *Picea abies* und *Pinus cembra-* Zweigen auf verschiedener Seehöhe und ihre Bedeutung für die winterliche Austrocknung der Bäume an der alpinen Waldgrenze. Centralblatt für das Gesamte Forstwesen 4:195–211

Bauer H, Nagele M, Comploj M, Galler V, Mair M, Unterpertinger E (1994) Photosynthesis in cold acclimated leaves of plants with various degrees of frost tolerance. Physiol Plant 91:403–412

Boyce RL, Lucero SA (1999) Role of roots in winter water relations of Engelmann spruce saplings. Tree Physiol 19:893–898

Boyce RL, Saunders GP (2000) Dependence of winter water relations of mature high-elevation *Picea engelmannii* and *Abies lasiocarpa* on summer climate. Tree Physiol 20:1077–1086

Christersson L (1972) The transpiration rate of unhardened, hardened and dehardened seedlings of spruce and pine. Physiol Plant 26:258–263

Christmann A, Havranek WM, Wieser G (1999) Seasonal variation of abscisic acid in needles of *Pinus cembra* L. at the alpine timberline and possible relations to frost resistance and water status. Phyton 39:23–30

Cochard H, Tyree MT (1990) Xylem dysfunction in Quercus: vessel sizes, tyloses, cavitation and seasonal changes in embolism. Tree Physiol 6:393–407

Cochard H, Froux F, Mayr S, Coutand C (2004) Xylem wall collapse in water-stressed pine needles. Plant Physiol 134:401–408

Giger T, Leuschner C (2004) Altitudinal change in needle water relations of *Pinus canariensis* and possible evidence of a

drought-induced alpine timberline on Mt. Teide, Tenerife. Flora 199:100–109
Gindl W, Grabner M, Wimmer R (2001) Effects of altitude on tracheid differentiation and lignification of Norway spruce. Can J Bot 79:815–821
Goldstein GH, Brubaker LB, Hinckley TM (1985) Water relations of white spruce (*Picea glauca* (Moench) Voss) at tree line in north central Alaska. Can J For Res 15:1080–1087
Groß M, Rainer I, Tranquillini W (1991) Über die Frostresistenz der Fichte mit besonderer Berücksichtigung der Zahl der Gefrierzyklen und der Geschwindigkeit der Temperaturänderung beim Frieren und Auftauen. Forstwissenschaftliches Centralblatt 110:207–217
Günthardt MS, Wanner H (1982) Veränderungen der Spaltöffnungen und der Wachsstruktur mit zunehmendem Nadelalter bei *Pinus cembra* L. und *Picea abies* (L.) Karsten an der Waldgrenze. Bot Helvet 92:47–60
Hacke UG, Sperry JS (2001) Functional and ecological xylem anatomy. Perspect Plant Ecol Evol Syst 4:97–115
Hacke UG, Sperry JS, Ewers BE, Ellsworth DS, Schäfer KVR, Oren R (2000) Influence of soil porosity on water use in *Pinus taeda*. Oecologia 124:495–505
Hadley JL, Smith WK (1983) Influence of wind exposure on needle desiccation and mortality for timberline conifers in Wyoming, U.S.A. Arct Alp Res 15:127–135
Hadley JL, Smith WK (1986) Wind effects on needles of timberline conifers: seasonal influence on mortality. Ecology 67:12–19
Hadley JL, Smith WK (1989) Wind erosion of leaf surface wax in timberline conifers. Arct Alp Res 21:392–398
Häsler R, Streule A, Turner H (1999) Shoot and root growth of young *Larix decidua* in contrasting microenvironments near the alpine treeline. Phyton 39:47–52
Havranek WM (1972) Über die Bedeutung der Bodentemperatur für die Photosynthese und Transpiration junger Forstpflanzen und die Stoffproduktion an der Waldgrenze. Angew Bot 46:101–116
Holbrook MN (1995) Stem water storage. In: Gartner BL (ed) Plant stems: physiology and functional morphology. Academic, New York
Holzer K (1959) Winterliche Schäden an Zirben nahe der alpinen Baumgrenze. Centralblatt für das Gesamte Forstwesen 76:232–244
Katz C, Oren R, Schulze E-D, Milburn JA (1989) Uptake of water and solutes through twigs of *Picea abies* (L.) Karst. Trees 3:33–37
Kavanagh KL, Bond BJ, Aitken SN, Gartner BL, Knowe S (1999) Shoot and root vulnerability to xylem cavitation in four populations of Douglas-fir seedlings. Tree Physiol 19:31–37
Kikuta S, Richter H (2003) Ultrasound acoustic emissions from freezing xylem. Plant Cell Environ 26:383–388
Körner C (1998) A re-assessment of high elevation treeline positions and their explanation. Oecologia 115:445–459
Körner C (2003) Alpine plant life: functional plant ecology of high mountain ecosystems. Springer Berlin, Heidelberg
Körner C (2007) The use of 'altitude' in ecological research. Trends Ecol Evol 22:569–574
Körner C, Bannister P, Mark AF (1986) Altitudinal variation in stomatal conductance, nitrogen content and leaf anatomy in different plant life forms in New Zealand. Oecologia 69:577–588
Larcher W (1957) Frosttrocknis an der Waldgrenze und in der alpinen Zwergstrauchheide auf dem Patscherkofel bei Innsbruck. Veröff Ferdinandeum Innsbruck 37:49–81
Larcher W (1963) Zur spätwinterlichen Erschwerung der Wasserbilanz von Holzpflanzen an der Waldgrenze. Berichte des naturwissenschaftlich medizinischen Vereins Innsbruck 53:125–137
Larcher W (1972) Der Wasserhaushalt immergrüner Pflanzen im Winter. Berichte der Deutschen Botanischen Gesellschaft 85:315–327
Martinez-Vilalta J, Sala A, Pinol J (2004) The hydraulic architecture of pinaceae – a review. Plant Ecol 171:3–13
Mattes H (1982) Die Lebensgemeinschaft von Tannenhäher und Arve, vol 241. Eidgenössische Anstalt für Forstliches Versuchswesen, Berlin, pp 1–74
Mayr S (2007) Limits in water relations. In: Wieser G, Tausz M (eds) Trees at their upper limit. Treelife limitation at the alpine timberline. Springer, Berlin, pp 145–162
Mayr S, Charra-Vaskou K (2007) Winter at the alpine timberline causes complex within-tree patterns of water potential and embolism in *Picea abies*. Physiol Plant 131:131–139
Mayr S, Sperry JS (2010) Freeze-thaw induced embolism in *Pinus contorta*: centrifuge experiments validate the "thaw-expansion hypothesis" but conflict with ultrasonic emission data. New Phytol 185:1016–1024
Mayr S, Zublasing V (2010) Ultrasonic emissions from conifer xylem exposed to repeated freezing. J Plant Physiol 167:34–40
Mayr S, Wolfschwenger M, Bauer H (2002) Winter-drought induced embolism in Norway spruce (*Picea abies*) at the alpine timberline. Physiol Plant 115:74–80
Mayr S, Gruber A, Bauer H (2003a) Repeated freeze-thaw cycles induce embolism in drought stressed conifers (Norway spruce, stone pine). Planta 217:436–441
Mayr S, Gruber A, Schwienbacher F, Dämon B (2003b) Winter-embolism in a "Krummholz"-shrub (*Pinus mugo*) growing at the alpine timberline. Aust J For Sci 120:29–38
Mayr S, Schwienbacher F, Bauer H (2003c) Winter at the alpine timberline: why does embolism occur in Norway spruce but not in stone pine? Plant Physiol 131:780–792
Mayr S, Hacke U, Schmid P, Schwienbacher F, Gruber A (2006a) Frost drought in conifers at the alpine timberline: xylem dysfunction and adaptations. Ecology 87:3175–3185
Mayr S, Wieser G, Bauer H (2006b) Xylem temperatures during winter in conifers at the alpine timberline. Agric For Meteorol 137:81–88
Mayr S, Cochard H, Ameglio T, Kikuta S (2007) Embolism formation during freezing in the wood of *Picea abies*. Plant Physiol 143:60–67
Mayr S, Beikircher B, Obkircher M-A, Schmid P (2010a) Hydraulic plasticity and limitations of alpine Rhododendron species. Oecologia. doi:DOI 10.1007/s00442-010-1648-7
Mayr S, Schwienbacher F, Dämon B (2010b) Increased winter transpiration of Norway spruce needles after infection by the rust *Chrysomyxa rhododendri*. Protoplasma 243:137–143
Michaelis P (1934a) Ökologische Studien an der Baumgrenze IV. Zur Kenntnis des winterlichen Wasserhaushaltes. Jahrbuch für wissenschaftliche Botanik 80:169–247

Michaelis P (1934b) Ökologische Studien an der Baumgrenze V. Osmotischer Wert und Wassergehalt während des Winters in den verschiedenen Höhenlagen. Jahrbuch für wissenschaftliche Botanik 80:337–362

Nardini A, Salleo S, LoGullo MA, Pitt F (2000) Different responses to drought and freeze stress of *Quercus ilex* L. growing along a latitudinal gradient. Plant Ecol 148:139–147

Neuwinger I (1980) Erwärmung, Wasserrückhalt und Erosionsbereitschaft subalpiner Böden. Mitt Forstl Bundes Versuchsanstalt (Wien) 129:113–144

Neuwinger-Raschendorfer I (1963) Bodenfeuchtemessungen. Mitteilungen der Forstlichen Bundesversuchsanstalt Mariabrunn 59:257–264

Pittermann J, Sperry JS (2003) Tracheid diameter is the key trait determining the extent of freezing-induced embolism in conifers. Tree Physiol 23:907–914

Pittermann J, Sperry JS (2006) Analysis of freeze-thaw embolism in conifers. The interaction between cavitation pressure and tracheid size. Plant Physiol 140:374–382

Platter W (1976) Wasserhaushalt, cuticuläres Transpirationsvermögen und Dicke der Cutinschichten einiger Nadelholzarten in verschiedenen Höhenlagen und nach experimenteller Verkürzung der Vegetationsperiode. Dissertation Universität Innsbruck

Richards JH, Bliss LC (1986) Winter water relations of a deciduous timberline conifer, *Larix lyallii* Parl. Oecologia 69:16–24

Schwienbacher F (2008) Winterembolien an der alpinen Waldgrenze: vergleichende Untersuchungen an Fichte. Dissertation Universität Innsbruck, Zirbe und Lärche

Smith WK, Young DR, Carter GA, Hadley JL, McNaughton GM (1984) Autumn stomatal closure in six conifer species of the Central Rocky Mountains. Oecologia 63:237–242

Sowell JB, Kouitnik DL, Lansing AJ (1982) Cuticular transpiration of whitebark pine (*Pinus albicaulis*) within a Sierra Nevadan timberline ecotone, USA. Arct Alp Res 14:97–103

Sparks JP, Black RA (1998) Winter hydraulic conductivity and xylem cavitation in coniferous trees from upper and lower treeline. Arct Antarct Alp Res 32:397–403

Sparks JP, Campbell GS, Black RA (2001) Water content, hydraulic conductivity, and ice formation in winter stems of *Pinus contorta*: a TDR case study. Oecologia 127:468–475

Sperry JS, Ikeda T (1997) Xylem cavitation in roots and stems of Douglas-fir and white fir. Tree Physiol 17:275–280

Sperry JS, Sullivan JEM (1992) Xylem embolism in response to freeze-thaw cycles and water stress in ring-porous, diffuse-porous and conifer species. Plant Physiol 100:605–613

Sperry JS, Nichols KL, Sullivan JEM, Eastlack SE (1994) Xylem embolism in ring-porous, diffuse-porous, and coniferous trees of northern Utah and interior Alaska. Ecology 75:1736–1752

Tranquillini W (1957) Standortsklima, Wasserbilanz und $CO_2$-Gaswechsel junger Zirben (*Pinus cembra* L.) an der alpinen Waldgrenze. Planta 49:612–661

Tranquillini W (1973) Der Wasserhaushalt junger Forstpflanzen nach dem Versetzen und seine Beeinflussbarkeit. Centralblatt für das gesamte Forstwesen 90:46–52

Tranquillini W (1974) Der Einfluß von Seehöhe und Länge der Vegetationszeit auf das cuticuläre Transpirationsvermögen von Fichtensämlingen im Winter. Berichte der Deutschen Botanischen Gesellschaft 87:175–184

Tranquillini W (1976) Water relations and alpine timberline. In: Lange OL, Kappen L, Schulze E-D (eds) Water and plant life, Ecological studies, Vol 19. Springer, Berlin, pp 473–491

Tranquillini W (1979) Physiological ecology of the alpine timberline. Tree existence at high altitudes with special reference to the European Alps. Ecological studies, Vol 31. Springer, Berlin

Tranquillini W, Platter W (1983) Der winterliche Wasserhaushalt der Lärche (*Larix decidua* Mill.) an der alpinen Waldgrenze. Verhandlungen Gesellschaft für Ökologie 9:433–443

Turner H (1961) Jahresgang und biologische Wirkung der Sonnen- und Himmelsstrahlung an der Waldgrenze der Ötztaler Alpen. Wetter Leben 13:93–113

Tyree MT, Davis SD, Cochard H (1994) Biophysical perspectives of xylem evolution: Is there a tradeoff of hydraulic efficiency for vulnerability to dysfunction? IAWA J 15:335–360

Van Gradingen P, Grace J, Jeffree CE (1991) Abrasive damage by wind to the needle surface of *Pinus sylvestris* L. and *Picea sitchensis* (Bong.) Carr. plant. Cell Env 14:185–193

Wieser G (2000) Seasonal variation of leaf conductance in a subalpine *Pinus cembra* during the winter months. Phyton 40:185–190

Zimmermann MH (1983) Xylem structure and the ascent of sap. Springer Verlag, Berlin

Zweifel R, Häsler R (2000) Frost-induced reversible shrinkage of bark of mature subalpine conifers. Agricultural and Forest Meteorology 102:213–222

Zwieniecki MA, Holbrook NM (2009) Confronting maxwell's demon: biophysics of xylem embolism repair. Trends in Plant Science 14:530–534

# Ice Formation and Propagation in Alpine Plants

# 12

Gilbert Neuner and Jürgen Hacker

## Abbreviations

IDTA infrared differential thermal analysis
INA ice nucleation active

## 12.1 Occurrence of Freezing Stress in Alpine Environments

### 12.1.1 Low Temperature Strain at High Altitudes

#### 12.1.1.1 Atmospheric Temperature Minima

Low atmospheric temperatures are among the well-known common features of the alpine macroclimate (see Körner 2003). Absolute low temperature extremes at high altitude sites are no greater than at low altitude sites. The lowest absolute air temperature minimum at high altitude in the Austrian Alps was −37.4°C (Mt. Sonnblick, 3,105 m, 1 Jan. 1905). This temperature minimum is only marginally lower than the absolute air temperature minimum recorded at low altitude sites in Austria, which was −36.6°C (Zwettl, 520 m, 11 Feb. 1929). Strikingly, the lowest air temperature record in Austria originates from the bottom of a doline at an altitude of 1,270 m (Grünloch, Lunz am See, Lower Austria, 19 Feb. 1932), where due to temperature inversion an absolute minimum of −52.6°C was recorded (Aigner 1952). Absolute seasonal low temperature extremes may not be a striking environmental feature of high altitudes. However, there is a low temperature dose-effect: the duration and intensity of freezing temperatures increases distinctly with altitude, e.g. the mean minimum temperature of the coldest month on Mt. Sonnblick was −14.4°C, while in Zwettl it was −5.8°C (1971–2000, ZAMG).

Furthermore, at high altitudes the probability and occurrence of low and freezing temperatures is not restricted to the winter period but can occur throughout the whole year. During the usually snow-free growing period in summer, ambient temperatures may have their full impact on plants, and freezing temperatures can catch plants in a physiologically active state with a generally reduced frost resistance (Larcher 1985, Sakai and Larcher 1987, Körner 2003, Taschler and Neuner 2004). The severity of night frosts during the growing season increases significantly with increasing altitude and depends on the respective month (Taschler and Neuner 2004, Larcher and Wagner 2009). Absolute air temperature minima (30 years records) in June range from −6°C (1,938 m) to −14°C (3,105 m) and in July and August from −2°C at 1,938 m to −10.1°C at 3,105 m. At 3,450 m, which is around the altitudinal limit of closed vegetation in the Alps (Grabherr et al. 1995), air temperature minima for June of −16°C and for July and August of −10°C are reported from the meteorological station on Mt. Brunnenkogel. Absolute air temperature minima during the growing period are low enough to explain the occurrence of frost damage in alpine plants (see Körner 2003) as the air temperature minima can exceed the frost resistance of vegetative parts of most alpine species (Taschler and Neuner 2004).

G. Neuner (✉) • J. Hacker
Institute of Botany, University of Innsbruck, Innsbruck, Austria
e-mail: Gilbert.neuner@uibk.ac.at

C. Lütz (ed.), *Plants in Alpine Regions*,
DOI 10.1007/978-3-7091-0136-0_12, © Springer-Verlag/Wien 2012

#### 12.1.1.2 Minimum Plant Temperature

Radiative cooling may result in even lower night time leaf temperatures than one would expect from atmospheric air temperature records. Radiative cooling occurs during clear nights with calm winds and is more pronounced at high altitudes. This can expose alpine plants to leaf temperatures distinctly lower than atmospheric air temperatures. In subalpine environments leaves were found to be up to 8 K colder than air (Jordan and Smith 1994).

On the other hand snow cover is an important factor mitigating low air temperature extremes. With increasing altitude the probability of sufficient snow cover increases, particularly during winter. In a survey of boundary layer temperatures of microsites in the Tyrolean Central Alps, it was shown that the temperatures under the snow cover at the subalpine timberline (1,950 m a.s.l.), in the upper-alpine vegetation belt (2,200 m–2,230 m a.s.l.), and at a subnival (2,880 m a.s.l.) and a nival (3,450 m a.s.l.) site remained constant at 0°C or a few degrees below (Larcher and Wagner 2009). Sufficient snow cover not only thermally insulates plants from low air temperature extremes but reduces the daily leaf temperature amplitudes and the number of freeze-thaw cycles they experience (Neuner et al. 1999). Although locally, e.g. on wind-exposed ridges, alpine plants may remain snow-free and exposed for extended periods even during winter time, the lowest air temperatures recorded by the weather service may not be relevant for the majority of alpine plants which will usually be sufficiently protected by snow during winter.

Bioclimate records indicate that the frequency of frost events increases significantly with altitude – while at 1,950 m temperatures lower than 0°C occurred on 20% of the 174 snow-free days, at 3,450 m the frequency of frosts increased to 69% of a total of 74 snow-free days (Larcher and Wagner 2009).

Long-term leaf temperature records are scarce, and little information exists about the frequency of leaf freezing events. Comparing minimum leaf temperatures of six alpine plant species of diverse growth forms (woody shrubs, cushions, rosette and herbaceous plants) determined at three alpine sites (1950, 2200 and 2600 m) throughout three successive growing periods (1998, 1999, 2000) may give an indication of the approximate frequency of plant freezing events. In June, July and August mean night time minimum leaf temperatures decreased from +6°C at 1,950 m to +3.3°C at 2,600 m. Night time leaf temperature minima decreased with altitude with a rate of 0.4°C/100 m (Fig. 12.1). From this a mean night time minimum temperature of 0°C can be expected to occur at an altitude of 3,425 m. Leaf temperatures lower than 0°C occurred at a frequency of 7.2% at 1,950 m, at a frequency of 12.6% at 2,200 m and at a frequency of 24.8% at 2,600 m. Absolute leaf temperature minima recorded during this period were around −5°C. Natural cooling rates of leaves in the subzero temperature range in the alpine environment were at a mean of −0.3°C/h and did not exceed −2°C/h (see Neuner and Buchner, Chapter 6).

#### 12.1.1.3 At What Temperature Does Ice Form in Plants in Nature?

In summer in leaves of alpine plants, measurements of freezing exotherms during freezing treatments with field portable frost chambers revealed an initial

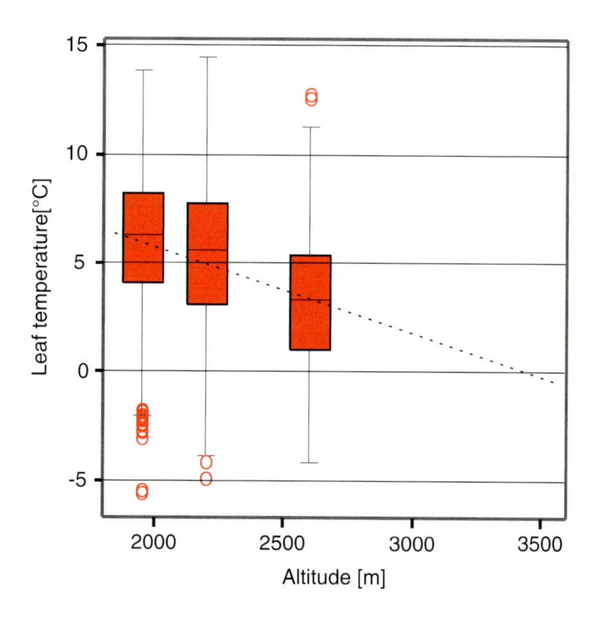


**Fig. 12.1** Box plot of leaf temperature minima of six alpine plant species of diverse growth forms (woody shrubs, cushions, rosette and herbaceous plants) determined at three alpine sites (1,950, 2,200 and 2,600 m) throughout three successive growing periods (1998, 1999, 2000) during June, July and August showing that mean night time minimum leaf temperatures decreased from +6°C at 1,950 m to +3.3°C at 2,600 m at a rate of 0.4°C/100 m (*dotted line*). The box plots show the median (line inside the box) and the 25 and 75 percentiles. The whiskers are Tukey's hinges. Values more than 1.5 interquartile range's (IQR's) but less than 3 IQR's from the end of the box are labelled as outliers (O)

intrinsic freezing event which was usually non-lethal at an average temperature of −1.9°C. In high alpine plants this could occur at between −2.6°C and −3.5°C when thermally insulated thermocouples were used (Taschler and Neuner 2004), and at between −5.2 and −6.3°C in *Loiseleuria procumbens* and *Rhododendron ferrugineum*, respectively, when thermocouples were not insulated (Hacker and Neuner, unpublished). These observations seem to overestimate the supercooling capacity of plants, as much milder subfreezing air temperatures between −0.6 and −2.6°C are reported to result in ice nucleation in plants in the field (Pearce 2001). This corresponds well with field observations where, during a mild late frost event of −2°C in April 2010 at low altitude, most meadow plants were found to be frozen stiff (Neuner, unpublished). Leaf temperatures lower than −1°C were recorded in leaves in the above mentioned study at a frequency of 1.4% (1,950 m), 2.9% (2,200 m) and 5.7% (2,600 m) of summer days. On these days ice formation in these leaves must be considered to be highly probable.

However, the measurement of realistic ice nucleation temperatures is somewhat tricky. Firstly, until now it has not been possible to simulate the environmental conditions that prevail during naturally occurring radiative cooling situations in the field, where the plant is always the coldest part. Secondly, detachment is known to provoke supercooling. When we try to avoid detachment of plant parts and treat attached plant parts in cooling chambers – even if it is done in situ – we need to consider that in a whole plant a higher number of ice nucleators are active at higher freezing temperatures than in a plant part, as the sample size significantly influences the supercooling capacity (Ashworth et al. 1985). Measurements on plant parts may, hence, overestimate the supercooling capability. Thirdly, when freezing exotherms are recorded by the use of thermocouples, freezing is recorded on a single spot. Ice can nucleate anywhere in the plant, but this will usually take place in the coldest part of the plant which need not be at the measurement spot of the thermocouple. Ice spreads rapidly at rates of up to 27 cm $s^{-1}$ (Hacker and Neuner 2007, Hacker and Neuner 2008, Hacker et al. 2008), and the ice wave will reach all parts that are colder than 0°C immediately. Hence, the temperature at which ice nucleation occurs in the plant is always a range of temperatures rather than one single temperature. Small temperature gradients within the cooled plant part of an average of 1.6 K were found to occur in plants cooled in our freezing chambers. This means that the recorded ice nucleation temperatures at a certain position can deviate by an average of 1.6 K from the actual ice nucleation temperature. This would have a great effect on the deduced supercooling capability (Fig. 12.2). Nevertheless, this may mirror the natural situation, as during natural frosts plants are exposed to temperature gradients and after ice nucleation in the coldest part this will also result in ice spreading throughout all parts colder than 0°C. Fourthly, thermal insulation of the thermocouple is usually used to increase the probability that small freezing exotherms can be detected. But this thermal insulation will protect the investigated part from fast cooling and hence it will not be the coldest part during freezing. Again such an experimental setup will cause a slight underestimation of the supercooling capacity.

Additionally, at least for some species it cannot be excluded that ice nucleation can also occur extrinsically, i.e. surface ice may grow into the plant. Under the current experimental procedures the plant is not the coldest part inside cooling chambers, and hence no surface ice is formed. Ventilation for uniform temperature distribution to prevent temperature gradients inside the frost chamber and within the sample will also prevent condensation of water on the surface of the plant. If a species is nucleated extrinsically, a result deviating from the natural situation will be obtained. External ice does not necessarily induce intrinsic freezing as the external ice has first to propagate

**Fig. 12.2** Temperature gradient observed within a single twig and attached leaves of *R. ferrugineum* during a freezing treatment in a controlled freezing chamber. Mean temperature gradients of 1.6 K were usually observed in twigs of woody species during frost treatment

into the plant (Wisniewski et al. 2002). This can occur through stomates in some species (*Phaseolus vulgaris, Prunus domestica, Malus domestica:* Wisniewski et al. 1997, Wisniewski and Fuller 1999), but in other species this is only possible via a damaged cuticle or other lesions (*Rhododendron sp.* (Wisniewski et al. 1997), *Lolium perenne* and *Poa supina* (Stier et al. 2003), *P. alpina* (Hacker and Neuner 2008)). The role of extrinsic ice nucleation is still not fully understood. Potentially, surface properties may play a decisive role for surface ice formation and ice growth into the plant. Plants from high altitudes and from open places have lower leaf wettability (Aryal and Neuner 2010), and lowered leaf wettability has been shown to prevent extrinsic ice nucleation in certain plants (Fuller et al. 2003).

## 12.2 Ice Propagation

### 12.2.1 Pattern of Whole Plant Freezing

Patterns of ice propagation were studied in whole plants of various typical alpine plant growth forms by Hacker and Neuner (2008) using infrared differential thermal analysis. Under the experimental conditions patterns of whole plant freezing were growth form specific. In woody dwarf shrubs, herbs, rosette and cushion plants a single ice nucleation anywhere in the plant was usually sufficient to produce freezing of the whole individual. This indicates that no anatomical ice barriers exist. The time necessary from initial ice formation until the whole plant was frozen

**Fig. 12.3** Digital image and sequence of IDTA images showing a typical pattern of ice propagation in a reproductive shoot of *R. ferrugineum* in the flower bud stage. Intrinsic ice nucleation occurred at −6.9°C in the stem, which caused immediate freezing of its vascular tissue and the veins in the adjacent leaves. Only in a second step of freezing did the whole leaf area turn white during extracellular mesophyll freezing. In the late flower bud stage ice propagates unhindered into the reproductive structures. The time specifications indicate the elapsed time after ice nucleation

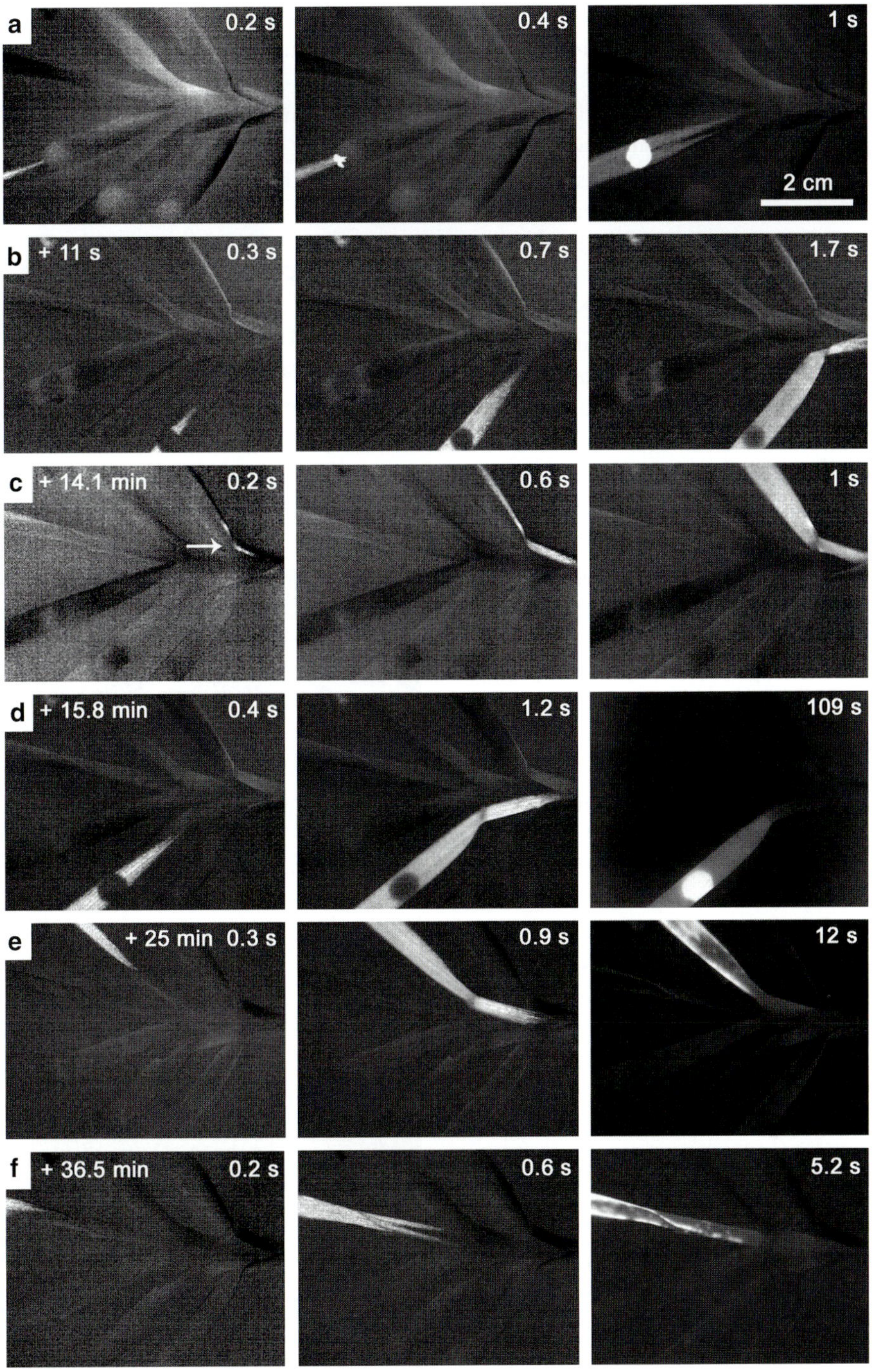

**Fig. 12.4** Sequence of IDTA images illustrating a typical freezing pattern in a tussock of *Poa alpina*. Each leaf needs a separate ice nucleation event to initiate freezing, as ice propagation from one leaf to another via the stem is blocked by an anatomical ice barrier. Ice nucleation occurred either in the leaf sheath (C, indicated by a *white arrow*) or in the leaf lamina outside the measuring frame (**a**, **b**, **d**–**g**); freezing temperatures of the leaves: **a**, **b**: −3.6°C, **c**: −5.2°C, **d**: −5.4°C, **e**: −6.3°C, **f**: −7.4°C. For the lower three leaves water droplets were applied to the leaf blade with a (**a**) damaged and (**b**, **d**) intact leaf surface, respectively. When the leaf surface was damaged, ice propagated unhindered from the leaf into the water droplet (**a**) and vice versa (data not shown). On the other hand with an intact leaf surface freezing of the water droplets was independent from leaf freezing. Time specifications marked with a "+" indicate elapsed time since freezing of the first leaf. Time specifications in each line indicate the elapsed time after ice nucleation or detection of freezing in the measuring frame of the respective leaf

was 0.2–7.2 min in woody plants and herbs and could last up to about 1 h in cushions (Hacker and Neuner 2008).

The principle pattern of ice propagation was usually found to be independent from the location of initial ice nucleation. Initially ice spreads rapidly at a rate of up to 27 cm $s^{-1}$ throughout the vascular tissue as soon as ice has contact with the water in the xylem. Only in a second step does freezing of extracellular water in other tissues occur (Hacker and Neuner 2007, 2008, Hacker et al. 2008). This freezing pattern is exemplarily shown for a reproductive shoot of *Rhododendron ferrugineum* (Fig. 12.3) where after freezing of the water in the xylem tissue, ice encroaches into the mesophyll and other adjacent tissues. In mesophyll cells this second freezing step induces functional disturbances in the thylakoids, however, in most plants they are recognizable only after a significant time lapse, i.e. 20–30 min after detection of initial ice formation in the mesophyll, and are very likely to be caused by freeze dehydration of mesophyll cells (Neuner and Pramsohler 2006, Hacker et al. 2008). In the late flower bud stage, ice propagates unhindered into the reproductive structures.

In contrast to most other alpine plant growth forms, graminoids show a distinctly different whole plant freezing pattern (Hacker and Neuner 2008). In graminoids each leaf requires a separate ice nucleation event, as there is no connection of the vascular tissue between the various leaves of the tussock (Fig. 12.4). This prolongs the time necessary for freezing of the whole tussock. Under the experimental setup whole tussock freezing took between 60 and 80 min. This slow freezing could have ecological significance for frost survival. In frost dehardened tussocks, leaf supercooling could also be important for frost survival, given that this supercooling also persists in nature when under radiation frost surface ice forms (Hacker and Neuner 2008). Controlled ice seeding experiments with INA bacteria dissolved in water droplets on grass leaves showed that ice propagation into the leaf tissue from the surface was inhibited as long as leaf surfaces were undamaged. This indicates that grass leaf supercooling is possible under natural conditions where formation of surface ice due to radiative cooling conditions is likely. During a late spring frost at a low altitude site (680 m) it was observed that this holds true for natural night frost conditions, as within a tussock of *Dactylis glomerata* the older leaves were found frozen stiff in the morning, but the youngest leaves in the centre of the same tussock must have supercooled during a night frost of −2°C, as they did not show any intrinsic ice (Fig. 12.5; unpublished observation, G. Neuner); this despite frozen rain droplets on the leaf surface.

## 12.2.2 Occurrence of Ice Barriers

### 12.2.2.1 Structural Ice Barriers

In the vegetative tissue of alpine species other than graminoids, structural ice barriers are also known to be present. For instance, in buds of some cold hardy

**Fig. 12.5** Older leaves of a *Dactylis glomerata* tussock growing in a meadow at 650 m show intrinsic ice, while the young leaves had no intrinsic ice in the morning after −2°C night frost on 2 April 2010. This indicates an efficient ice barrier between the leaves in the tussock. Young leaves remain unfrozen despite frozen water droplets on the leaf surface

**Fig. 12.6** Translocated ice masses can be seen in the cavity below the crown of a *P. abies* bud whilst the unfrozen apical meristem remains ice-free (photo Sonja Zimmermann)

**Fig. 12.7** Digital image and sequence of IDTA images showing a typical pattern of ice propagation in a reproductive shoot of *R. ferrugineum* in winter. Intrinsic ice nucleation occurred in the veins of a single leaf at –4.6°C, from where ice spreads immediately throughout the vascular tissue into the shoot and the adjacent leaves. In a second step of freezing, ice encroaches the extracellular spaces of the mesophyll tissue. Then the bud scales of the flower bud froze, and only after further lowering of the temperature (–11.6°C to –15.9°C) sudden single freezing events could be detected inside the flower buds that corresponded to intracellular lethal freezing of single florets in the bud. The time specifications indicate the elapsed time after ice nucleation

conifers, the spread of ice into the meristematic tissue is efficiently prevented (for review see Zwiazek et al. 2001). These ice barriers are also present in cold hardy buds of timberline conifers such as *P. abies*. There spreading of ice into the bud meristem is prevented during formation of extraorgan ice masses in the cavity beneath the crown (Fig. 12.6). The same frost survival mechanism was found for *L. decidua* but not for *P. cembra* buds whose survival mechanism, and that for other species from the genus *Pinus*, is still under discussion (Zwiazek et al. 2001).

In reproductive tissues of some alpine species during certain developmental stages, ice propagation barriers appear to be present that prevent the propagation of ice from the vegetative tissue into the flower or flower primordial tissue. In the flowering stage of *R. ferrugineum*, for example, ice can propagate nearly unhindered into the florets (see Fig. 12.3), however,

**Fig. 12.8** Digital image of a flowering cushion of *Silene acaulis* and sequence of infrared images obtained on this part of the cushion during a controlled freezing treatment. Freezing flowers are marked with black circles. All flowers froze separately within a temperature range of −8.6°C to −17.4°C. Each flower needed an autonomous ice nucleation event to initiate freezing, as ice propagation between flowers is blocked by a thermal ice barrier

this is not the case in the flower bud during winter (Fig. 12.7). Spreading of ice into the florets of the flower bud of *Rhododendron* species must somehow be prevented, probably by structural ice barriers (Ishikawa and Sakai 1981). Similar observations were made for *Prunus* and *Forsythia* flower buds,

**Fig. 12.9** Digital image of a flowering cushion of *Saxifraga moschata* during the flower bud stage and sequence of IDTA images obtained on this part of the cushion during a controlled freezing treatment. All flowers and vegetative shoots froze separately within a temperature range of −7°C to −14.3°C. Each reproductive unit needed an autonomous ice nucleation event to initiate freezing due to a thermal ice barrier formed by the cushion. If time specifications are given they indicate the elapsed time to the previous image

where the supercooling ability of floral primordia depended on the existence of a functional xylem (Ashworth 1984, Ashworth et al. 1992). When xylem continuity between the floral bud and the subtending stem was established, ice could propagate from the frozen stem via the vascular system and nucleate the water within the primordia.

A similar ice barrier preventing the spreading of ice into the flower bud tissue was found in other alpine woody species such as *Loiseleuria procumbens* and *Calluna vulgaris* during winter time (Hacker et al. unpublished). However, the occurrence of ice barriers between vegetative and reproductive organs may not be a distinctive feature of alpine plants. Preliminary results obtained for spring-flowering lowland plants show that ice barriers between shoots and reproductive organs can also be found in lowland species such *Corylus avellana* and even in herbs such as *Anemone nemorosa* (Neuner et al. unpublished). As these species flower during seasonal periods with a high frost probability at low altitudes, this adaptation appears to be crucial for survival of the reproductive organs. In fruits of *S. acaulis* single ovules freeze independently from each other and at much lower temperatures than the fruit, indicating the existence of an ice barrier (Hacker et al. 2011). The structural nature of these ice barriers is still under investigation.

#### 12.2.2.2 Thermal Ice Barriers

For alpine cushion plants a thermal ice barrier has been detected (Hacker et al. 2011). In cushion plants a thermal gradient builds up during freezing, where the flower heads are coldest and the vegetative parts and the shoots inside the compact cushion body remain significantly warmer. This gradient may be even more pronounced under natural radiative cooling conditions and may allow these inner shoots to maintain a non-freezing temperature. If ice nucleation occurs in a single flower, this thermal ice barrier prevents the spreading of ice from a frozen flower into the others, which means that for each single flower a separate ice nucleation event is necessary (Hacker et al. 2011). The typical freezing pattern, as characteristically observed in cushion plants, showing these thermally functioning ice barriers is shown exemplarily for cushions of *Silene acaulis* (Fig. 12.8) and *Saxifraga moschata* (Fig. 12.9).

A lack of sufficient ice barriers against extrinsic ice nucleation could be involved in the particularly low frost resistance of early developmental stages of alpine plants such as germinating seeds and seedlings (Marcante et al. unpublished). The rhizodermis (root epidermis) seems to be unable to protect the seedling from ice entrance (Hacker and Neuner, unpublished), which is in contrast to the epidermis of shoots and cotyledons where ice cannot enter. In this way, these early stages will be ice-nucleated as soon as the rhizodermis has contact with extrinsic ice masses. During germination at the soil surface, the rhizodermis may come into contact with frozen soil water, the risk will decrease with protrusion of the root into the soil.

**Conclusions**

In summer the frost resistance of most alpine plant species appeared to be insufficient to protect them from occasional frost damage at altitudes higher than 2,500 m (Taschler et al. 2004, Taschler and Neuner 2004), and severe frost damage to alpine plants can occur naturally (Körner 2003). In winter frost survival of alpine plants is strictly intermeshed with snow cover and micro-site exposure (Sakai and Larcher 1987).

The risk of frost extinction of a species will depend on the leeway between the prevailing low temperature extremes and the temperature of frost damage: both carry uncertainties. We still lack relevant reports on plant body temperatures from various altitudes, and snow cover, which often coincides with cold spells in summer, will significantly co-determine low temperature extremes. Additionally, frost resistance differs significantly between life forms (Taschler and Neuner 2004) and developmental stages (Marcante et al. unpublished), plant organs (leaves, shoots, roots, buds, flowers and fruits) and tissues (Larcher et al. 2010, Ladinig et al. unpublished). Frost survival may furthermore depend on recuperation capability, which has been shown to occur rapidly in species that maintain a low frost resistance during summer, such as *Oxyria digyna* (Taschler and Neuner 2004). Even *Primula minima* was capable of recuperating after loss of the total above-ground tissue (Fig.12.10). Ice propagation barriers paired with a high supercooling capacity are another important feature allowing frost survival of at least some alpine species, organs and tissues.

**Fig. 12.10** A *Primula minima* cushion (**a**) was exposed to artificial night frost in the leaf layer down to −10°C in situ at 1,950 m on 19 June 2007. The freezing treatment resulted in 100% frost damage to the above-ground tissue (**b**), but 3 months afterwards regeneration of leaf rosettes via buds from the below-ground undamaged rhizomes became visible (**c**; Neuner and Kirschner, unpublished)

However, their supercooling capability in nature is difficult to simulate in laboratory tests due to the problems discussed above. Particularly for meristematic tissues in buds and reproductive organs, which do not tolerate ice formation at all during some developmental stages, ice protection via ice barriers seems to be the only mechanism to ensure survival and the reproductive success, respectively (Hacker et al. 2011, Ladinig et al. unpublished).

## References


Aigner S (1952) Die Temperaturminima im Gstettnerboden bei Lunz am See. Niederösterreich. Wetter und Leben, Sonderheft, pp 34–37

Aryal B, Neuner G (2010) Leaf wettability decreases along an extreme altitudinal gradient. Oecologia 162:1–9

Ashworth EN (1984) Xylem development in Prunus flower buds and the relationship to deep supercooling. Plant Physiol 74:862–865

Ashworth EN, Davis GA, Anderson JA (1985) Factors affecting ice nucleation in plant tissues. Plant Physiol 79:1033–1037

Ashworth EN, Willard TJ, Malone SR (1992) The relationship between vascular differentiation and the distribution of ice within Forsythia flower buds. Plant Cell Environ 15:607–612

Fuller M, Hamed F, Wisniewski M, Glenn DM (2003) Protection of plants from frost using hydrophobic particle film and acrylic polymer. Ann Appl Biol 143:93–97

Grabherr G, Gottfried M, Gruber A, Pauli H (1995) Patterns and current changes in alpine plant diversity. In: Chapin FS, Körner C (eds) Arctic and alpine plant biodiversity, Ecological studies 113. Springer, Berlin, Heidelberg, pp 167–181

Hacker J, Neuner G (2007) Ice propagation in plants visualized at the tissue level by IDTA (infrared differential thermal analysis). Tree Physiol 27:1661–1670

Hacker J, Neuner G (2008) Ice propagation in dehardened alpine plant species studied by infrared differential thermal analysis (IDTA). Arc Antarc Alp Res 40:660–670

Hacker J, Spindelböck J, Neuner G (2008) Mesophyll freezing and effects of freeze dehydration visualized by simultaneous measurement of IDTA and differential imaging chlorophyll fluorescence. Plant Cell Environ 31:1725–1733

Hacker J, Ladinig U, Wagner J, Neuner G (2011) Inflorescences of alpine cushion plants freeze autonomously and may survive subzero temperatures by supercooling. Plant Sci 180:149–156

Ishikawa M, Sakai A (1981) Freezing avoidance mechanisms by supercooling in some Rhododendron flower buds with reference to water relations. Plant Cell Physiol 22:953–967

Jordan DN, Smith WK (1994) Energy balance analysis of night time leaf temperatures and frost formation in a subalpine environment. Agr For Meteorol 71:359–372

Körner C (2003) Alpine plant life: functional plant ecology of high mountain ecosystems. Springer, Berlin

Larcher W (1985) Kälte und Frost. In: Sorauer P (found) Handbuch der Pflanzenkrankheiten, 7th edn., part V. Parey, Berlin, pp 107–320

Larcher W, Wagner J (2009) High mountain bioclimate: temperatures near the ground recorded from the timberline to the nival zone in the Central Alps. Contrib Nat Hist 12:857–874

Larcher W, Kainmüller C, Wagner J (2010) Survival types of high mountain plants under extreme temperatures. Flora 205:3–18

Neuner G, Pramsohler M (2006) Freezing and high temperature thresholds of photosystem 2 compared to ice nucleation, frost and heat damage in evergreen subalpine plants. Physiol Plant 126:196–204

Neuner G, Ambach D, Aichner K (1999) Impact of snow cover on photoinhibition and winter desiccation in evergreen Rhododendron ferrugineum leaves during subalpine winter. Tree Physiol 19:725–732

Pearce RS (2001) Plant freezing and damage. Ann Bot 87:417–424

Sakai A, Larcher W (1987) Frost survival of plants. Responses and adaptation to freezing stress. In: Billings WD, Golley F,

Lange OL, Olson JS, Remmert H (eds) Ecological studies, vol 62. Springer, Berlin

Stier JC, Filiault DL, Wisniewski M, Palta JP (2003) Visualization of freezing progression in turfgrasses using infrared video thermography. Crop Sci 43:415–420

Taschler D, Neuner G (2004) Summer frost resistance and freezing patterns measured in situ in leaves of major alpine plant growth forms in relation to their upper distribution boundary. Plant Cell Environ 27:737–746

Taschler D, Beikircher B, Neuner G (2004) Frost Resistance and ice nucleation in leaves of five woody timberline species measured in situ during shoot expansion. Tree Physiol 24:331–337

Wisniewski M, Fuller M (1999) Ice nucleation and deep supercooling in plants: new insights using infrared thermography. In: Margesin R, Schinner F (eds) Cold adapted organisms. Ecology, physiology, enzymology and molecular biology. Springer, Berlin, pp 105–118

Wisniewski M, Lindow SE, Ashworth EN (1997) Observations of ice nucleation and propagation in plants using infrared video thermography. Plant Physiol 113:327–334

Wisniewski M, Fuller M, Glenn D, Gusta L, Duman J, Griffith M (2002) Extrinsic ice nucleation in plants: what are the factors involved and can they be manipulated? In: Li PH, Tapio P (eds) Plant cold hardiness. Gene regulation and genetic engineering. Kluwer Academic Publishers, New York, pp 223–236

Zwiazek JJ, Renault S, Croser C, Hansen J, Beck E (2001) Biochemical and biophysical changes in relation to cold hardiness. In: Bigras FJ, Colombo SJ (eds) Conifer cold hardiness. Kluwer Academic Publisher, Dordrecht, pp 165–186

# Cell Structure and Physiology of Alpine Snow and Ice Algae

# 13

Daniel Remias

## 13.1 "Lower Plants" in High Alpine Regions

Due to climatic and orographic reasons, the occurrence of vascular plants in high alpine regions is limited. At locations that are not suitable for the establishment of higher plants because of exposure, substrate or other abiotic factors, cryptogams can be the dominant life forms. Mosses, lichens and algae particularly thrive on places such as bare rocks, permafrost soils or, exceptionally, even in melting snow and permanent ice. Since these lower plants are poikilohydric and lack complex morphological tissues like the cormophytes, unfavourable conditions (like drought) can be overcome by physiological inactivity, and structural damage is not the critical issue for these poikilohydric organisms. The vegetation period of cryptogams can be very short (from days to a few weeks per year), and growth and reproduction have to be adapted to limiting factors such as low temperatures, limited water-availability or irradiation stress.

While mosses and lichens are prominent representatives in places with sparse or almost no other vegetation, the virtual dominance of microalgae is somehow "cryptic" to the human eye, not only because of their size but also because of inconspicuous macroscopic signs of their occurrence in many cases. Alpine algae are neither limited to more familiar habitats such as permanent water bodies (lakes or streams), nor to lichens as embedded phycobiont. In fact, many species are specialists in growing on bare surfaces such as stones (epilithic), in microcavities within rocks (endochasmolithic), at lower mountain ranges on tree bark (aerophytic), superficially in open habitats (e.g. soil crusts), subterrestrially in the first centimetres of the ground layer (soil algae) or – as is the main subject of this chapter – in wet snow and ice.

All these niches are also frequently populated by cyanobacteria, which are typical primary successors in still ungrown places. However, in melting snow of the European Alps, these photoautotrophic procaryota have so far not been reported to generate populations. Generally, both cyanobacteria and green algae produce the first organic matter and form primary soils with nutrients that are available for mosses, higher plants or fungi after decomposition (Elster and Benson 2004).

Although non-vascular alpine plants are less well investigated concerning their taxonomy and ecophysiology compared to vascular plants, it is assumed that they have also evolved a distinct speciation including ecophysiological strategies, which are necessary for survival in alpine (and in many cases similar polar) regions due to the adaptation to high altitude (or high latitude) ecosystems.

## 13.2 Snow and Ice as a Habitat

The "Red Snow" phenomenon is the probably best-known case where macroscopic visible mass accumulations of microalgae attract the viewers' attention. A historical background about discoveries and investigations of red colourations caused by algae is given by Werner (2007). Snow coloured by algae

D. Remias (✉)
Institute of Pharmacy/Pharmacognosy, University of Innsbruck, Innsbruck, Austria
e-mail: Daniel.Remias@uibk.ac.at

C. Lütz (ed.), *Plants in Alpine Regions*,
DOI 10.1007/978-3-7091-0136-0_13, © Springer-Verlag/Wien 2012

has been reported for many polar and alpine places in the world and can be regarded as cosmopolitan (Kol 1968). Red Snow from the Tyrolean Alps has been reported by Ettl (1968), by Kol (1961, 1970) or by Remias et al. (2005).

The circumstances that make such blooms possible depend on the availability of liquid water. Consequently, only wet snow that lasts long enough is suitable for the development of an algal population. These conditions are given during summer at higher elevations (in central Europe typically from 1,800 m a.s.l. on), where snow from the previous winter persists for several weeks in a water-logged stage before completely melting away or even remains until the beginning of the next winter. Snow algae thrive neither in dry winter snow nor in fast-melting slush of the lowlands. Old summer snow, however, has a soft and grainy consistency due to numerous freeze-thaw-cycles and thus supports the establishment of a continuous water film surrounding macroscopic, roundish snow crystals (typical sizes from 1 to 3 mm; Jones et al. 2001). Unicellular algae with distinctive sizes of 10–30 μm have plenty of space to flourish in this microhabitat as long as they are capable of sustaining several kinds of abiotic stresses that are common in this harsh ecosystem. The environmental conditions include constant temperatures around 0–1°C in wet snow, but they can be significantly deeper at the snow surface during an alpine cold snap. Moreover, the irradiation at the snow surface can be higher than full sunlight because of local reflections (high albedo of snow and ice). On the other hand, irradiation decreases logarithmically with snow depth (Gorton et al. 2001), thus algae situated deeper in the snow may be subject to very low irradiation. Overviews about the taxonomy, ecology and physiology of snow algae are given by Hoham and Duval (2001) and Komárek and Nedbalová (2007).

The kind of macroscopic colouration and its intensity depends on the individual pigment composition of an algal species, on the phase in the life cycle of the population and on the cell concentration per snow volume. On a cellular basis, the primary and secondary pigments, like chlorophylls and carotenoids, can undergo drastic changes in quantity and quality during the season and thus cause very different *in situ* colourations (Fig. 13.1).

Alpine snow and ice colourations can be roughly grouped into:

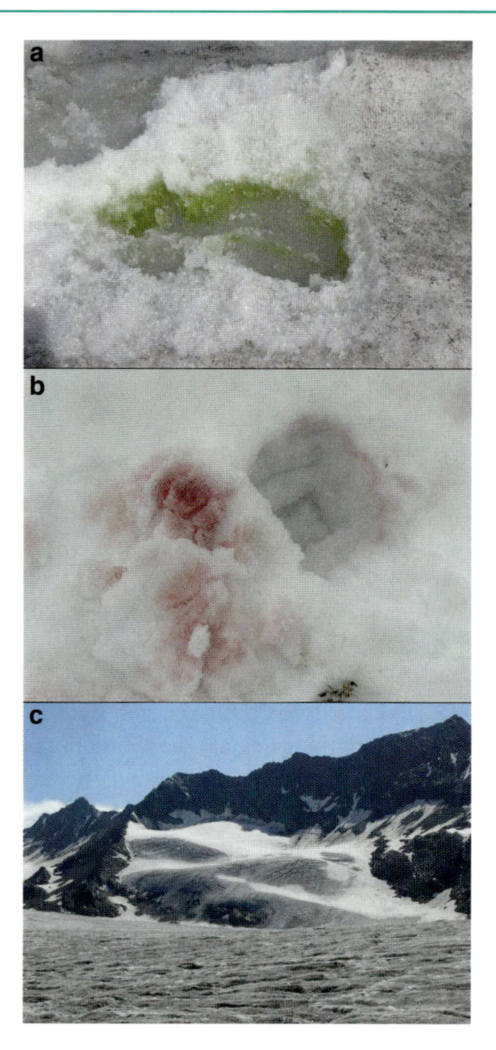

**Fig. 13.1** Typical cryosestic habitats. (**a**) Green snow caused by *Chloromonas* sp. at Hallsteiner Glacier, Upper Austria, 2,650 m. Note that the green slush layer rests on the glacier surface and was covered by approx. 1–2 cm white snow. (**b**) Red snow caused by *Chlamydomonas nivalis*, Kühtai (Tyrol), 2,300 m. The crimson colour usually reaches several centimetres below the surface. (**c**) Grey ice at Gurgler Glacier (Tyrol), 3,000 m. The dark colour of the ice surface is caused by *Mesotaenium berggrenii* as well as by cryoconite particles

*Green Snow* (Fig. 13.1a): Usually caused by green algae (Chlorophyta s.str.) of genera such as *Chloromonas* or *Chlamydomonas* (order Chlamydomonadales), where chlorophylls dominate and that (still) lack secondary pigments. Green snow is hard to find in the Alps, because it is mostly a transient stage in early summer (April, May) and still covered under white, algae-free snow. In most cases, green populations turn into red snow during the later season.

*Red Snow* (Fig. 13.1b): Also caused by green algae, as above. However, in this case, high amounts of secondary carotenoids dominate the green chlorophylls. Depending on the species and the cell concentration in the field, all transitions in terms of colour between dark-yellow, orange, pinkish to red or even crimson can be found. In this chapter, all such variations are summarised as Red Snow. It can be regarded as the "climax stage" in terms of seasonal development (e.g. originating from Green Snow). In the Alps, it can be found quite often in middle and late summer (June to September), and it usually develops above the timberline either in seasonal snowfields (bedded on soil or on permanent ice) or in the snow slush of the melting ice cover of high alpine lakes.

*Grey Ice* (Fig. 13.1c): Green algae with microscopically green/brownish appearance that exclusively thrive on glacier surfaces. The main representative in the Alps is *Mesotaenium berggrenii* var. *alaskana* (Zygnematales, Streptophyta). A relative, however filamentous and mostly polar species is *Ancylonema nordenskiöldii*. The dark colour of the habitat is also caused by cryoconite particles that commonly cover old ice. Consequently, these algae are difficult or even impossible to recognize in the field without a microscope. Sometimes, when growing in large amounts, blooms can cause faint dark-violet reflections when the ice surface is illuminated by bright sunlight. In this circumstance, it is important to stress that glacier ice is somehow a physically different habitat compared to snow. Since freshwater ice algae such as *Mesotaenium* have no moving stages in their life cycles, they depend on being permanently attached to a cold surface.

The classification of these three groups is based on the situation in the European Alps. In other cold regions, there can be several further types of coloured snow, mostly due to different climatic conditions. For example, long lasting Green Snow caused by *Chloromonas* ssp. can be found in shaded places below the timberline in North America (Hoham and Duval 2001) or in low mountain ranges of Eastern Europe (Komárek and Nedbalová 2007). Also, Yellow Snow caused by Chrysophyceae (*Ochromonas* sp.) is a frequent phenomenon in coastal regions of maritime Antarctica, but has not been found in the Alps yet.

By definition, the snow and freshwater ice algae dealt with in this chapter are species which proliferate exclusively in snow or on glacier surfaces. Vegetative and generative reproduction is limited to temperatures around 0–1°C. These "true" snow and ice species may be passively removed from their cold habitat by meltwater streams or by winds. At warmer places, however, they survive only temporarily as inactive cysts, alternatively they are subject to lethal stress.

By contrast, the group of "non-true" snow and ice algae may survive in the short term because of transient cold-tolerance, however progeny on snow or ice is not possible, consequently a lasting population cannot be established. Typical members of these "guests" on snow are alpine or arctic soil algae such as *Raphidonema* or *Koliella*. They are short-filamented green alga which may be blown onto snow by strong winds (Stibal and Elster 2005). Similar effects may explain the occurrence of soil algae such as *Xanthonama* sp. (Tartari and Forlani 2008) or "*Chlorella*" (Vona et al. 2004), where cells were isolated and cultivated from snow, but in fact had never been observed to cause true populations in these habitats.

Snow and ice algae are the primary producers in a relatively simple ecosystem poor in species numbers and trophic levels. They have virtually no competition in their habitat, and typically one single species forms monospecific blooms at a given location. Consequently, the same species can be found at the same place in the next year again, indicating the distinct phytogeographical distributions, also at the local level. Other organisms that benefit from algal blooms are psychrophilic bacteria (Thomas and Duval 1995), snow fungi or small animals such as ciliates or copepods. The latter (e.g. *Isotoma saltans*) are typical for glaciers, yet it has not been proven yet that they feed on snow algae. Snow worms, as observed in Alaska (Shain et al. 2001), are still unknown in the Alps.

## 13.3 Cell Forms in an Ecophysiological Context

Snow and ice algae are single-celled organisms or, more uncommonly, short, unbranched filaments. There is no individual differentiation of cells to fulfil a certain role e.g. within a tissue. Consequently, each cell must be able to sustain its vitality and reproductivity itself. This paragraph describes the strategies of algae living in snow and ice at a cellular level. There are several differences to higher plants, because the latter lack morphological and physiological flexibility at a single-cell level compared to microalgae. Figure 13.2 presents light micrographs of typical alpine snow algae.

One key factor in understanding the success of snow algae is their life cycle which includes different cell stages with different physiological and morphological attitudes. It enables them to adapt quickly to changing environmental conditions. The vegetative (asexual) and generative (sexual) life cycle of many Chlamydomonadales (from which most snow algae are derived) is complex and allows several kinds of cell reproduction strategies. An overview of the taxonomy of this group of algae is given by Ettl (1983), and principal seasonal life cycles in snow are described in Hoham and Duval (2001) and also in Sattler et al. (2010). Figure 13.3 depicts the main physiological and cytological differences between

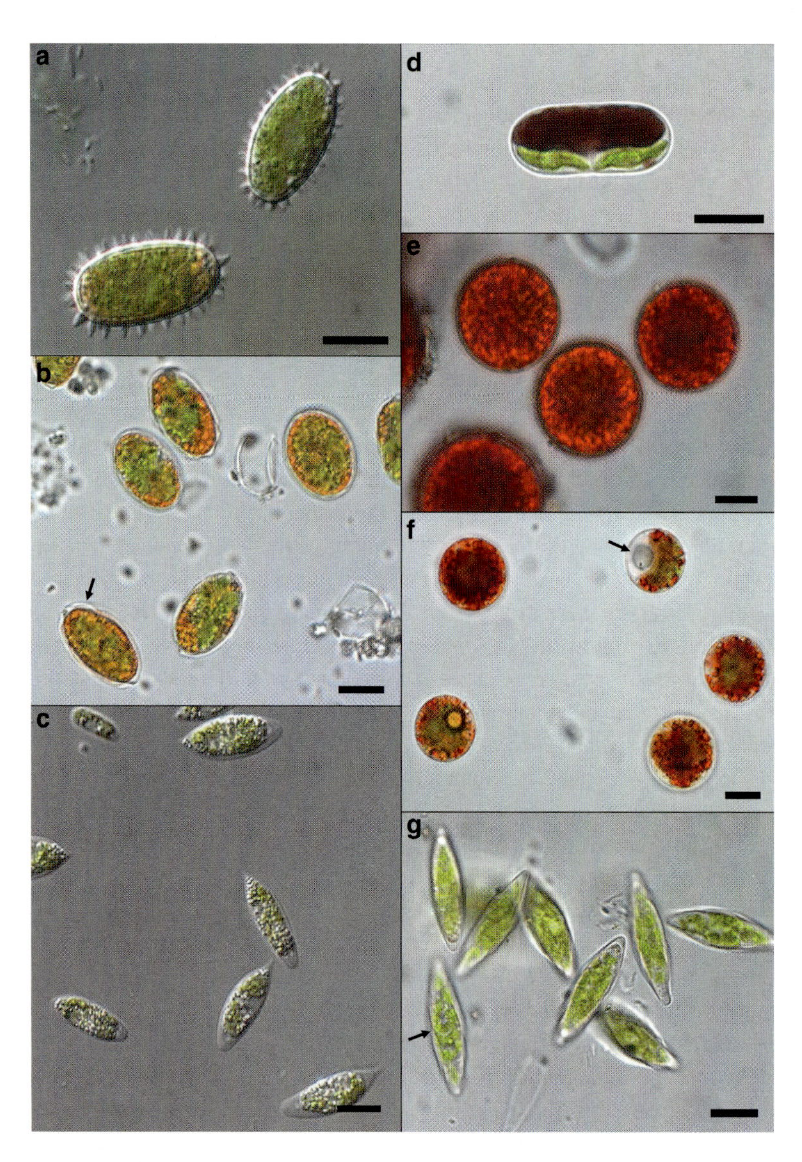


**Fig. 13.2** Microphotographs of snow and ice algae. (**a**) *Cr. brevispina* from Kühtai, cysts with typical coned spikes; (**b**) *Cr. nivalis* from Kühtai (Tyrol), cysts with green chloroplasts and orange-red lipid globules. Note the secondary cell wall structures; (**c**) cells of a green snow species, *Chloromonas* sp. from Hallstein Glacier (location also shown in Fig. 1c); (**d**) the ice alga *M. berggrenii* from Tiefenbach Glacier (Tyrol); (**e**) mature cysts of *Cd. nivalis* from red snow at Tiefenbach glacier (Tyrol). Note the central dark red part representing the (masked) chloroplast; (**f**) *Cd. nivalis* from Kühtai, typical transient stage of a young cyst still with visible green parts of the chloroplast and one, large unpigmented vacuole (*arrow*). (**g**) *Cr. rosae* var. *psychrophila* (Kühtai), fusiform cysts with ridges (*arrow*)

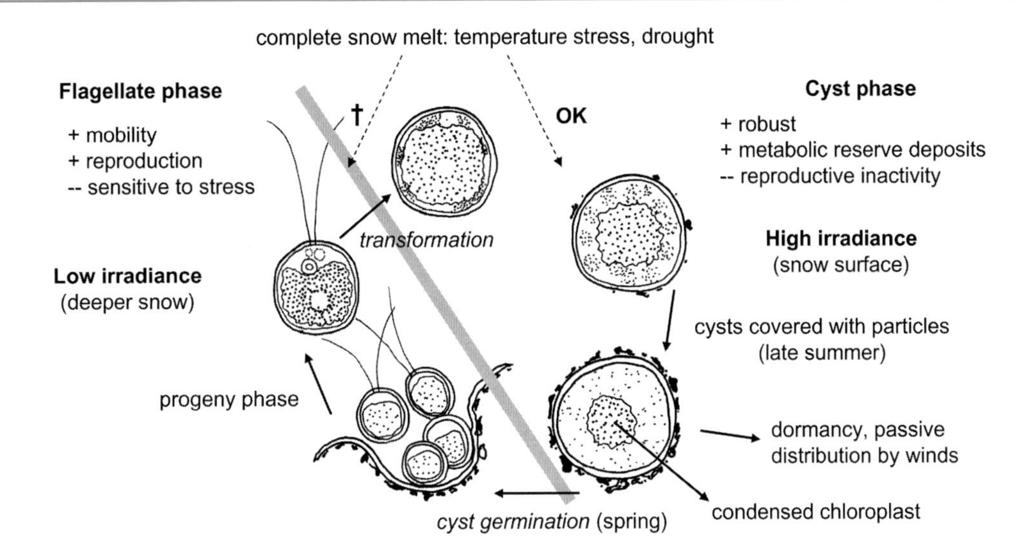


Fig. 13.3 Scheme of the main ecophysiological differences between motile-vegetative (*green*) and cyst-like (*red*) stages in the seasonal life cycle of a Chlamydomonadales snow alga (e.g. *Chlamydomonas nivalis*)

the green cells (flagellate phase) and red stages (immotile cyst phase) of a Chlamydomonadacean snow alga in relation to major abiotic factors. It demonstrates that flagellates, which are able to reproduce, are much more sensitive to freezing and high irradiation than cysts of the same species.

### 13.3.1 Life Cycle of Chlamydomonadales Cells in Snow

In early summer, when the snow starts to be waterlogged, flagellates emerge from cysts (as shown in Fig. 13.3). They can actively swim upwards (from places where the cysts had overwintered, e.g. soil surface or old ice layer) until they reach adequate light conditions for performing vegetative and/or sexual reproduction. Thus, they are responsible for vertical migration in the snow (to a limited extent) and the growth of the population. There, temperatures are constantly around 0–1°C, and cells are still protected from excessive irradiation due to the remaining snow pack above them. Moreover, they are protected against harsh freeze-thaw-cycles at this location because of the stabilising self-isolation of the snow cover, regardless of any late frost events in early summer. Endogenous or external triggers (e.g. habitat conditions getting harder) can transform these flagellates directly into vegetative cysts (hypnoblasts) or, by sexual fusion of two gametes, into generative zygotes. Both loose motility, cease any cleavages and become very resistant against environmental stress due to intracellular reorganisation (chloroplast rearrangement), lipid production (less water content of the cytoplasm), accumulation of secondary metabolites such as pigments or the reinforcement of cell walls (Remias et al. 2010a). The secondary surface structures of these cyst cell walls can be very individual and species-characteristic (Fig. 13.4). Vegetative and cyst stages of genera like *Chloromonas* and *Chlamydomonas* can be so dissimilar in their morphology that they were frequently described as separate taxa until they were recognized to be just different parts of one life cycle. For example, the fusiform cysts of *Chloromonas nivalis* (Figs. 13.2c and 13.4d) were initially referred to as *Scotiella nivalis* (Hoham and Duval 2001; Remias et al. 2010a).

During the ongoing season, snow melt continues and the nutrients in the snow may become exhausted because of cellular absorption, too. Consequently, the population has to transform into cysts on time, because the vegetative cells would neither withstand a complete snowmelt (high temperatures) nor longer exposure on the snow surface (excessive irradiation). Later, the population passively accumulates on the snow surface, where solar irradiation (UV and VIS) can be very high and diurnal freeze-thaw-cycles (which result in water stress) may take place.

The cysts germinate after a period of dormancy or when favourable conditions return in the next season. In this case, they transmute into

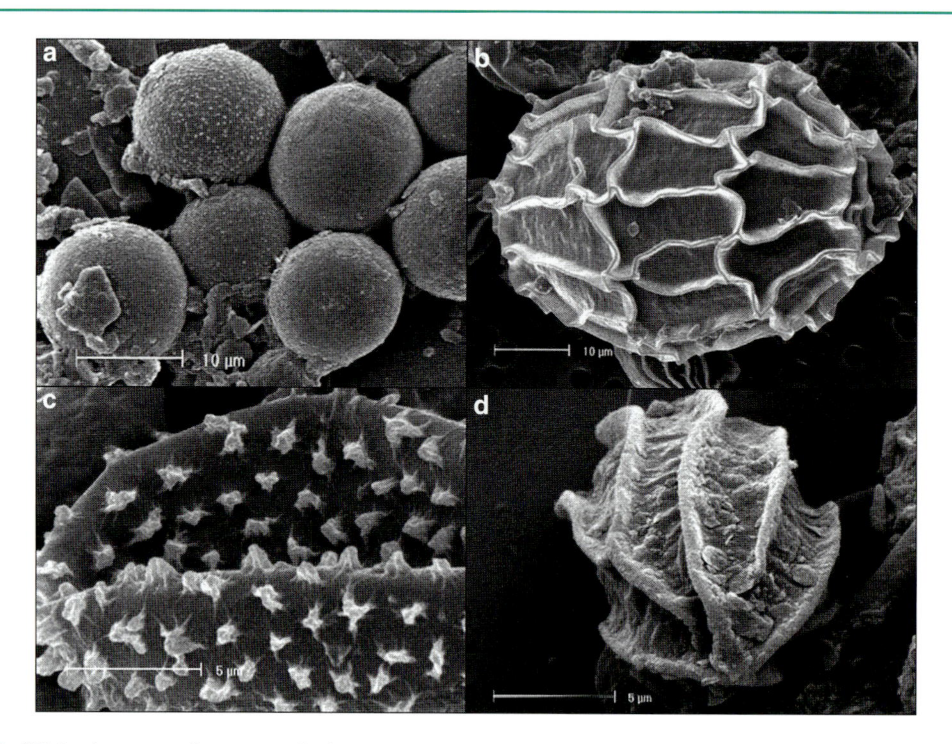


**Fig. 13.4** SEM images of snow algal cysts. (**a**) *Chlamydomonas nivalis* (Tiefenbach Glacier, Tyrol) round cysts with smooth or slightly ornamented cell wall. (**b**) Unknown algal cysts from red snow on a high alpine lake ice-cover, probably *Chlainomonas* sp. (Kühtai, Tyrol). (**c**) *Cr. brevispina*, spikes have a length of approx. 1 µm. (**d**) *Cr. nivalis* (Kühtai) cysts with undulated ridges, mostly reaching from cell pole to cell pole

autosporangia by making internal cleavages, resulting in several daughter cells (typically four to eight individuals), which are released after mother-wall rupture. Usually, flagellated snow algae can be found only for a few weeks in the early mountain summer when snow melt has begun, whereas non-dividing cysts dominate during the rest of the year.

#### 13.3.1.1 Cell Structures of Chlamydomonadales in Snow

The morphological flexibility of Chlamydomonadales is also depicted in the TEM images in Fig. 13.5. Motile cells have a typical cup-shaped chloroplast, and in this stage much starch and an eyespot are still present (Fig. 13.5a). During cyst formation, lipid bodies (which store secondary carotenoids) appear, and the cytoplasmic volume is reduced (Fig. 13.5b). Moreover, the chloroplast contains hardly any starch grains anymore and can be more irregularly in shape. The number of plastoglobules increases. Sometimes it is structured into several smaller parts for surface enhancement. Furthermore, small vesicles with crystalline content can be observed (Fig. 13.5c). The ornamented, secondary cell walls of *Chloromonas* cysts develop below the smooth primary wall (Fig. 13.5d). A somewhat different cytoarchitecture can be found in *Chlainomonas* sp. (corresponding to *Chloromonas* sp. GU117574 in Remias et al. 2010a), a species that thrives only on snow slush on the melting ice cover of high alpine lakes: the cells possess numerous small peripheral chloroplasts, the lipid bodies are more central and surround the nucleus (Fig. 13.5e).

### 13.3.2 Life Cycle of Mesotaeniaceae Cells at Ice

Freshwater ice algae on glacier surfaces have a different life cycle and other ecophysiological strategies compared to the previous group. They have a prominent brownish secondary pigmentation, which resembles ice algae in marine brine, namely diatom assemblages at the bottom of ice floes in polar oceans, but are not taxonomically related to them. Glacier dwelling algae

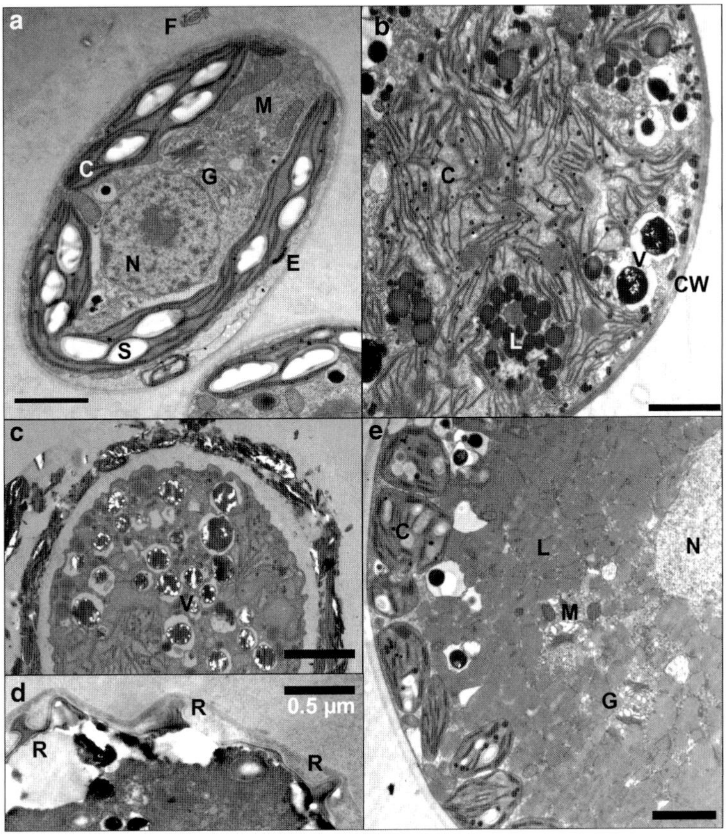

Fig. 13.5 TEM images of snow algae. (**a**) Flagellate of *Cr. nivalis* with flagella (F; one of two), cup-shaped chloroplast (C) with starch (S) and eyespot (E), nucleus (N), mitochondria (M) and Golgi stacks (G). (**b**) Young cyst of *Cd. nivalis* with smooth secondary cell wall (CW), irregularly shaped chloroplast (C), initially pigment-loaded lipid bodies (L) and several crystal-vacuoles (V). (**c**) Cyst of *Chlamydomonas* sp. with thickened cell wall and a layer of inorganic particles on its surface. Note the large number of crystal containing vesicles (V) in the cytoplasm. (**d**) Secondary cell wall of *Cr. nivalis* with ridges (R), still under a smooth primary wall layer (compare with Figs. 2c and 4d). (**e**) *Chlainomonas rubra* cysts possess peripherical positioned small chloroplasts (C) with starch grains. Only a small volume of the cytoplasm with Golgi stacks (G) and mitochondria (M) is left. The central nucleus (N) is surrounded by a large number of astaxanthin-containing lipid bodies (L). Bar: 1 µm

belong to the Zygnematophyceae (Streptophyta), which means that they lack all flagellated/motile stages. As a consequence, they cannot migrate into snow like the Chlamydomonadales (Remias et al. 2009). They are restricted to permanent ice surfaces, but airborne long distance dispersal may occur in autumn, when strong winds abrade the glacier. Remarkably, the cells stay in the vegetative stage and constantly grow and divide throughout the season. They typically grow about one and a half time in length before cleavage. They possess smooth cell walls, vacuoles filled with the water-soluble brownish pigment (a putative tannin-like phenolic; Remias et al., unpublished) and one or two discus-like chloroplasts with pyrenoids and starch sheaths (Remias et al. 2009). Cysts or sporangia do not play a significant role in the population and occur only occasionally as reproductive zygotes. Their most prominent members are *Mesotaenium berggrenii* (Fig. 13.2b) in the Alps and the filamentous *Ancylonema nordenskiöldii* mainly at polar sites.

## 13.4 Physiology

The main physiological needs of snow and ice algae in their special habitat include cold- and high-irradiation adaptation of the metabolism, avoidance of intracellular freezing, coping with a short vegetation period accompanied by a poor nutritional situation and persisting as dormant stages during the long alpine winter. Margesin (Chapter 14, this book) summarizes the different metabolic adaptation strategies of microorganisms.

Remias et al. (2005) performed photosynthesis measurements to show that red snow caused by mature cysts of *Cd. nivalis* from the European Alps has active, oxygen-producing chloroplasts, which contradicts the notion of inactive dormant stages as believed earlier. Moreover, temperatures of 10 or 20°C are not lethal for these cysts, seemingly a pre-adaptation to the conditions that will prevail after snowmelt. Similar physiological trends were also shown for cysts of *Cr. nivalis* from the Tyrolean Alps by Remias et al. (2010a; Austria), for *Cr. brevispina* by Kvíderova et al. (2005; Czechia) and for *Cr. nivalis* from the Giant Mts. by Stibal (2003). The photosynthetic optimum for the latter species turned out to be clearly below 10°C.

Chlorophyll fluorescence measurements ("Kautsky effect") provide information about the vitality state of the photosystem II of snow algal cysts. The common indicator $F_v/F_m$ has been used to compare different samples of *Cd. nivalis* from the Tyrolean Alps: young cysts (visually not yet fully red) have an $F_v/F_m$ of 0.683 ± 0.013. After keeping this sample in an illuminated chamber for 11 days (approx. 100 μmol $m^{-2}$ $s^{-1}$ PAR, 4°C), the value rose to 0.716 ± 0.05, which is quite high, with such values never having been measured in the field. Older cysts generally decrease the activity of their photosystem II: mature cysts (harvested in July) had 0.517 ± 0.044, and old cysts in the late season (August) had only 0.479 ± 0.012. $F_v/F_m$ can also be different within a given population: in June cysts in rather dry and hard snow had 0.509 ± 0.05, whereas cells of the same developmental stage in the direct proximity, which were bedded in very wet snow above a brook, reached 0.603 ± 0.023. Consequently, the chlorophyll fluorescence decreases during the season because of cyst maturation and preparation for overwintering. This is also in accordance with cytological findings of morphological chloroplast reduction (Fig. 13.5; e.g. less chloroplast volume with less thylakoid membranes but more plastoglobules). Furthermore, the harshness of local field conditions (e.g. water availability in the snow) can cause variations in the vitality of the photosystem II during the season. Comparing these $F_v/F_m$ values with those common in higher plants (around 0.8), it has to be considered that microalgae have a generally lower $F_v/F_m$ possibly due to a differently structured photosystems.

The role of secondary pigments in the ecophysiology of snow algae, which usually occur in high abundances, is still under investigation. Their production can be regarded in several helpful aspects, e.g. as a physiological "sink" for excessive photosynthetic energy, as a metabolic reserve depot, as photosystem protection by shading the chloroplast against high irradiation or as a pool of antioxidative compounds. Remias et al. (2005) applied HPLC to confirm that the secondary carotenoid astaxanthin is the main pigment causing the red colour of *Cd. nivalis* in cytoplasmic lipid bodies from snow fields in Austria, as similarly depicted by Bidigare et al. (1993) with samples from North America. Astaxanthin is also the secondary carotenoid of *Cr. nivalis*, however in much lower relative quantity (Remias et al. 2010a), causing the cysts to be less red. It was suggested that in this species compensation occurs partly as a result of the presence of a larger and more active (= potentially higher deepoxidised) pool of xanthophyll cycle pigments, which is able to buffer high light stress at the chloroplast level, and which seems much less developed in *Cd. nivalis*. The absorption spectrum of astaxanthin remains unchanged regardless of esterification or glycosilation, and it is a compound with powerful VIS absorption capacities, thus shielding the chloroplast against excessive irradiation. Many snow algae also contain significant amounts of a *cis*-isomer of astaxanthin (13*Z* astaxanthin), which provides advanced protection in the UVA region (Fig. 13.6; Remias and Lütz 2007). It is responsible for irradiation tolerance in high light exposed habitats such as snow surfaces. However, astaxanthin is located in extraplastidal lipid bodies and therefore only indirectly involved in the protection of photosynthetic processes.

Řezanka et al. (2008a) discovered that cysts of Red Snow from Bulgaria deposit large amounts of astaxanthin-diglucoside-diesters, which combine the three different classes of a carotenoid, a sugar and fatty acids. Řezanka et al. (2008b) also reported that *Cr. brevispina*, collected from patches of green snow in the Bohemian Forest of Czechia, produces unusual medium-chained polyunsaturated fatty acids under a total number of 43 different fatty acids. While carotenoid synthesis in the chloroplasts of *Chlamydomonas*

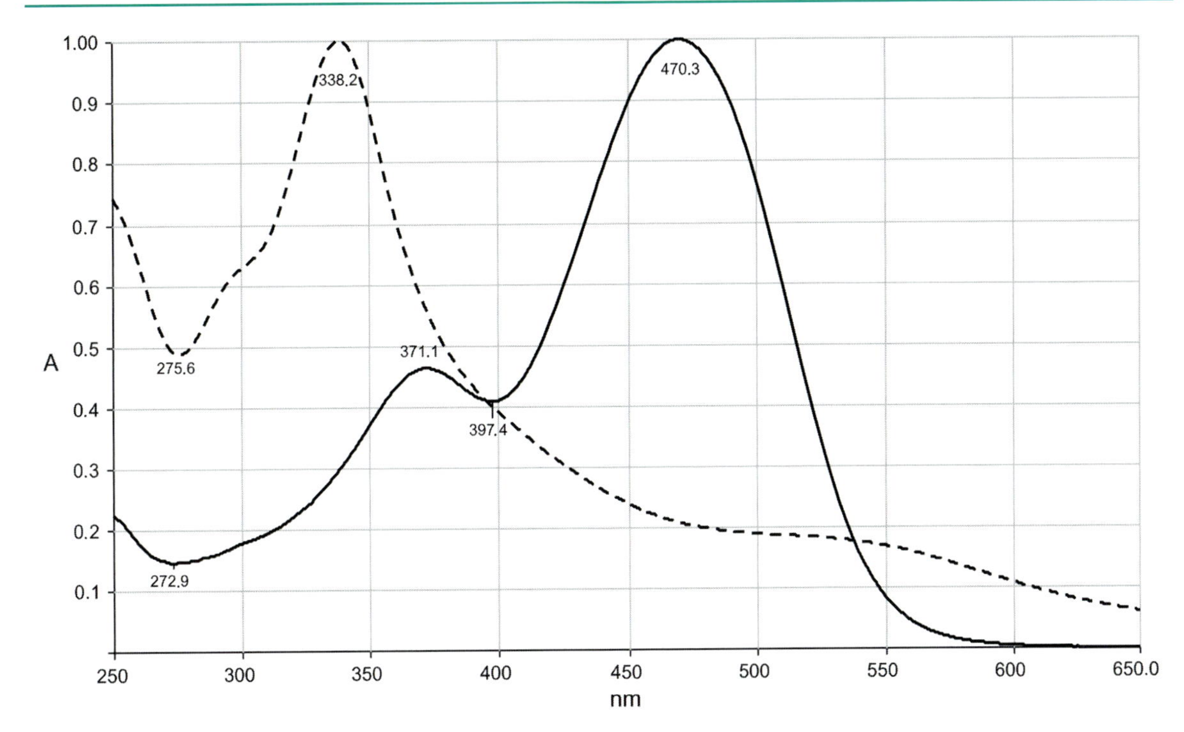


**Fig. 13.6** Spectral absorbance of two main secondary pigments in psychrophilic algae. The ice alga *Mesotaenium berggrenii* (dashed) has a water-soluble putative phenolic pigment; its brown colour is caused by a broad absorbance in the VIS region. Moreover, there is a large potential against harmful UV irradiation ($\lambda_{max}$ 338 nm). By contrast, the apolar keto-carotenoid isomer 13*Z* *cis*-astaxanthin, typical for *Chlamydomonas nivalis* and other relatives, has its maximum absorbance in the VIS region ($\lambda_{max}$ 470 nm) but still a significant capability of UV protection (shoulder at 371 nm)

is well-known (Lohr 2009), the information about regulation and the procedures of decomposition in case of depletion (e.g. regreening during cyst germination in spring) in katabolic pathways is scarce. The general existence of carotenoid cleavage dioxygenases in microorganisms is established (Marasco et al. 2006), and the cleavage products are volatile short-chained carbohydrate bodies, which may cause a scent due to their aromatic ring. These smelling catabolites may be the origin of a faint "watermelon" aroma that is noticed when coming close to masses of ripe red algal cysts in the field. Since young red cysts do not release a scent, it can be assumed that the volatile cleavage products are released only when cells are already oversaturated with astaxanthinesters in their lipid bodies.

Remias et al. (2010b) tested, for first time with a snow alga, the reaction of *Cd. nivalis* cysts collected from the field to realistically elevated UVB irradiation in an indoor sun simulator. After a 3-day scenario of increased UVB, the cells reacted with an increase of carotenoids and chlorophylls per dry weight; however, photosynthesis declined down to 40%. This can be interpreted as a short-time decrease in photosynthetic performance due to excessive UV irradiation, while the alga reacts with the production of additional pigments.

In an extended lab culture study, Leya et al. (2009) tested ten strains of arctic snow- and soil algae for their ability of secondary carotenoid production when exposed to less nitrogen and elevated irradiation. They found out that two species of *Raphidonema* were not able to produce any secondary carotenoids under the given environmental parameters (the authors suggested that this may be a trend in all members of the Trebouxiophyceae), whereas starving and light-stressed Chloromonad species can be valuable sources of canthaxanthin and astaxanthin, but also of alpha-tocopherol. The lack of secondary carotenoids in Trebouxiophyceae on snow can be also explained by the observations of Stibal and Elster (2005) and Stibal et al. (2007) in Svalbard that *Raphidonema* (and some similar *Koliella* species) is more likely a cold-tolerant soil alga. Moreover, several Trebouxiophyceae microalgae contain MAAs (Karsten et al. 2005); this points to an alternative UV-protective pigment in substitution of secondary *cis*-carotenoids. In agreement with this, no MAAs were detected in the non-Trebouxiophyceae *M. berggrenii* (Remias et al. 2009) and *Cr. nivalis* (Remias et al. 2010a).

A third group of constituents relevant to irradiation stress are phenolics, a group of mainly water-soluble pigments. These are well-known for seaweeds or vascular plants, but there are only few studies concerning microalgae. Duval et al. (2000) tested cysts of *Cd. nivalis* for changes in total phenolic contents after exposure to harsh UVC irradiation. Their spectrophotometric assays showed semiquantitatively that, with their colourimetric method, the cells had augmented phenolic content after stress exposure; however, no structural information concerning the constituent could be given. Also, Li et al. (2007) semiquantitatively measured a high antioxidant capacity of raw extracts of *Cd. nivalis* (the geographical origin of the material was not stated), concluding that putative phenolics are the responsible antioxidants.

Remias et al. (2009) collected and described populations of *M. berggrenii* from glaciers in the Tyrolean Alps and found high amounts of a brownish pigment located in cytoplasmic vacuoles. A similar pigmentation has also been found in other members of the Zygnematophyceae, like in the polar ice alga *Ancylonema nordenskiöldii* (Leya 2004) or in an alpine soil crust alga, *Zygogonium ericetorum* (Holzinger et al. 2010). To date, there are no known reports about the structure of these putative phenolics occurring in sarcoderm desmids that live in cold habitats, and an earlier suggestion of Ling and Seppelt (1990) that it may be ferric tannin has not been confirmed yet. Figure 13.6 shows the spectral absorption of an aqueous extract of *M. berggrenii*. Due to remarkable absorption in the UV A and B region, it can be assumed that this pigment plays a role in the protection against harmful irradiation.

### Conclusion

Snow and freshwater ice algae thrive exclusively in summer in old snow and wet glacier surfaces. They have adapted to harsh environmental conditions such as excessive irradiation, low temperatures or short vegetation period. Mass accumulations can cause alpine phenomena like Red Snow (*Chlamydomonas nivalis*) or Grey Ice (*Mesotaenium berggrenii*). The cellular survival strategies of snow algae, besides a psychrophilic metabolism, can be seen in a flexible life cycle comprising flagellated cells for reproduction during good conditions and specialized, immotile cyst stages for persistence. The latter possess secondary cell walls, frequently with fortifications such as spikes or flanges, and intracellular depots of lipids and secondary metabolites like astaxanthin. For unknown reasons cysts do not play a significant role in populations of ice algae, and the cell walls show no secondary structures. Instead, they have vacuoles filled with a brownish secondary pigment (a putative phenolic), which has a high spectral absorbance in the UV range.

**Acknowledgement** This research was supported by a grant of the Austrian FWF (200810) to C. Lütz.

## References

Bidigare RR, Ondrusek ME, Kennicutt MC II, Iturriaga R, Harvey HR, Hoham RW, Macko SA (1993) Evidence for a photoprotective function for secondary carotenoids of snow algae. J Phycol 29:427–434

Duval B, Shetty K, Thomas WH (2000) Phenolic compounds and antioxidant properties in the snow alga *Chlamydomonas nivalis* after exposure to UV light. J Appl Phycol 11:559–566

Elster J, Benson EE (2004) Life in the polar terrestrial environment – a focus on algae and cyanobacteria. In: Fuller B, Lane N, Benson EE (eds) Life in the frozen state. Taylor & Francis, London, pp 111–149

Ettl H (1968) Ein Beitrag zur Kenntnis der Algenflora Tirols. Ber Nat Med Ver Innsbruck 56:177–354

Ettl H (1983) Chlorophyta I. Phytomonadina. In: Ettl H, Gerloff J, Heynig H, Mollenhauer D (eds) Süßwasserflora von Mitteleuropa, vol. 9. Gustav Fischer, Stuttgart, p. 807

Gorton HL, Williams WE, Vogelmann TC (2001) The light environment and cellular optics of the snow alga Chlamydomonas nivalis (Bauer) Wille. Photochem Photobiol 73:611–620

Hoham RW, Duval B (2001) Microbial ecology of snow and freshwater ice with emphasis on snow algae. In: Jones HG et al (eds) Snow ecology. Cambridge University Press, New York, pp 168–228

Holzinger A, Tschaikner A, Remias D (2010) Cytoarchitecture of the desiccation-tolerant green alga *Zygogonium ericetorum*. Protoplamsa 243:15–24

Jones HG, Pomeroy JW, Walker DA, Hoham RW (2001) Snow ecology. Cambridge University Press, Cambridge, 378pp

Karsten U, Friedl T, Schumann R, Hoyer K, Lembcke S (2005) Mycosporine like amino acids (MAAs) and phylogenies in green algae: *Prasiola* and its relatives from the Trebouxiophyceae (Chlorophyta). J Phycol 41:557–566

Kol E (1961) Über den roten und grünen Schnee der Alpen. Verh Internat Verein Limnol 14:912–917

Kol E (1968) Kryobiologie. Biologie und Limnologie des Schnees und Eises. I. Kryovegetation. In: Elster HJ, Ohle W (eds) Die Binnengewässer, band XXIV. Schweizerbart'sche Verlagsbuchhandlung, Stuttgart, p 216

Kol E (1970) Vom roten Schnee der Tiroler Alpen. Annals Hist Nat Mus Nat Hung 62:129–136

Komárek J, Nedbalová L (2007) Green cryosestic algae. In: Seckbach J (ed) Cellular origin, life in extreme habitats and astrobiology (volume 11): algae and cyanobacteria in extreme environments, part 4: phototrophs in cold environments. Springer, Dordrecht, pp 323–344

Kvíderova J, Stibal M, Nedbalová L, Kaštovská K (2005) The first record of snow algae vitality in situ by variable fluorescence of chlorophyll. Fottea (Czech Phycol) 5:69–77

Leya T (2004) Feldstudien und genetische Untersuchungen zur Kryophilie der Schneealgen Nordwestspitzbergens. Dissertation. Shaker, Aachen, 145pp

Leya T, Rahn A, Lütz C, Remias D (2009) Response of arctic snow and permafrost algae to high light and nitrogen stress by changes in pigment composition and applied aspects for biotechnology. FEMS Microbiol Ecol 67:432–443

Li HB, Cheng KW, Wong CC, Fan KW, Chen F, Jiang Y (2007) Evaluation of antioxidant capacity and total phenolic content of different fractions of selected microalgae. Food Chem 102:771–776

Ling HU, Seppelt RD (1990) Snow algae of the Windmill Islands, continental Antarctica. *Mesotaenium berggrenii* (Zygnematales, Chlorophyta) the alga of grey snow. Antarct Sci 2:143–148

Lohr M (2009) Carotenoids. In: Stern DB (ed) Chlamydomonas sourcebook, vol 2, Organellar and metabolic processes. Academic Press, Oxford, pp 799–819, Chapter 21

Marasco EK, Vay K, Schmidz-Dannert C (2006) Identification of carotenoid cleavage dioxygenases from Nostoc sp. PCC 7120 with different cleavage activities. J Biol Chem 281:31583–31593

Remias D, Lütz C (2007) Characterisation of esterified secondary carotenoids and of their isomers in green algae: a HPLC approach. Algolog Stud 124:85–94

Remias D, Lütz-Meindl U, Lütz C (2005) Photosynthesis, pigments and ultrastructure of the alpine snow alga *Chlamydomonas nivalis*. Eur J Phycol 40:259–268

Remias D, Holzinger A, Lütz C (2009) Physiology, ultrastructure and habitat of the ice alga *Mesotaenium berggrenii* (Zygnemaphyceae, Chlorophyta) from glaciers in the European Alps. Phycologia 48:302–312

Remias D, Albert A, Lütz C (2010a) Effects of realistically simulated, elevated UV irradiation on photosynthesis and pigment composition of the alpine snow alga *Chlamydomonas nivalis* and the arctic soil alga *Tetracystis* sp. (Chlorophyceae). Photosynthetica 48:269–277

Remias D, Karsten U, Lütz C, Leya T (2010b) Physiological and morphological processes in the Alpine snow alga *Chloromonas nivalis* (Chlorophyceae) during cyst formation. Protoplasma 243:73–86

Řezanka T, Nedbalová L, Sigler K, Cepák V (2008a) Identification of astaxanthin diglucoside diesters from snow alga *Chlamydomonas nivalis* by liquid chromatography–atmospheric pressure chemical ionization mass spectrometry. Phytochemistry 69:479–490

Řezanka T, Nedbalová L, Sigler K (2008b) Unusual medium-chain polyunsaturated fatty acids from the snow alga *Chloromonas brevispina*. Microbiol Res 163:373–379

Sattler B, Remias D, Lütz C, Dastych H, Psenner R (2010) Glaziale und periglaziale Lebensräume im Raum Obergurgl. In: Koch EM, Erschbamer B (eds) Leben auf Schnee und Eis. Innsbruck University Press, Innsbruck, pp 229–249

Shain D, Mason T, Farrell A, Michalewicz L (2001) Distribution and behaviour of ice worms (*Mesenchytraeus solifugus*) in south-central Alaska. Can J Zool 79:1813–1821

Stibal M (2003) Ecological and physiological characteristics of snow algae from Czech and Slovak mountains. Fottea (Czech Phycol) 3:141–152

Stibal M, Elster J (2005) Growth and morphology variation as a response to changing environmental factors in two Arctic species of *Raphidonema* (Trebouxiophyceae) from snow and soil. Polar Biol 28:558–567

Stibal M, Elster J, Šabacká M, Kaštovská K (2007) Seasonal and diel changes in photosynthetic activity of the snow alga Chlamydomonas nivalis (Chlorophyceae) from Svalbard determined by pulse amplitude modulation fluorometry. FEMS Microbiol Ecol 59:265–273

Tartari A, Forlani G (2008) Osmotic adjustments in a psychrophilic alga, Xanthonema sp. (Xanthophyceae). Environ Exp Bot 63:342–350

Thomas WH, Duval B (1995) Sierra Nevada, California, USA, snow algae: snow albedo changes, algal-bacterial interrelationships, and ultraviolet radiation effects. Arct Alp Res 27:389–399

Vona V, Di Martino Rigano V, Lobosco O, Carfagna S, Esposito S, Rigano C (2004) Temperature responses of growth, photosynthesis, respiration and NADH: nitrate reductase in cryophilic and mesophilic algae. New Phytol 163:325–331

Werner P (2007) Roter Schnee oder Die Suche nach dem färbenden Prinzip. Akademie Verlag, Berlin, p 190

# Psychrophilic Microorganisms in Alpine Soils

# 14

Rosa Margesin

## 14.1 Introduction

The Earth is a cold planet. About 85% of the biosphere is exposed to temperatures below 5°C throughout the year. Cold habitats span from the Arctic to the Antarctic, from high mountain range environments to the deep ocean. The major fraction of this low-temperature environment is represented by the deep sea (nearly 75% of the Earth is covered by oceans and 90% of the ocean volume is below 5°C), followed by snow (35% of land surface), permafrost (24% of land surface), sea ice (13% of the Earth's surface) and glaciers (10% of land surface). Psychrophilic microorganisms, including bacteria, archaea, yeasts, filamentous fungi and algae, have successfully colonized these cold environments, because they evolved special mechanisms to overcome the life-endangering influence of low temperature. This chapter describes mechanisms of microbial cold adaptation and aspects of microbial activity and biodiversity in cold alpine soils.

## 14.2 Mechanisms of Microbial Adaptation to Cold

A change in temperature has an immediate effect on all cellular processes of microorganisms since they are too small to insulate themselves or to use avoidance strategies by moving away from thermal extremes (Russell 2008). To survive and grow successfully in cold environments, psychrophilic microorganisms have evolved a complex range of adaptations of all their cellular constituents, which enable them to compensate for the negative effects of low temperatures on biochemical reactions. These adaptation mechanisms are summarized below. Survival strategies of algae in ice and snow have been described by Remias (see chapter in this book).

### 14.2.1 Growth Characteristics

#### 14.2.1.1 Arrhenius Law and Growth

When the environmental temperature of a population of microorganisms drops, the growth rate decreases until a point is reached when one or more critical functions proceed so slowly that they are insufficient to support cellular requirements, and cell growth ceases. The effect of temperature on microbial growth is described by the Arrhenius law relating the exponential rise of the reaction rate to the temperature increase:

$$K = A\ e^{-Ea/RT}$$

where A is a constant (relating to steric factors and molecular collision frequency), Ea is the activation energy, R is the gas constant, and T is the absolute temperature.

According to this equation, any decrease in temperature causes an exponential decrease of the reaction rate, the magnitude of which depends on the value of

R. Margesin (✉)
Institute of Microbiology, University of Innsbruck, Innsbruck, Austria
e-mail: rosa.margesin@uibk.ac.at

C. Lütz (ed.), *Plants in Alpine Regions*,
DOI 10.1007/978-3-7091-0136-0_14, © Springer-Verlag/Wien 2012

the activation energy. The linear range of the Arrhenius plot (the logarithmic value of the growth rate is plotted as the reaction rate constant versus the reciprocal of the absolute temperature) corresponds to a physically "normal" temperature for growth, whereas the plot derives from linearity at temperatures near the upper or lower growth limits. Temperatures outside the linear range are stress-inducing temperatures, as shown by decreased microbial activity (e.g., enzyme production, degradation activities), protein synthesis, membrane permeability, and increased cellular stress (Feller and Gerday 2003; Jaouen et al. 2004; Margesin et al. 2005; D'Amico et al. 2006; Feller 2007). For psychrophiles, Arrhenius plots remain linear down to 0°C, while plots for mesophiles deviate from linearity at about 20°C (Gounot and Russell 1999).

At low temperatures, growth rates of psychrophiles are higher than those of mesophiles. While growth and enzyme production of mesophilic microorganisms is stopped in a refrigerator, psychrophiles actively divide and secrete enzymes under such conditions (see below, Fig. 14.1). Some wild-type psychrophilic bacteria display doubling times at 4°C comparable to that of fast-growing *E. coli* laboratory strains grown at 37°C. The latter fail to grow exponentially below 8°C, whereas psychrophilic bacteria maintain doubling times as low as 2–3 h at 4°C (Margesin and Feller 2010). Both for psychrophiles and mesophiles, the temperature for maximum biomass formation is well below the maximum temperature for growth. Psychrophilic bacteria and yeasts produced the highest amounts of cells per dry mass at 1°C, while cell numbers of mesophiles were highest at 20°C (Margesin 2009).

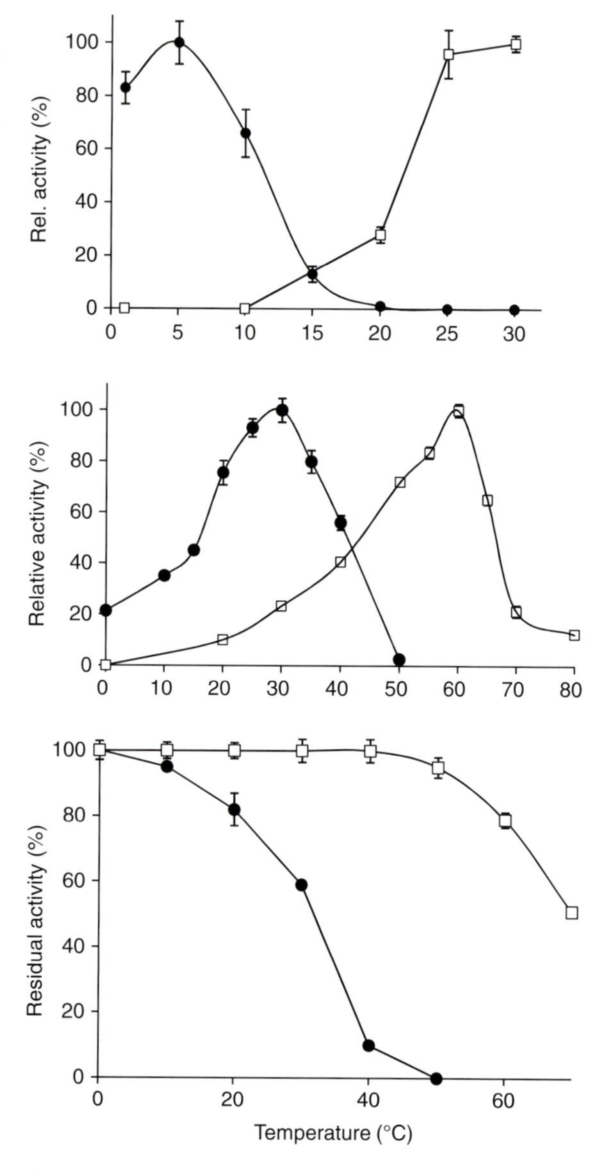


**Fig. 14.1** Effect of temperature on enzyme production (*top*), on activity (*middle*) and on stability (*bottom*; residual activity after 15 min of incubation at 25°C) of the cold-active pectate lyase produced by the alpine *Mrakia frigida* strain A15 (●) and its mesophilic counterpart produced by *Bacillus subtilis* (□). Modified from Margesin et al. (2005)

#### 14.2.1.2 Upper and Lower Temperature Limits for Growth

The slope of a microbial growth curve is usually greater at the high temperature compared to the low temperature end of the scale. The reason lies in the different mechanisms that are responsible for setting the upper and lower limits of growth, particularly for psychrophilic microorganisms.

The upper temperature limit for growth results from heat denaturation of cellular proteins. Psychrophiles have lower upper growth temperature limits than mesophiles because of the particular thermolability of one or more of their proteins (e.g. enzymes such as aminoacyl-tRNA synthetases) which are essential for growth/survival of the microorganisms. Other factors that are responsible for the comparatively low upper temperature limit for growth of psychrophiles include the inability to synthesize RNA at superoptimum temperatures, a reduced capacity of ribosomes to bind tRNA, a lower precision of translation, as well as alteration of the cell morphology and inhibition of cell division (Margesin and Schinner 1994).

The lower temperature limit for growth of psychrophiles is usually below 0°C, and its determination in practice is very difficult because of very slow growth rates and the need to include antifreeze in the culture medium which may further reduce the growth rate. The theoretical minimum for psychrophiles may be as low as −26.5°C (Nichols et al. 1997). This estimation assumes that there has been no phase change in the system. However, there are a number of potential phase changes which make such calculated lower temperature limits an overestimate of the true biological value (Margesin et al. 2002). Cold denaturation of proteins generally occurs at a temperature below −15°C (Franks 1995). The lower growth temperature limit is fixed by the physical properties of aqueous solvent systems inside and outside the cell. The major growth-limiting factor at subzero temperatures appears to be the availability of liquid water. Below -10 to -15°C, the cell water begins to freeze and intracellular salt concentrations increase due to the progressive removal of water into ice crystals. The resulting ionic imbalances, lowered water activity, and desiccation have a toxic effect on cells (Ingraham and Stokes 1959; Russell 1990). However, liquid water has been shown to exist at grain contacts as low as -20°C (Jakosky et al. 2003).

Currently the functional low-temperature limits of psychrophiles are -12°C for reproduction and -20°C for metabolism (Bakermans 2008). *Psychromonas ingrahamii* grows exponentially at -12°C with a doubling time of 240 h (Riley et al. 2008). Microbial activity at temperatures ranging from -9°C to -20°C and even below has been convincingly demonstrated by several laboratories and different techniques. Such activities include DNA/protein synthesis at -15°C (Christner 2002) and protein synthesis at -20°C (Junge et al. 2006) by laboratory cultures, as well as $CH_4$ production at -16.5°C (Rivkina et al. 2002) or glucose oxidation at -20°C (Panikov et al. 2006) in permafrost soil. Permafrost isolates have even been shown to grow and to be active at temperatures as low as -35°C (Panikov et al. 2006; Panikov and Sizova 2007).

### 14.2.2 Cold Sensing, Lipids and Membrane Fluidity

The ability to adapt to low temperatures depends on the ability to sense changes in temperature. One of the primary cold sensors is the cell membrane that acts as an interface between external and internal environments (Rowbury 2003). At cold temperatures, the membrane becomes more rigid, which activates a membrane-associated sensor. The sensor transduces the signal to a response regulator, which induces up-regulation of genes involved in membrane fluidity modulation, and ultimately results in up-regulation of a number of genes involved in cold adaptation of bacteria, such as genes for fatty acid desaturases, genes that serve as RNA chaperones similar to cold-shock proteins, genes involved in replication, transcription and translation, and genes that encode a number of enzymes (Shivaji and Prakash 2010).

The membranes of microorganisms, like other organisms, contain a lipid bilayer that is essential for many of the major cellular functions, including passive and active permeability, nutrient uptake and electron transport, environmental sensing, photosynthesis and recognition processes. All of these functions demand the maintenance of membrane stability. Lipid fluidity is most influenced by the fatty acyl moieties, whereas lipid phase depends more on the nature of the head-group of the membrane lipids. Both the gel to liquid-crystalline transition as well as the bilayer (lamellar) to non-bilayer phase transition are influenced by growth temperature. However, changes in microbial culture temperature usually lead to greater modifications in the fatty-acyl composition than the head-group composition of membrane lipids, and so the focus of attention has been on fluidity effects (Margesin et al. 2002).

To increase membrane fluidity, microorganisms apply various strategies. When growth temperature is lowered, the most frequently observed change in fatty acid composition is in the extent of unsaturation; increased fatty acid unsaturation has been observed with bacteria, archaea, fungi, and algae. Other bacterial strategies include an increased content in methyl-branched fatty acids, changes in fatty acid isomerization, and an increase in the ratio of anteiso/iso-branched fatty acids. A decrease in the average chain length of fatty acids (only possible in growing cells) as well as in the ratio of sterol/phospholipids has been detected with bacteria, fungi and algae (Robinson 2001; Russell 2008). A further mode of modulation of membrane fluidity includes changes in the composition of carotenoids; polar carotenoids stabilize the membrane to a greater extent than non-polar ones (Chintalapati et al. 2004; Russell 2008; Shivaji and Prakash 2010). Snow algae also produce large amounts of carotenoids in response to environmental conditions (see chapter by Remias).

Among fatty acid changes in response to temperature, two categories can be distinguished: (1) Alteration of the existing membrane ("modification synthesis"; resulting in fatty acid unsaturation by desaturases) is a more rapid process, especially in response to sudden temperature decrease. (2) Some fatty acid changes ("addition synthesis", such as methyl branching, altered chain length, ratio of sterol/phospholipids) require de novo biosynthesis. In general, after a temperature decrease, modification synthesis takes place to restore membrane fluidity, and later addition synthesis takes over (Russell 2008).

### 14.2.3 Cold-Active Enzymes

Psychrophiles produce cold-active enzymes. These enzymes can be up to ten times more active at low and moderate temperatures than their mesophilic homologues (D'Amico et al. 2006). Furthermore, psychrophilic enzymes are heat-labile and are frequently inactivated at temperatures that are not detrimental to their mesophilic counterparts (see below, Fig. 14.1).

The conformation and 3D structures of psychrophilic proteins are not markedly different from their mesophilic homologues, and, furthermore, all amino acid side chains that are essential for the catalytic mechanism are strictly identical. It was found, however, that cold-active enzymes maintain the appropriate flexibility and dynamics of the active site at temperatures at which their mesophilic and thermophilic counterparts have severely restricted molecular motions (Feller and Gerday 2003; D'Amico et al. 2006). Thus, cold-active enzymes have a higher structural flexibility in order to compensate for the freezing effect of their cold habitats (Feller 2007). This is achieved by the disappearance of discrete stabilizing interactions either in the whole molecule or at least in structures adjacent to the active site. Amongst these destabilizing factors, the most relevant include a reduced number of proline residues and of electrostatic interactions (ion pairs, H-bonds, aromatic interactions), a weakening of the hydrophobic effect, the strategic location of glycine residues, an improved interaction of surface side chains with the solvent or an improved charge-induced interaction with substrates and cofactors (Siddiqui and Cavicchioli 2006). This adaptive destabilization of psychrophilic enzymes has been demonstrated to be responsible for both cold-activity and low thermal stability (D'Amico et al. 2003; Feller 2007).

### 14.2.4 Cold-Shock Proteins and Cold-Acclimation Proteins

As a response to sudden temperature changes, representatives of all thermal classes of bacteria (psychro-, meso- and thermophilic) display cold-shock responses. Mesophilic bacteria react with a transient overexpression of cold-shock proteins (CSPs) that are involved in a number of cellular processes, e.g., transcription, translation, protein folding, regulation of membrane fluidity, general metabolism, and chemotaxis (Phadtare 2004; Phadtare and Inoue 2008). The basic principles of cold-shock response are similar in psychrophiles and mesophiles. However, the cold-shock response in psychrophiles differs from that in mesophiles or thermophiles bacteria in two major aspects: cold shock does not repress the synthesis of housekeeping proteins, and the number of CSPs is higher and increases with the severity of the cold shock. In addition, psychrophiles permanently produce one set of proteins (cold-acclimation proteins, CAPs) during growth at low temperature and increase the steady-state level of CAPs when the temperature is lowered. These CAPs are mostly constitutively (rather than transiently) expressed at low temperatures and may be fundamental to life in the cold and ensure improved protein synthesis at low temperature (Gounot and Russell 1999; Margesin et al. 2002; Phadtare and Inoue 2008).

### 14.2.5 Cryoprotectants and Ice-Binding Proteins

In frozen environments, bacteria are exposed to conditions that require the partial removal of water from the intracellular space to maintain the structure and function of the cell. Since water is essential for the functioning of macromolecular structures, any significant deviation in the accessibility of water, such as the physical state (alteration from the aqueous phase to an ice crystal), poses a severe threat to the survival of organisms (Beall 1983). Psychrophilic microorganisms produce various compounds to protect themselves or the extracellular environment against intracellular freezing or to minimize the deleterious effects of ice crystal formation (Kawahara 2002).

#### 14.2.5.1 Low-Molecular Mass Cryoprotectants

Freezing results in an osmotic shock. Osmoprotection of bacterial and fungal cells is achieved by the accumulation of compatible solutes (low molecular mass compounds) after cold shock in bacteria and fungi (Gounot and Russell 1999; Robinson 2001; Kawahara 2008; Shivaji and Prakash 2010). These compounds include polyamines, sugars (e.g., glucose, fructose, sucrose, trehalose, ribose), polyols (a class of alcohols derived from sugar; e.g., glycerol, sorbitol, mannitol), and amino acids (e.g., alanine, proline).

Ribose-1-phosphate acts as cryoprotectant of enzymes (observed with *Pantoea agglomerans*), while the accumulation of glucose results in the depression of freezing points (observed with *Pantoea ananatis*) (Kawahara 2008). Trehalose accumulation in bacteria plays a role in preventing protein denaturation and aggregation (Phadtare 2004). This sugar is also accumulated in alpine mycorrhizal roots and in fungal hyphae in response to low temperatures (Niederer et al. 1992; Weinstein et al. 2000). For example, trehalose accumulation in *Mortierella elongata* at 5°C increased by 75% compared to the accumulation at 15°C. Polyols (e.g. glycerol, mannitol) act as cryoprotectants in fungi (Robinson 2001).

Glycine betaine aids to maintain optimum membrane fluidity at low temperatures by preventing cold-induced aggregation of proteins; this compound has been shown to enhance growth of *Listeria monocytogenes* at low temperatures (Chattopadhyay 2002). *Colwellia psychrerythraea* (Methé et al. 2005) and *Psychromonas ingrahmii* (Riley et al. 2008) have genes for the production of compatible solutes, such as glycine betaine and betaine cholin, which may balance the osmotic pressure under freezing conditions.

#### 14.2.5.2 Ice-Nucleation Proteins

Some bacteria (at least six Gram-negative and epiphytic species of the genera *Pseudomonas*, *Pantoea* and *Xanthomonas*) and fungi (e.g., *Fusarium* and related genera) produce proteins that can induce ice-nucleation at temperatures higher than -3°C. Ice-nucleating agents serve as templates for ice crystallization and provide resistance to desiccation. The induction of frost damage in plants by bacteria that produce ice-nucleating agents can be an adaptive advantage to get access to nutrients from plants (Lundheim 2002).

According to the sequences of genes conferring ice-nucleating activity in six bacterial strains, all strains encode ice-nucleating proteins with a molecular mass of 120–150 kDa and similar primary structures. The ice-nucleating proteins contain three domains: the N-domain is responsible for the binding of lipids, polysaccharides and ice-nucleating proteins; the R-domain acts as a template for ice formation, and their length (ca. 800–1,300 amino acids) is correlated with the amplitude of ice nucleation activity; the C-terminal domain is required for ice-nucleation activity as demonstrated with mutants (Kawahara 2008).

#### 14.2.5.3 Antifreeze Proteins

Antifreeze proteins (AFPs) are ice-binding proteins that have the ability to modify the ice crystal structure and inhibit the growth of ice in two ways. (1) Prior to freezing, they lower the freezing point of water without altering the melting point (thermal hysteresis activity). (2) In the frozen state, AFPs show ice recrystallization inhibition activity, whereby the proteins inhibit the growth of large crystals at the expense of small crystals at subzero temperatures (Gilbert et al. 2004).

AFPs have been detected in bacteria, fungi, plants and animals (Margesin et al. 2007). Bacterial and plant AFPs generally show substantially lower thermal hysteresis compared to AFPs from animals. Insects and fish have up to 2°C and 5°C of thermal hysteresis, respectively, while bacterial and fungal representatives show values of ≤0.1°C (Gilbert et al. 2005; Hoshino et al. 2009). Bacteria that produce AFPs with a low thermal hysteresis activity, however, use the recrystallization inhibition activity of the AFPs (Xu et al. 1998; Yamashita et al. 2002; Gilbert et al. 2005). As an exception, the AFP produced by the Antarctic lake-ice bacterium *Marimonas primoryensis* has a thermal hysteresis activity (lowers the freezing point of water) of more than 2°C, which is higher than the maximum activity of most fish AFPs. The protein is $Ca^{2+}$-dependent and located in the periplasmic space, while bacterial and fungal AFPs are generally secreted extracellularly (Gilbert et al. 2004; Gilbert et al. 2005).

Thus, bacteria may apply different strategies: freeze tolerance can be obtained with low levels of thermal hysteresis activity but high recrystallization inhibition activity, a strategy similar to the one employed by some plants (rye grass, carrot, winter rye) (Griffith et al. 1992; Worrall et al. 1998; Sidebottom et al. 2000). On the other hand, freeze avoidance by high thermal hysteresis activity can inhibit the growth of ice crystals before they propagate into the bacterium (Gilbert et al. 2005).

Among AFP-producing bacteria from Antarctic lakes, members of the *Gammaproteobacteria* dominated (Gilbert et al. 2004). The AFP produced by an Arctic plant growth promoting rhizobacterium (*Pseudomonas putida*) is an extracellular glycolipoprotein that also has ice-nucleating activity (Xu et al. 1998).

In fungi, extracellular AFPs are assumed not only to prevent hyphae from freezing, but also to ensure substrate availability by preventing nutrients from freezing at subzero temperatures (Robinson 2001). AFPs have been detected in psychrophilic phythopathogenic fungi causing snow molds. These fungi belong to various taxa (*Oomycetes*, *Ascomycetes* and *Basidiomycetes*), grow at temperatures as low as $< -7°C$, and can grow and attack dormant plants (crops, winter cereals and conifer seedlings) at low temperatures under snow cover (Hoshino et al. 2009). Basidiomycetous snow molds produce extracellular AFPs to keep the extracellular environment unfrozen, which, however, does not support mycelial growth. In contrast, the ascomycete *Sclerotia borealis* does not produce extracellular AFPs but grows at subzero temperatures due to osmotic stress tolerance; its mycelial growth is even higher under frozen conditions compared to unfrozen conditions (Hoshino et al. 2009).

#### 14.2.5.4 Exopolymers

Exopolymeric substances (EPS) are complex organic materials composed primarily of high-molecular mass exopolysaccharides. Exopolysaccharides contain major amounts of hexose and pentose. Contrary to intracellular adjustments to cold stress, EPS are secreted as mucous slime by many aquatic microorganisms. Key functions of EPS include the mediation of adhesion to wet surfaces and the formation of the biofilm matrix, which traps nutrients, protects the cell against unfavorable environmental conditions and mediates biochemical interaction (Mancuso Nichols et al. 2005). EPS production is high in bacteria living in aquatic environments; high EPS abundance has been found in Antarctic and Arctic sea ice (Krembs et al. 2002; Mancuso Nichols et al. 2005).

### 14.2.6 Antioxidant Defense

Protection against reactive oxygen species (ROS) is important for survival at low temperatures where the solubility of gases is increased. ROS can result in significant damage to cell structures. Bacterial strategies for the detoxification of ROS include the production of high amounts of antioxidant enzymes (catalase, superoxide dismutases, dioxygen-consuming lipid desaturases) or the absence of ROS-producing pathways. *Pseudoalteromonas haloplanktis* employs both strategies; it entirely lacks the ROS-producing molybdopterin metabolism. In addition, the bacterium produces dioxygen-consuming lipid desaturases in order to obtain protection against oxygen and to maintain membrane fluidity at the same time. By contrast, *Colwellia psychrerythraea* achieves an enhanced antioxidant capacity through the presence of catalase and superoxide dismutases (Medigue et al. 2005; Methé et al. 2005).

### 14.2.7 Genomic and Proteomic Insights into Microbial Cold Adaptation

Microbial adaptation to low temperatures requires a vast array of metabolic and structural adjustments at nearly all organization levels of the cell, which are gradually being understood thanks to the availability of genome sequences and proteomic studies of a number of psychrophilic bacteria. A survey of these data shows that the main up-regulated functions for growth at low temperatures are protein synthesis (transcription, translation), RNA and protein folding (adaptation of the molecular structure of proteins to ensure increased flexibility at low temperatures), maintenance of membrane fluidity, production and uptake of compounds for cryoprotection (extracellular polysaccharides, compatible solutes), antioxidant activities and regulation of specific metabolic pathways. However, only few features are commonly shared by all psychrophilic genomes and proteomes, which suggests that cold adaptation superimposes on pre-existing cellular organization and, accordingly, the strategies to cope with cold environments may differ among psychrophiles (Medigue et al. 2005; Methé et al. 2005; Kurihara and Esaki 2008; Riley et al. 2008; Bakermans et al. 2009; Qiu et al. 2009).

## 14.3 Microbial Activity and Biodiversity in Alpine Soils

Compared to the Arctic, the European Alpine region is characterized by higher maximum temperatures, lower minimum temperatures, large and frequent (diurnal) temperature fluctuations and freeze-thaw events, higher precipitation (up to 2,000–3,000 mm per year) and air humidity, lower atmospheric pressure, and higher intensity of solar radiation.

Alpine microorganisms are equally well-adapted to low temperatures as polar microorganisms. The comparison of cold-active enzymes (pectate lyase) from alpine and Siberian psychrophilic yeasts (*Mrakia frigida*) clearly showed that the enzymes produced by these strains had an almost identical activity and stability pattern (Fig. 14.1). Both enzymes were thermolabile, but resistant to repeated freezing and thawing (Margesin et al. 2005). The two strains had almost identical growth characteristics (high cell densities at 1–15°C, no growth above 20°C), yet their enzyme production patterns were completely different. The Siberian strain produced pectate lyase over the entire growth temperature range, with a maximum at 1°C, whereas enzyme production by the alpine strain was highest at 5°C, very low at 15°C and absent at 20°C. Enzyme production patterns may be related to the natural environmental conditions of the strains.

### 14.3.1 Soil Microbial Activity at Low Temperatures

Soil microorganisms play an essential role in soil organic matter turnover and biogeochemical cycling. Soil microbial activity and community composition are influenced by a number of biotic and abiotic factors, such as vegetation type, soil type, and a range of environmental conditions including temperature. Low temperature is not a limiting factor for microbial activity in cold soils. There is evidence of a wide range of metabolic activities in all cold ecosystems; microbial activity in soil has been reported to occur at subzero temperatures down to -20°C (Lipson and Schmidt 2004; Panikov and Sizova 2007) and substantial carbon mineralization has been described to occur in cold soils during winter months (Clein and Schimel 1995).

A change in temperature affects soil microbial communities and nutrient cycling (Uchida et al. 2000; Hart 2006). Microbial activities in cold soils respond quickly to seasonal changes (Lipson 2007; Edwards and Jefferies 2010). In seasonally frozen soils from some alpine and arctic sites, microorganisms metabolize slowly at subzero temperatures, presumably in contact with unfrozen water. However, microbial biomass declines in late winter (at the winter-spring transition), before the soil temperature rises above 0°C. This decline in biomass has been attributed to low levels of available nutrients, rupture of cell membranes due to repeated freeze-thaw cycles, and the loss of compatible solutes from viable cells due to an abrupt change in osmotic potential (Jefferies et al. 2010).

Like polar microorganisms, psychrophilic alpine microorganisms, able to grow and to be active at low temperatures, play a key ecological role in their natural habitats. Measurement of microbial activities in the Austrian Central Alps at altitudes of 2,300–2,500 m a.s.l. included litter decomposition, $CO_2$-release and enzyme activities (phosphatase, urease, xylanase, cellulase). Soil activities were generally lower on wind-exposed sites and were low in poorly drained soils of the snowbed, which was explained by a deficiency of substrates and frequent drought stress during the vegetation period. Irrespectively of the site, soil microbial activities increased immediately after the frozen topsoils thawed, when bacterial and fungal populations increased (Schinner 1982a, 1983). Another factor influencing soil microbial respiration and enzyme activities is soil depth; activities were considerably higher in surface layers of alpine soils and sharply decreased with depth. Soil microbial activities are further influenced by vegetation. For example, activities in soils with alpine dwarf shrubs (*Loiseleuria procumbens*) were higher by a factor of five compared to activities in *Carex curvula* grassland soils (Schinner 1982a).

The degradation of xylan, the major polysaccharide in plant cell walls, occurs mainly by microbial xylanases. Recently it has been shown that xylanase activities in cold alpine tundra soil are very diverse and widely distributed among soil bacteria; they could be clustered into six groups and were related to xylanases from *Actinobacteria*, *Proteobacteria*, *Verrucomicrobia*, *Bacteroidetes*, *Firmicutes* and *Acidobacteria* (Wang et al. 2010).

Soil microbial respiration is a critical component of the global carbon cycle. In subalpine coniferous forest soil, microbial communities isolated from under-snow soil were characterized by high biomass-specific respiration rates, i.e. higher growth rates and lower growth yields. Bacteria may contribute to soil heterotrophic respiration to a greater extent, as demonstrated by higher bacterial growth rates and lower growth yields compared to those from fungi. In winter, psychrophilic bacteria of the genus *Janthinobacterium* dominated (Lipson et al. 2009).

Litter decomposition is an important factor in nutrient cycling. Microorganisms decomposing plant litter belong to phylogenetically diverse taxa. In cold Arctic and Antarctic ecosystems, wood decomposition appears to proceed via "soft rot" by anamorphous ascomycetes, rather than by "white rot" or "brown rot" basidiomycetes (Ludley and Robinson 2008).

Cold periods during the growing season can significantly limit the symbiotic association of legumes with rhizobia. Cold-adapted rhizobia, isolated from alpine or arctic legumes, are useful to improve the symbiosis under cold stress. Arctic rhizobia increased the production of legumes by 30% through improved nitrogen fixation (Prevost et al. 2003).

### 14.3.2 Microbial Activity and Biodiversity Related to Altitude

The change of temperature and other environmental conditions with altitude in mid-latitude mountains has often been compared to their change with latitude: a 1,000 m higher altitude in the Alps may roughly be equivalent to a 1,000 km move northward (Kuhn 2008). Thus, temperature gradients in mountains can be similar to those relating to latitude; the altitude-controlled vegetation belts on mountain slopes represent an analogue to the different latitudinally controlled climatic zones. The annual average temperature decreases with increasing latitude; in mountain areas the temperature decreases with increasing altitude. While climate changes (e.g. temperature decrease) are spread over thousands of kilometres along latitude gradients, they occur on a comparatively small scale along altitude gradients, which makes mountain regions useful for climate change studies (Diaz et al. 2003).

Altitudinally defined climatic conditions, soil properties, and vegetation regulate microbial community structures and metabolic rates in mountain soils (Whittaker 1975; Schinner and Gstraunthaler 1981). An increase in altitude, and thus in environmental harshness (lower annual temperature, lower soil nutrient contents), generally results in a decrease in microbial abundance and activity (respiration rate, microbial biomass, litter degradation, enzyme activities), as well as in shifts in microbial (bacterial and fungal) community composition (Schinner and Gstraunthaler 1981; Schinner 1982b; Väre et al. 1997; Ma et al. 2004; Giri et al. 2007; Lipson 2007; Niklinska and Klimek 2007). With increasing altitude, and thus colder climate conditions (lower air and soil temperatures, more ice and frost days, higher precipitation) over a gradient ranging from 1,500 to 2,530 m in the Austrian Central Alps, a number of significant changes were observed: an increase in altitude resulted in a significant decrease of bacterial and fungal biomass, on one hand, and in a significant increase in the relative amounts of psychrophilic heterotrophic bacteria and fungal populations, on the other hand. Gram-negative bacteria detected by FISH (fluorescence in situ hybridization) increased with altitude. Since FISH is based on the detection of rRNA, and the rRNA content is associated with the metabolic state of microbial cells, FISH-detected cells represent the active, ecologically relevant part of the microbial community (Wagner et al. 2003). *Proteobacteria* dominated at high altitudes, while the amount of members of the *Cytophaga-Flavobacterium-Bacteroides* group decreased with altitude (Margesin et al. 2009).

With increasing altitude, and thus colder climate conditions, microorganisms are better adapted to the cold. Microbial activity (soil dehydrogenase) decreased with altitude, yet relative activities at low temperatures were significantly higher in alpine than

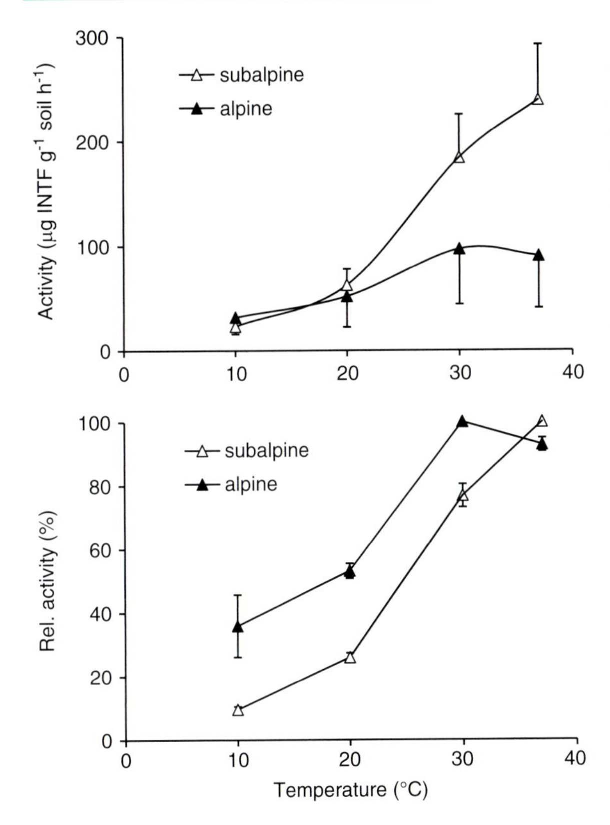


**Fig. 14.2** Effect of temperature on soil dehydrogenase activity (*top*) and on the relative enzyme activity (*bottom*; maximum activity as determined in the figure on *top* = 100%) in subalpine (1,500–1,900 m) and alpine (2,100–2,530 m, above the forest line) soils (Margesin et al. 2009). INTF = iodonitrotetrazoium formazan

in subalpine soils (Fig. 14.2), which means that enzymes from soils from higher altitudes are better adapted to the prevailing cold climate conditions. This can be attributed partly to the influence of altitude on physicochemical properties; e.g., lower contents of clay, humus and nitrogen due to unfavorable conditions for soil formation with increasing altitude; slower nutrient cycling at high altitudes due to cold temperatures could possibly affect organic matter structure and quality (Margesin et al. 2009).

Similarly to European alpine soils, the diversity of the psychrophilic bacterial community in high altitude cold soils of the Himalayan mountains decreased with increasing altitude. The culture-independent approach revealed a dominance of *Preoteobacteria*. However, viable bacteria consisted of almost equal amounts of Gram-negative bacteria (with a dominance of *Gammaproteobacteria* and a low amount of *Bacteroidetes*) and Gram-positive bacteria (with a dominance of *Firmicutes*). Isolates produced a number of hydrolytic enzymes; the most frequently observed enzyme was lipase (Gangwar et al. 2009). The abundance of ammonia-oxidizing bacteria and archaea in high-altitude soils (4,000–6,500 m) of Mt. Everest was also influenced by altitude. Archaeal ammonia oxidizers were more abundant than bacterial ones at altitudes below 5,400 m, while the situation was reversed at higher altitudes (Zhang et al. 2009).

Soils at high altitudes (3,000–5,400 m) in Annapurna Mountains, Nepal, are characterized by low water activity due to dry climate, and consequently these soils contained psychrophilic fungi with xerophilic characteristics; the most extreme xerophiles belonged to the ascomycetous genera *Eurotium* and *Aspergillus* (Petrovic et al. 2000). *Chytridiomycota* dominated fungal diversity in periglacial soils at high altitudes in the Himalayans and Rockies, which can be attributed to the high abundance of carbon sources that support chytrid growth (eolian deposited pollen and microbial phototrophs) as well to the saturation of soils with water under snow (Freeman et al. 2009).

## Conclusions

A change in temperature has an immediate effect on all cellular processes of microorganisms, since they are too small to insulate themselves from the cold or to use avoidance strategies such as moving away from thermal extremes. Therefore, they alter their cellular composition. To survive and grow successfully in cold environments, psychrophilic microorganisms have therefore evolved a complex range of adaptations of all their cellular constituents, which enable them to compensate for the negative effects of low temperatures on biochemical reactions. The main up-regulated functions for growth at low temperatures are protein synthesis (transcription, translation), RNA and protein folding, maintenance of membrane fluidity, production and uptake of compounds for cryoprotection (extracellular polysaccharides, compatible solutes), antioxidant activities and regulation of specific metabolic pathways. The emerging fields of genome and proteome analyses will give further new insights into the psychrophilic lifestyle.

Microorganisms in cold soils play an essential role in organic matter turnover and biogeochemical cycling. Like polar microorganisms, psychrophilic alpine microorganisms, able to grow and to be active at low temperatures, play a key ecological role in their natural habitats. An increase in environmental harshness (e.g. lower air and soil temperatures, more frost and ice days, higher precipitation at higher altitudes) generally results in a decrease in microbial abundance and activity, as well as in shifts in microbial community composition. On the other hand, microorganisms living in colder climate conditions are better adapted to the cold, as shown by higher relative amounts of psychrophilic bacterial and fungal populations and higher relative enzyme activities.

## References


Bakermans C (2008) Limits for microbial life at subzero temperatures. In: Margesin R, Schinner F, Marx JC, Gerday C (eds) Psychrophiles: from biodiversity to biotechnology. Springer, Berlin, pp 17–28

Bakermans C, Bergholz PW, Ayala-del-Río H, Tiedje J (2009) Genomic insights into cold adaption of permafrost bacteria. In: Margesin R (ed) Permafrost soils, vol 16, Soil biology. Springer, Berlin, pp 159–168

Beall PT (1983) States of water in biological systems. Cryobiology 20:324–443

Chattopadhyay MK (2002) The cryoprotective effects of glycine betaine on bacteria. Trends Microbiol 10:311

Chintalapati S, Kiran MD, Shivaji S (2004) Role of membrane lipid fatty acids in cold adaptation. Cell Mol Biol 50:631–642

Christner BC (2002) Incorporation of DNA and protein precursors into macromolecules by bacteria at -15°C. Appl Environ Microbiol 68:6435–6438

Clein JS, Schimel JP (1995) Microbial activity of tundra and taiga soils at sub-zero temperatures. Soil Biol Biochem 27:1231–1234

D'Amico S, Gerday C, Feller G (2003) Temperature adaptation of proteins: engineering mesophilic-like activity and stability in a cold-adapted alpha-amylase. J Mol Biol 332:981–988

D'Amico S, Collins T, Marx JC, Feller G, Gerday C (2006) Psychrophilic microorganisms: challenges for life. EMBO Rep 7:385–389

Diaz HF, Grosjean M, Graumlich L (2003) Climate variability and change in high elevation regions: past, present and future. Clim Change 59:1–4

Edwards KA, Jefferies RL (2010) Nitrogen uptake by *Carex aquatilis* during the winter-spring transition in a low Arctic wet meadow. J Ecol 98:737–744

Feller G (2007) Life at low temperatures: is disorder the driving force? Extremophiles 11(2):11–216

Feller G, Gerday C (2003) Psychrophilic enzymes: hot topics in cold adaptation. Nat Rev Microbiol 1:200–208

Franks F (1995) Protein destabilization at low temperatures. Adv Protein Chem 46:105–139

Freeman KR, Martin AP, Karki D, Lynch RC, Mitter MS, Meyer AF, Longcore JE, Simmons DR, Schmidt SK (2009) Evidence that chytrids dominate fungal communities in high-elevation soils. Proc Natl Acad Sci USA 106:18315–18320

Gangwar P, Alam SI, Bansod S, Singh L (2009) Bacterial diversity of soil samples from the western Himalayas, India. Can J Microbiol 55:564–577

Gilbert JA, Hill PJ, Dodd CER, Laybourn-Parry J (2004) Demonstration of antifreeze protein activity in Antarctic lake bacteria. Microbiology 150:171–180

Gilbert JA, Davies PL, Laybourn-Parry J (2005) A hyperactive, $Ca^{2+}$-dependent antifreeze protein in an Antarctic bacterium. FEMS Microbiol Lett 245:67–72

Giri DD, Shukla PN, Kashyap S, Singh P, Kashyap AK, Pandey KD (2007) Variation in methanotrophic bacterial population along an altitude gradient at two slopes in tropical dry deciduous forest. Soil Biol Biochem 39:2424–2426

Gounot AM, Russell NJ (1999) Physiology of cold-adapted microorganisms. In: Margesin R, Schinner F (eds) Cold-adapted organisms. Springer, Berlin, pp 33–55

Griffith M, Ala P, Yang DS, Hon WC, Moffat BA (1992) Antifreeze protein produced endogenously in winter rye leaves. Plant Physiol 100:593–596

Hart SC (2006) Potential impacts of climate change on nitrogen transformations and greenhouse gas fluxes in forests: a soil transfer study. Global Change Biol 12:1032–1046

Hoshino T, Xiao N, Tkachenko OB (2009) Cold adaptation in the phythopathogenic fungi causing snow molds. Mycoscience 50:26–38

Ingraham JL, Stokes JL (1959) Psychrophilic bacteria. Bacteriol Rev 23:97–108

Jakosky BM, Nealson KH, Bakermans C, Ley RE, Mellon MT (2003) Subfreezing activity of microorganisms and the potential habitability of Mars' polar regions. Astrobiology 3:343–350

Jaouen T, De E, Chevalier S, Orange N (2004) Size dependence on growth temperature is a common characteristic of the major outer membrane protein OprF in psychrotrophic and mesophilic *Pseudomonas* species. Appl Environ Microbiol 70:6665–6669

Jefferies JL, Walker NA, Edwards KA, Dainty J (2010) Is the decline of soil microbial biomass in late winter coupled to changes in the physical status of cold soils? Soil Biol Biochem 42:129–135

Junge K, Eicken H, Swanson BD, Deming JW (2006) Bacterial incorporation of leucine into protein down to -20°C with evidence for potential activity in sub-eutectic saline ice formations. Cryobiology 52:417–429

Kawahara H (2002) The structure and function of ice crystal-controlling proteins from bacteria. J Biosci Bioeng 94:492–496

Kawahara H (2008) Cryoprotection and ice-binding proteins. In: Margesin R, Schinner F, Marx JC, Gerday C (eds) Psychrophiles: from biodiversity to biotechnology. Springer, Berlin, pp 229–246

Krembs C, Eicken H, Junge K, Deming JW (2002) High concentrations of exopolymeric substances in Arctic winter sea ice: Implications for the polar ocean carbon cycle and cryoprotection of diatoms. Deep Sea Res 49:2163–2181

Kuhn M (2008) The climate of snow and ice as boundary condition for microbial life. In: Margesin R, Schinner F, Marx JC, Gerday C (eds) Psychrophiles: from biodiversity to biotechnology. Springer, Berlin, pp 3–15

Kurihara T, Esaki N (2008) Proteomic studies of psychrophilic microorganisms. In: Margesin R, Schinner F, Marx JC, Gerday C (eds) Psychrophiles: from biodiversity to biotechnology. Springer, Berlin, pp 333–344

Lipson DA (2007) Relationships between temperature responses and bacterial community structure along seasonal and altitudinal gradients. FEMS Microbiol Ecol 59:418–427

Lipson DA, Schmidt SK (2004) Seasonal changes in an alpine soil bacterial community in the Colorado Rocky Mountains. Appl Environ Microbiol 70:2867–2879

Lipson DA, Monson RK, Schmidt SK, Weintraub MN (2009) The trade-off between growth rate and yield in microbial communities and the consequences for under-snow soil respiration in a high elevation coniferous forest. Biogeochemistry 95:23–35

Ludley KE, Robinson CH (2008) Decomposer Basidiomycota in Arctica and Antarctic ecosystems. Soil Biol Biochem 40:11–29

Lundheim R (2002) Physiological and ecological significance of biological ice nucleators. Phil Trans R Soc Lond B 357:937–943

Ma X, Chen T, Zhang G, Wang R (2004) Microbial community structure along an altitude gradient in three different localities. Folia Microbiol 49:105–111

Mancuso Nichols CA, Guezennec J, Bowman JP (2005) Bacterial exopolysaccharides from extreme marine environments with special consideration of the southern ocean, sea ice, and deep-sea hydrothermal vents: A review. Marine Biotechnol 7:253–271

Margesin R (2009) Effect of temperature on growth parameters of psychrophilic bacteria and yeasts. Extremophiles 13:257–262

Margesin R, Feller G (2010) Biotechnological applications of psychrophiles. Environ Technol 31:844–845

Margesin R, Schinner F (1994) Properties of cold-adapted microorganisms and their potential role in biotechnology. J Biotechnol 33:1–14

Margesin R, Feller G, Gerday C, Russell NJ (2002) Cold-adapted microorganisms: adaptation strategies and biotechnological potential. In: Bitton G (ed) The encyclopedia of environmental microbiology, vol 2. John Wiley & Sons Inc., New York, pp 871–885

Margesin R, Fauster V, Fonteyne PA (2005) Characterization of cold-active pectate lyases from psychrophilic *Mrakia frigida*. Lett Appl Microbiol 40:453–459

Margesin R, Neuner G, Storey KB (2007) Cold-loving microbes, plants and animals – fundamental and applied aspects. Naturwissenschaften 94:77–99

Margesin R, Jud M, Tscherko D, Schinner F (2009) Microbial communities and activities in alpine and subalpine soils. FEMS Microbiol Ecol 67:208–218

Medigue C, Krin E, Pascal G, Barbe V, Bernsel A, Bertin PN, Cheung F, Cruveiller S, D'Amico S, Duilio A, Fang G, Feller G, Ho C, Mangenot S, Marino G, Nilsson J, Parrilli E, Rocha EPC, Rouy Z, Sekowska A, Tutino ML, Vallenet D, von Heijne G, Danchin A (2005) Coping with cold: the genome of the versatile marine Antarctica bacterium *Pseudoalteromonas haloplanktis* TAC125. Genome Res 15:1325–1335

Methé BA, Nelson KE, Deming JW, Momen B, Melamud E, Zhang X, Moult J, Madupa R, Nelson WC, Dodson RJ, Brinkac LM, Daugherty SC, Durkin AS, DeBoy RT, Kolonay JF, Sullivan SA, Zhou L, Davidsen TM, Wu M, Huston AL, Lewis M, Weaver B, Weidman JF, Khouri H, Utterback TR, Feldblyum TV, Fraser CM (2005) The psychrophilic lifestyle as revealed by the genome sequence of *Colwellia psychrerythraea* 34 H through genomic and proteomic analyses. Proc Natl Acad Sci USA 102(31): 10913–10918

Nichols DS, Nichols PD, Russell NJ, Davies NW, McMeekin TA (1997) Polyunsaturated fatty acids in the psychrophilic bacterium *Shewanelle gelidimarina* ACAM456T: molecular species analysis of major phospholipids and biosynthesis of eicosapentaenoic acid. Biochim Biophys Acta 1347:164–176

Niederer M, Pankow W, Wiemken A (1992) Seasonal changes of soluble carbohydrates in mycorrhizas of Norway spruce and changes induced by exposure to frost desiccation. Eur J For Pathol 22:291–299

Niklinska M, Klimek B (2007) Effect of temperature on the respiration rate of forest soil organic layer along an elevation gradient in the Polish Carpathians. Biol Fertil Soil 43:511–518

Panikov NS, Sizova MV (2007) Growth kinetics of microorganisms isolated from Alaskan soil and permafrost in solid media frozen down to -35°C. FEMS Microbiol Ecol 59:500–512

Panikov NS, Flanaganb PW, Oechelc WC, Mastepanovd MA, Christensend TR (2006) Microbial activity in soils frozen to below -39°C. Soil Biol Biochem 38:785–794

Petrovic U, Gunde-Cimerman N, Zalar P (2000) Xerotolerant mycobiota from high altitude Anapurna soils, Nepal. FEMS Microbiol Lett 182:339–342

Phadtare S (2004) Recent developments in bacterial cold-shock response. Curr Issues Mol Biol 6:125–136

Phadtare S, Inoue M (2008) Cold-shock proteins. In: Margesin R, Schinner F, Marx JC, Gerday C (eds) Psychrophiles: from biodiversity to biotechnology. Springer, Berlin, pp 191–209

Prevost D, Drouin P, Laberge S, Bertrand A, Cloutier J, Levesque G (2003) Cold-adapted rhizobia for nitrogen fixation in temperate regions. Can J Bot Rev Can Bot 81:1153–1161

Qiu Y, Vishnivetskaya A, Lubman DM (2009) Proteomic insights: cryoadaptation of permafrost bacteria. In: Margesin R (ed) Permafrost soils, vol 16, Soil biology. Springer, Berlin, pp 169–181

Riley M, Staley JT, Danchin A, Wang TZ, Brettin TS, Hauser LJ, Land ML, Thompson LS (2008) Genomics of an extreme psychrophile *Psychromonas ingrahamii*. BMC Genom 9:210

Rivkina EM, Laurinavichus KS, Gilichinsky DA, Shcherbakova VA (2002) Methane generation in permafrost sediments. Dokl Biol Sci V383:179–181

Robinson CH (2001) Cold adaptation in Arctic and Antarctic fungi. New Phytol 151:341–353

Rowbury RJ (2003) Temperature effects on biological systems: introduction. Sci Prog 86:1–8

Russell NJ (1990) Cold adaptation of microorganisms. Phil Trans R Soc Lond B 329:595–611

Russell NJ (2008) Membrane components and cold sensing. In: Margesin R, Schinner F, Marx JC, Gerday C (eds) Psychrophiles: from biodiversity to biotechnology. Springer, Berlin, pp 177–190

Schinner F (1982a) $CO_2$-Freisetzung, Enzymaktivitäten und Bakteriendichte von Böden unter Spaliersträuchern und Polsterpflanzen in der alpinen Stufe. Ecol Plant 3:49–58

Schinner F (1982b) Soil microbial activities and litter decomposition related to altitude. Plant Soil 65:87–94

Schinner F (1983) Litter decomposition, $CO_2$-release and enzyme activities in a snowbed and on a windswept ridge in an alpine environment. Oecologia 59:288–291

Schinner F, Gstraunthaler G (1981) Adaptation of microbial communities to the environmental conditions in alpine soils. Oecologia 50:113–116

Shivaji S, Prakash JSS (2010) How do bacteria sense and respond to low temperatures? Arch Microbiol 192:85–95

Siddiqui KS, Cavicchioli R (2006) Cold-adapted enzymes. Ann Rev Biochem 75:403–433

Sidebottom C, Buckley S, Pudney P, Twigg S, Jarman C, Holt C, Telford J, McArthur A, Worrall D, Hubbard R, Lillford P (2000) Heat-stable antifreeze protein from grass. Nature 406:256

Uchida M, Nakatsubo T, Kasai Y, Nakane K, Horikoshi T (2000) Altitudinal differences in organic matter mass loss and fungal biomass in a subalpine coniferous forest, Mt. Fuji, Japan. Arct Antarct Alp Res 32:262–269

Väre H, Vestberg M, Ohtonen R (1997) Shifts in mycorrhiza and microbial activity along an oroarctic altitudinal gradient in Northern Fennoscandia. Arct Alp Res 29:93–104

Wagner M, Horn M, Daims H (2003) Fluorescence in situ hybridisation for the identification and characterisation of prokaryotes. Curr Opin Microbiol 6:302–309

Wang GZ, Wang YR, Yang PL, Luo HY, Huang HQ, Shi PJ, Meng K, Yao B (2010) Molecular detection and diversity of xylanase genes in alpine tundra soil. Appl Microbiol Biotechnol 87:1383–1393

Weinstein RN, Montiel PO, Johnstone K (2000) Influence of growth temperature on lipid and soluble carbohydrate synthesis by fungi isolated from fellfield soil in the maritime Antarctic. Mycologia 92:222–229

Whittaker RH (1975) Communities and ecosystems, 2nd edn. Mac Millan, New York

Worrall D, Elias L, Ashford D, Smallwood M, Sidebottom C, Lillford P, Telford J, Holt C, Bowles D (1998) A carrot leucine-rich-repeat protein that inhibits ice recrystallization. Science 282:115–117

Xu H, Griffith M, Patten CL, Glick BR (1998) Isolation and characterization of an antifreeze protein with ice nucleation activity from the plant growth promoting rhizobacterium *Pseudomonas putida* GR12-2. Can J Microbiol 44:64–73

Yamashita Y, Kawahara H, Obata H (2002) Identification of a novel anti-ice-nucleating polysaccharide from *Bacillus thuringiensis* YY529. Biosci Biotechnol Biochem 66: 948–954

Zhang LM, Wang M, Prosser JI, Zheng YM, He JZ (2009) Altitude ammonia-oxidizing bacteria and archaea in soils of Mount Everest. FEMS Microbiol Ecol 70:208–217

# Index




C. Lütz (ed.), *Plants in Alpine Regions*,
DOI 10.1007/978-3-7091-0136-0, © Springer-Verlag/Wien 2012

Printing and Binding: Stürtz GmbH, Würzburg